Electric Power Distribution: Systems and Engineering

Electric Power Distribution: Systems and Engineering

Editor: Linda Morand

NY RESEARCH
P R E S S

New York

Published by NY Research Press
118-35 Queens Blvd., Suite 400,
Forest Hills, NY 11375, USA
www.nyresearchpress.com

Electric Power Distribution: Systems and Engineering
Edited by Linda Morand

International Standard Book Number: 978-1-63238-561-1 (Hardback)

Cataloging-in-Publication Data

Electric power distribution : systems and engineering / edited by Linda Morand.
     p. cm.
Includes bibliographical references and index.
ISBN 978-1-63238-561-1
1. Electric power distribution. 2. Electric power systems. 3. Electrical engineering.
I. Morand, Linda.
TK3001 .E44 2017
621.31--dc23

Printed in the United States of America.

Contents

Preface

Every book is initially just a concept; it takes months of research and hard work to give it the final shape in which the readers receive it. In its early stages, this book also went through rigorous reviewing. The notable contributions made by experts from across the globe were first molded into patterned chapters and then arranged in a sensibly sequential manner to bring out the best results.

This book on electric power distribution discusses the various mechanisms that regulate power transmission. Power is generated across a complex set of pathways and networks. Electric power distribution is a field of study that focuses on the transmission of power from the source of its generation to the end users. The aim of this text is to present researches that have transformed this discipline and aided its advancement. Coherent flow of topics, student-friendly language and extensive use of examples make this book an invaluable source of knowledge. This book is a vital tool for all researching and studying this field. It includes some of the vital pieces of work being conducted across the world, on various topics related to electric power grids and power distribution.

It has been my immense pleasure to be a part of this project and to contribute my years of learning in such a meaningful form. I would like to take this opportunity to thank all the people who have been associated with the completion of this book at any step.

Editor

Hybrid Wind Speed Prediction Based on a Self-Adaptive ARIMAX Model with an Exogenous WRF Simulation

Erdong Zhao [1], Jing Zhao [2,*], Liwei Liu [1], Zhongyue Su [3] and Ning An [4]

Academic Editor: Frede Blaabjerg

[1] School of Economics and Management, North China Electric Power University, Beijing 102206, China; teacherzed@163.com (E.Z.); vliu2006@163.com (L.L.)
[2] School of Mathematics and Statistics, Lanzhou University, Lanzhou 730000, China
[3] College of Atmospheric Sciences, Lanzhou University, Lanzhou 730000, China; suzy10@lzu.edu.cn
[4] Gerontechnology Lab, School of Computer and Information, Hefei University of Technology, Hefei 230009, China; ning.an@hfut.edu.cn
* Correspondence: zhaoj13@lzu.edu.cn

Abstract: Wind speed forecasting is difficult not only because of the influence of atmospheric dynamics but also for the impossibility of providing an accurate prediction with traditional statistical forecasting models that work by discovering an inner relationship within historical records. This paper develops a self-adaptive (SA) auto-regressive integrated moving average with exogenous variables (ARIMAX) model that is optimized very-short-term by the chaotic particle swarm optimization (CPSO) algorithm, known as the SA-ARIMA-CPSO approach, for wind speed prediction. The ARIMAX model chooses the wind speed result from the Weather Research and Forecasting (WRF) simulation as an exogenous input variable. Further, an SA strategy is applied to the ARIMAX process. When new information is available, the model process can be updated adaptively with parameters optimized by the CPSO algorithm. The proposed SA-ARIMA-CPSO approach enables the forecasting process to update training information and model parameters intelligently and adaptively. As tested using the 15-min wind speed data collected from a wind farm in Northern China, the improved method has the best performance compared with several other models.

Keywords: wind speed; self-adaptive strategy; ARIMAX; WRF simulation

1. Introduction

1.1. Time Series Forecasting and Wind Energy

Time series forecasting plays an essential role in many fields, especially in meteorology, economics and energy. Time series models produce forecasts by discovering the inner relationships within historical records. This paper focuses on wind speed forecasting, which is crucial in the whole life-cycle of wind farm construction and operation and is also the basic technique to guarantee the grid security of a wind-connected system. Wind power is economic and ecologically friendly, which makes it one of the most popular and promising alternative energy sources. Wind power accounts for approximately 10% of the national power use in many European countries, and for more than 15% in Spain, Germany and the US [1]. However, the main obstacle for wind industry development is the variability of output power, which seriously prevents wind power penetration and threatens grid security. To guarantee the security of the grid system, the dispatching department have to balance the grid's consumption and production within very small time intervals [2]. Moreover, because the lack of accurate information on wind occurrence, the efficiency of wind turbine may also be limited [3]. In actual power generation, wind predictions—especially the short-term forecasts—are important for scheduling, controlling and dispatching the energy conversion systems [4]. However, as the most important characteristic of wind, speed can be easily influenced by other meteorological factors, such as air pressure, air temperature and terrain [5]. Thus, wind speed prediction is not easy to address. Moreover, wind speed modelling has become one of the most difficult problems [6,7].

1.2. Wind Speed Forecasting: Existing Works

Many methods have been attempted to forecast wind speed. In general, they can be classified into two categories: physical and statistical methods. Physical methods are always referred to as meteorological predictions of wind speed, including the numerical approximation of models that describe the state of the atmosphere [8], such as the Weather Research and Forecasting (WRF) model [9]. These models always choose physical data such as topography information, pressure and temperature to forecast wind speed in the future [10,11]. As one of the current-generation physical models, WRF [9] is widely used in both research [12–14] and operational forecasts. Reference [13] used the WRF model to manage ocean surface wind simulations forced by different initial and boundary conditions. Reference [14] compared WRF with the Wind Atlas Analysis and Application Program (WAsP) model, to test the performance in terms of flow characteristics and energy yields estimates. Considering numerical weather prediction (NWP) models, one important issue is downscaling. Generally, two categories are focused on: dynamic and statistical downscaling methods. Dynamical downscaling methods have clear physical meanings and are unaffected by the observation data. However, they require large computational costs. Being different, statistical downscaling—including transfer function method (TFM), weather pattern method (WPM) and stochastic weather generator (SWG)—is simple to establish and needs a small amount of calculation, but it may be influenced strongly by observations [15]. Recently, many new statistical downscaling techniques have been developed, such as the similarity method, hidden Markov model (HMM), generalized linear model (GLM) and others [15].

Unlike physical models, statistical methods make forecasts by discovering the relationships in historical wind speed data and sometimes other variables (e.g., wind direction or temperature). The data used is recorded at the observation site or other nearby locations where data are available. Moreover, many statistical methods have been applied, such as the auto-regressive integrated moving average (ARIMA) model, Kalman filters, and the generalized auto-regressive conditional heteroscedasticity (GARCH) model, *etc.* The statistical models can be used at any stage during modelling, and they often merge various methods into one. Physical and statistical models each have their own advantages for wind speed prediction, but few forecasts use only one of them. The physical prediction results are just the first step of wind forecast; then, the physically predicted wind speed can be regarded as an auxiliary input to other statistical models [16–18]. Currently, grey models (GM) [19,20] and models based on artificial intelligence (AI) techniques [21,22] have been developed for this area, containing the artificial neural networks (ANNs) of multi-layer perceptrons (MLP) [23], radial basis function (RBF) [24], recurrent neural networks [25,26], and fuzzy logic [27,28].

As one of the most widely used time series approach, ARIMA has been used as an effective and efficient forecasting technique in many fields, including traffic, energy, and the economy. Generally, ARIMA is a linear model that represents both stationary and non-stationary series [29] and uses historical time series patterns to make forecasts for the future data trend. In terms of the wind speed prediction problem, which is studied in this paper, the ARIMA models are effective and suitable for short-term and very-short-term predictions. References [30–32] applied the auto-regressive moving average (ARMA) model to wind speed predictions with different time horizons. Furthermore, because wind-related data always show obvious periodicity, a seasonal ARIMA model can be defined with the consociation of a seasonal difference process [33]. Later, the fractional-ARIMA model was proposed by Kavasser and Seetharaman [34], which assumes that the differencing parameter d of ARIMA (p, d, q) is a fractionally continuous value in the interval $(-0.5, 0.5)$. Their model was used for wind speed prediction on the day-ahead and two-day-ahead time horizons in North Dakota. When there is little knowledge available or there is no suitable model relating the predicted variables to other explanatory factors, the ARIMA model is particularly useful [35]. Some articles made a hybrid approach by combining the ARIMA model with other methods. Studies take ARIMA as the first step of a hybrid method, and then the residual series of ARIMA can be regarded as the nonlinear part of the original series. Reference [36] developed a hybrid ARIMA-ANN model for hourly wind speed prediction. In their method, the ARIMA model was first used for wind speed forecasting, while the ANN was chosen to reduce the errors from the ARIMA models. Later, a hybrid method combined the seasonal ARIMA, and the least square support vector machine (LSSVM) was developed in Reference [5] for monthly wind speed prediction in the Hexi Corridor of China. Here, both ANN and LSSVM are quite effective for addressing series within nonlinear signals.

Improvement made on ARIMA has enhanced the model performance substantially. However, by considering either the improved ARIMA model or the ARIMA-combined hybrid methods for wind forecast, most approaches employ only the historical observations but not the factors of atmospheric dynamics. Some studies claimed that an accurate wind prediction method must include a numerical weather prediction (NWP)-based process [37].

1.3. Original Contribution: Developed Self-Adaptive Wind Speed Forecasting Strategy

The original contribution of this paper is the development of a self-adaptive (SA) auto-regressive integrated moving average with exogenous variables (ARIMAX) model optimized by the chaotic particle swarm optimization (CPSO) algorithm called the SA-ARIMAX-CPSO approach, which is applied to wind speed prediction. Specifically, the applied ARIMAX model takes the WRF simulation as an exogenous part, which makes the forecasting model a combination of both statistical and physical information. Moreover, the CPSO-driven SA strategy enables the proposed method to syncretize the previous model and the recently updated information. In this paper, the self-adaptation contains two parts. The first one is new model fitting, when the recent measurements or WRF data are available. This paper updates the fitting coefficients every time-step, while the WRF model runs once a day. The second one is adaptation process, where the optimal adaptive weights are determined only based on the training set.

On the issue of very-short-term wind speed prediction, models were established generally based on a statistical process, while the NWP simulations were typically used for short-term predictions. This is mainly due to the model accuracy and calculation costs. This paper develops a hybrid approach for very-short-term wind speed prediction combining both statistical and physical models, which has an acceptable amount of calculation and effective model performance. Specifically, the WRF model is now the current generation physics-based atmospheric model, which is widely applied; the ARIMA process is the typical time series model, which emphasizes modelling the relationship among historical observations. Thus, the proposed ARIMAX model in this paper considers not only the statistical information from historical wind speed observations but also the physical process of atmospheric motion.

Furthermore, this paper develops a SA strategy to apply for the ARIMAX method. Model parameters are always fixed values that are determined by the training data set; this may be unreasonable in a dynamic process. When new information is obtained, the prediction system should be updated. In this paper, the new information includes two parts—the newly updated measurement records and the WRF simulation result. From this opinion, this paper develops a SA-ARIMAX model, which has adaptive model parameters when the new information is available. During this process, the CPSO algorithm is applied to obtain the optimized parameters. Simulation results show that the developed SA-ARIMAX-CPSO method in this paper performs considerably better than the original auto-regressive moving average with exogenous variables (ARMAX), ARIMAX, and adaptive ARMAX models.

1.4. Structure of This Paper

The rest of this paper is organized as follows: Section 2 reviews the original ARIMAX model. Section 3 introduces the improved SA-ARIMA optimized by the CPSO algorithm. Section 4 shows the available data sets and model measurements. Sections 5 and 6 display the experiments and analysis. Afterward, conclusions are discussed in Section 7. Finally, acknowledgements and references are given.

2. Original ARIMAX Model

The developed ARIMAX model in this paper is a single-input and single-output (SISO) system, which is defined as follows:

$$A\left(z^{-1}\right) y\left(t\right) = B\left(z^{-1}\right) u\left(t\right) + C\left(z^{-1}\right) e\left(t\right) \tag{1}$$

The input data passes through a difference filter D times, where:

$$A\left(z^{-1}\right) = 1 - a_1 z^{-1} - \cdots - a_p z^{-p} \tag{2}$$

$$B\left(z^{-1}\right) = b_1 + b_2 z^{-1} + \cdots + b_q z^{-q+1} \tag{3}$$

$$C\left(z^{-1}\right) = 1 + c_1 z^{-1} + \cdots + c_r z^{-r} \tag{4}$$

$y\left(t\right)$ is the output at time t, $u\left(t\right)$ is the exogenous variable at time t, $e\left(t\right)$ is the white noise, and p, q and r are the orders of auto-regressive (AR), moving average (MA) and exogenous (X), respectively. Moreover, z^{-1} represents the delay operator, and $A\left(z^{-1}\right)$, $B\left(z^{-1}\right)$ and $C\left(z^{-1}\right)$ are the parameters of AR, MA and X parts, respectively. It is assumed that the zero points of $A\left(z^{-1}\right)$ and $C\left(z^{-1}\right)$ are located in the unit circle. Equation (1) can be re-written as:

$$\begin{aligned}
y\left(t\right) = a_1 y\left(t-1\right) + \cdots + a_p y\left(t-p\right) + b_1 u\left(t\right) + b_2 u\left(t-1\right) + \cdots + b_q u\left(t-q+1\right) \\
+ e\left(t\right) + c_1 e\left(t-1\right) + \cdots + c_r e\left(t-r\right)
\end{aligned} \tag{5}$$

To determine the model order, the most popular one is the Bayesian Information Criterion (BIC) [38]. Reference [39] provides a detailed discussion on order determination for the ARIMAX model by using the BIC method.

The parameters $A\left(z^{-1}\right)$, $B\left(z^{-1}\right)$ and $C\left(z^{-1}\right)$ are obtained by the recursive maximum likelihood estimation method [40]. Thus:

$$\theta = \left[a_1, \ldots, a_p, b_1, \ldots, b_q, c_1, \ldots, c_r\right] \tag{6}$$

The recursive estimation of θ can be expressed as:

$$\theta\left(t+1\right) = \theta\left(t\right) + K\left(t\right)\left(y\left(t+1\right) - \varphi^T\left(t\right)\theta\left(t\right)\right) \tag{7}$$

where $\theta\left(0\right)$ can be any value, $\theta\left(i\right) = 0$ if $i < 0$, and:

$$K\left(t\right) = \frac{P\left(t\right)\varphi\left(t\right)}{1 + \varphi^T\left(t\right)P\left(t\right)\varphi\left(t\right)} \tag{8}$$

$$P\left(t+1\right) = P\left(t\right) - K\left(t\right)\varphi^T\left(t\right)P\left(t\right) \tag{9}$$

$$\varphi\left(t\right) = \left[y\left(t-1\right), \ldots, y\left(t-p\right), u\left(t\right), \ldots, u\left(t-q+1\right), y\left(t-1\right)\right.$$
$$\left. -\varphi^{T}\left(t-1\right)\theta\left(t\right), \ldots, y\left(t-r\right) - \varphi^{T}\left(t-r\right)\theta\left(t-r+\right)\right] \tag{10}$$

3. Self-Adaptive ARIMAX Optimized by CPSO Algorithm

3.1. Self-Adaptive ARIMAX (SA-ARIMAX) Method

In the original ARIMAX model introduced in Section 2, the model parameters $A\left(z^{-1}\right)$, $B\left(z^{-1}\right)$ and $C\left(z^{-1}\right)$ are fixed by the training data set. This is unreasonable in real applications. When new information is obtained, the forecast system should be updated. From this point of view, this paper develops a SA-ARIMAX model with adaptive model parameters.

The model parameters are denoted at time t as $A^{(t)}\left(z^{-1}\right)$, $B^{(t)}\left(z^{-1}\right)$, and $C^{(t)}\left(z^{-1}\right)$, as follows:

$$A^{(t)}\left(z^{-1}\right) = 1 - a_1^{(t)}z^{-1} - \cdots - a_p^{(t)}z^{-p} \tag{11}$$

$$B^{(t)}\left(z^{-1}\right) = b_1^{(t)} + b_2^{(t)}z^{-1} + \cdots + b_q^{(t)}z^{-q+1} \tag{12}$$

$$\varphi C^{(t)}\left(z^{-1}\right) = 1 + c_1^{(t)}z^{-1} + \cdots + c_r^{(t)}z^{-r} \tag{13}$$

Assuming that the model parameters at time t are estimated, Equation (5) can be re-rewritten as:

$$\hat{y}\left(t+1\right) = a_1^{(t)}y\left(t\right) + \ldots + a_p^{(t)}y\left(t-p+1\right) + b_1^{(t)}u\left(t+1\right) + b_2^{(t)}u\left(t\right) + \ldots$$
$$+ b_q^{(t)}u\left(t-p+2\right) + e\left(t+1\right) + c_1^{(t)}e\left(t\right) + \ldots + c_r^{(t)}e\left(t-r+1\right) \tag{14}$$

When the new information is obtained at time $(t + 1)$, the model parameters should be updated. As fitted by ARIMAX with the same model orders as previously stated, parameters are obtained and denoted as $\hat{A}^{(t)}\left(z^{-1}\right)$, $\hat{B}^{(t)}\left(z^{-1}\right)$, and $\hat{C}^{(t)}\left(z^{-1}\right)$. Then, at time $(t + 1)$, the parameters of the forecasting model should be influenced not only by the parameters at time t, $A^{(t)}\left(z^{-1}\right)$, $B^{(t)}\left(z^{-1}\right)$, and $C^{(t)}\left(z^{-1}\right)$ but also by the new information $\hat{A}^{(t)}\left(z^{-1}\right)$, $\hat{B}^{(t)}\left(z^{-1}\right)$, and $\hat{C}^{(t)}\left(z^{-1}\right)$. Thus, this paper takes a weighted average of the two aspects, as:

$$A^{(t+1)}\left(z^{-1}\right) = (1-\alpha)\hat{A}^{(t+1)}\left(z^{-1}\right) + \alpha A^{(t)}\left(z^{-1}\right) \tag{15}$$

$$B^{(t+1)}\left(z^{-1}\right) = (1-\beta)\hat{B}^{(t+1)}\left(z^{-1}\right) + \beta B^{(t)}\left(z^{-1}\right) \tag{16}$$

$$C^{(t+1)}\left(z^{-1}\right) = (1-\gamma)\hat{C}^{(t+1)}\left(z^{-1}\right) + \gamma C^{(t)}\left(z^{-1}\right) \tag{17}$$

where $0 < \alpha, \beta, \gamma < 1$ are three weights.

3.2. Parameters in the SA-ARIMAX Model

There are two categories of parameters in the SA-ARIMAX model. One category are the ARIMAX parameters, named $A\left(z^{-1}\right)$, $B\left(z^{-1}\right)$ and $C\left(z^{-1}\right)$ and defined by Equations (11)–(13). This set of parameters can always be obtained by the least square (LS) method during the model fitting process. The other category is the self-adaptive parameters, $\alpha, \beta,$ and γ, defined in Equations (15)–(17) when applying the SA strategy to an ARIMAX process. It can be easily found that parameters $\alpha, \beta,$ and γ represent a weighted average between the historical and the newly fitted model parameters. Larger $\alpha, \beta,$ and γ prove that the prediction model takes more information from the historical model parameters, while smaller values of $\alpha, \beta,$ and γ prove that the newly fitted model parameters cause more influence on the final forecasting results.

Values of $\alpha, \beta,$ and γ affect the model performance by constructing a different information balance between the historical and newly fitted model parameters. The determination of $\alpha, \beta,$ and γ is difficult but quite essential. To search for the optimized parameters α, β, and γ during the SA process, this paper applies the CPSO algorithm, which is a swarm intelligent method. The combination with the CPSO algorithm enables the developed SA-ARIMAX model to absorb the newly updated information with an optimized coefficient.

3.3. Model Optimization by CPSO Algorithm

3.3.1. Working Principle of CPSO Algorithm

Particle swarm optimization (PSO) simulates the social psychological metaphor based on swarm intelligence. Two best values exist in the simulation process of PSO. For each particle in the problem space, the best value obtained up to now is denoted as *pBest*. In terms of the global version, the overall best solution achieved up to now is called *gBest*. The procedure for PSO can be expressed as shown in Appendix A [41,42].

3.3.2. Developed Method: SA-ARIMAX Optimized by CPSO (SA-ARIMAX-CPSO)

In this paper, the three coefficients, α, β, and γ, are optimized by the CPSO algorithm introduced in Section 3.3.1. Then, the prediction value at time $(t + 1)$ can be calculated by Equation (14) using the optimized parameters. The developed self-adaptive ARIMAX method optimized by CPSO, called SA-ARIMAX-CPSO in this paper, can be divided into several steps as Appendix B shows. Figure 1 shows the flowchart.

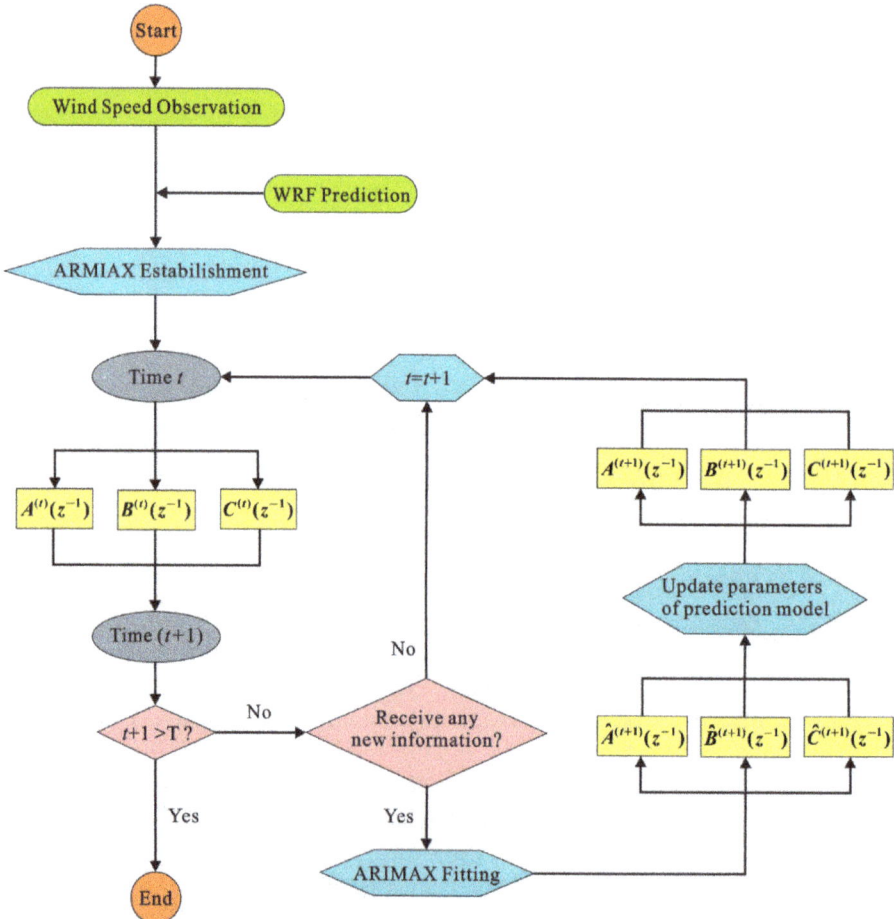

Figure 1. The flow chart of SA-ARIMAX-CPSO method.

In practice, the new information is not updated at every time t, thus the model parameters should be updated when the new information, newly-obtained observation or WRF simulation data, is available. This forecasting process could be concluded as follows: assuming that $t + 1 < T$ and the parameters at time t, $A^{(t)}(z^{-1})$, $B^{(t)}(z^{-1})$, and $C^{(t)}(z^{-1})$, are obtained. Thus, the parameters at time $(t + 1)$ can be calculated:

$$A^{(t+1)}\left(z^{-1}\right) = \begin{cases} A^{(t)}\left(z^{-1}\right), & \text{if no new information is received} \\ (1-\alpha)\,\hat{A}^{(t+1)}\left(z^{-1}\right) + \alpha A^{(t)}\left(z^{-1}\right), & \text{otherwise} \end{cases} \tag{18}$$

$$B^{(t+1)}\left(z^{-1}\right) = \begin{cases} B^{(t)}\left(z^{-1}\right), & \text{if no new information is received} \\ (1-\beta)\,\hat{B}^{(t+1)}\left(z^{-1}\right) + \beta B^{(t)}\left(z^{-1}\right), & \text{otherwise} \end{cases} \tag{19}$$

$$C^{(t+1)}\left(z^{-1}\right) = \begin{cases} C^{(t)}\left(z^{-1}\right), & \text{if no new information is received} \\ (1-\gamma)\,\hat{C}^{(t+1)}\left(z^{-1}\right) + \gamma C^{(t)}\left(z^{-1}\right), & \text{otherwise} \end{cases} \tag{20}$$

where $\hat{A}^{(t+1)}(z^{-1})$, $\hat{B}^{(t+1)}(z^{-1})$, and $\hat{C}^{(t+1)}(z^{-1})$ represent the new information.

3.3.3. How the CPSO Works: An AI-Based Optimization Process

As introduced in Section 3, the CPSO algorithm is employed as a parameter searching tool, optimizing the parameters α, β, and γ in Equations (15)–(17). This section aims to display how the parameters α, β, and γ are optimized during a CPSO-driven process. The maximum iteration is set as 100; Figure 2 displays the parameter values for each iteration.

The CPSO-driven optimization process is a parameter searching process, promoting the reduction of the fitness value. For each step, the best values, $pBest$ and $gBest$, will be updated if the fitness value meets a better value, which means a lower fitness value in this study. Figure 2 shows the parameter changing trace, where alpha, beta, gamma imply α, β, and γ, respectively. Denote the t-th iteration of α, β, and γ as $x(t) = (x_1(t), x_2(t), x_3(t))$. Then, the $(t + 1)$-th iteration can be expressed as:

$$v(t+1) = w \cdot v(t) + c_1 \cdot uD(t) \cdot [pBest(t) - x(t)] + c_2 \cdot UD(t) \cdot [gBest(t) - x(t)] \tag{21}$$

$$x(t+1) = x(t) + \Delta t \cdot v(t+1) \tag{22}$$

where w is the parameter called inertia weight, c_1 and c_2 are positive constants, and uD and UD are random figures uniformly distributed in [0, 1]. Thus, the vector v can be regarded as the velocity vector of the parameter iteration. Then, the position $x(t + 1)$ can be calculated by adding the velocity vector $v(t + 1)$ onto the previous position $x(t)$, where Δt means the step length.

To strengthen the randomness, different combination methods regarding parameters were given [43]. In this paper, w is iterated by the following Tent Map:

$$w(t+1) = \begin{cases} \dfrac{w(t)}{0.7}, & w(t) < 0.7 \\ \dfrac{10}{3w(t)(1-w(t))}, & w(t) \geqslant 0.7 \end{cases} \tag{23}$$

$$w(t+1) = w(t+1) + 0.5 \tag{24}$$

c_1 and c_2 are updated by the Logistic Map:

$$c_1(t+1) = a \cdot c_1(t)(1 - c_1(t)) \tag{25}$$

$$c_2(t+1) = a \cdot c_2(t)(1 - c_2(t)) \tag{26}$$

where, generally, $a = 4$, $c_1, c_2 \in (0, 1)$.

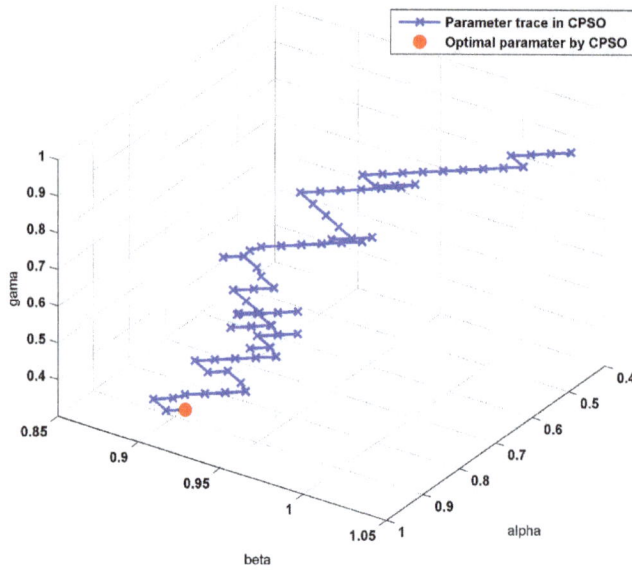

Figure 2. Trace of parameters during a CPSO-driven process.

4. Available Data Sets

This study site is a wind farm in Shandong Province in Northern China. Figure 3 shows the topography of Shandong Province and the location of the study site. The data set used in this paper was collected from an anemometer tower located in the range of this wind farm, which measuring height is 70 m. The available data are from 6:00, 2011-9-30 to 13:45, 2011-11-18, with a time interval of 15 min (Figure 4) There are 3.26% missing data, and the missing data are filled by linear interpolation. Table 1 shows the basic statistical description of the available data. Moreover, Figure 5 displays the frequency distribution of the available data and the probability distribution function (pdf) fitted by the two-parameter Weibull distribution:

$$f\left(v\right) = \frac{k}{c}\left(\frac{v}{c}\right)^{k-1} exp\left[-\left(\frac{v}{c}\right)^{k}\right], \ v \geqslant 0 \tag{27}$$

where k and c are the shape parameter and scale parameter, respectively, and v represents the wind speed records. Applying the maximum likelihood (ML) method, the shape and scale parameters are 5.0258 and 2.1490, respectively.

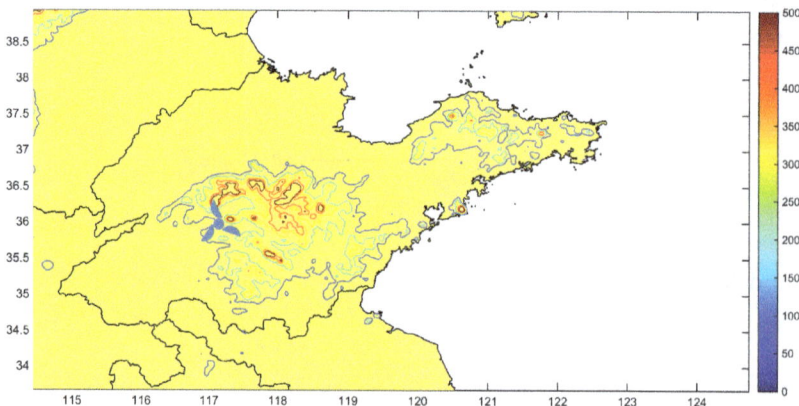

Figure 3. The topography of Shandong Province.

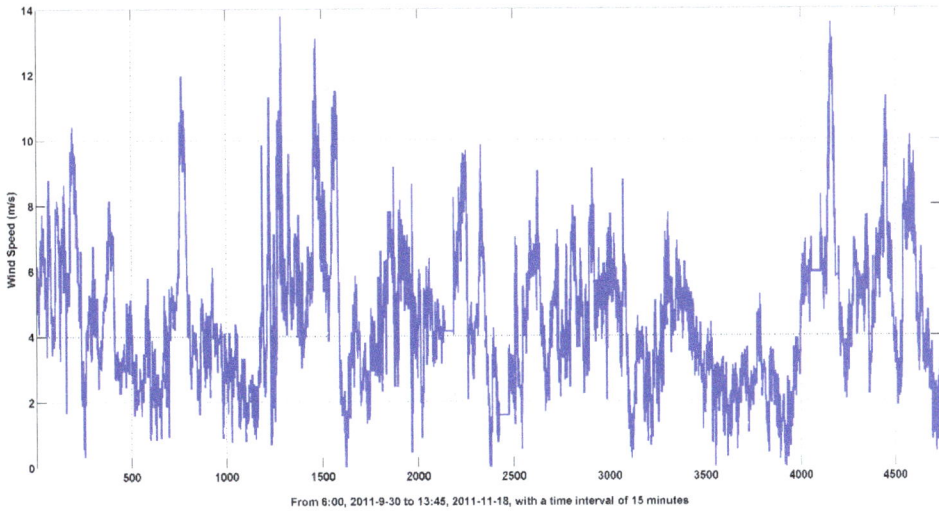

From 6:00, 2011-9-30 to 13:45, 2011-11-18, with a time interval of 15 minutes

Figure 4. The available data.

Table 1. Statistical description of available data.

Number	Minimum (m/s)	Maximum (m/s)	Mean (m/s)	Standard Deviation	Skewness	Kurtosis
5000	0.1	13.8	4.29	2.23	0.74	0.69

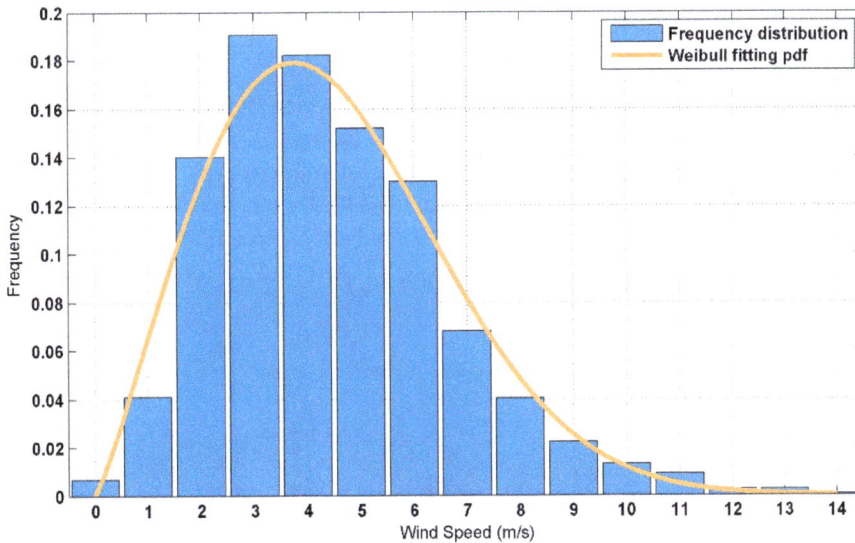

Figure 5. Frequency distribution of the available data.

5. Experiments and Analysis

Experiments in this paper are composed of three parts. The first part displays the WRF configuration and prediction, which will be used as an exogenous input in the following model construction. Then, the second part discusses the original ARMAX and ARIMAX predictions. This part aims to test the model performance without the SA strategy, which means the fixed model parameters will be used during the whole tested period. After that, the developed SA strategy is applied, with the same model setup as the original ARMAX and ARIMAX models.

5.1. WRF Meso-Scale Numerical Model Prediction

The WRF meso-scale numerical model is now the current generation physics-based atmospheric model, serving the needs of both atmospheric research and operational forecasting. Recently, the WRF model has become one of the most popular and widely used tools for numeric weather prediction. In this paper, the WRF model is selected as a representative for the physical models.

The WRF model domain has 150 by 120 horizontal grid points, spaced at 27 km, situated on 47 terrain following vertical levels. In the WRF model, a grid is defined as an integration of three dimensional points. It contains a set of weather data (wind speed, atmospheric pressure *etc.*). Physical equations are used to simulate the atmospheric state; this is based not only on the data on grid but also a specific physical model. Then, the simulations are calculated by discretized time-steps [44].

In the developed ARIMAX model, a wind speed prediction from the WRF simulation is adopted as an exogenous variable. First, this section provides the WRF results. The initial and boundary conditions of the WRF simulation are extracted from the National Centres for Environmental Prediction (NCEP, http://www.ncep.noaa.gov/) reanalysis data ($1° \times 1°$); the time resolution is 15 min, and the spatial resolution is 27 km. The physical options of the WRF model are described in Table 2. The WRF calculation discussed in this paper is a one-day simulation, which starts at 8:00 am (China Standard Time, CST) on the first day to 8:00 am (CST) on the second day.

Table 2. Model configuration of WRF simulation.

	Physical Options
Cumulus parameterization	Grell 3d ensemble cumulus scheme
Longwave/Shortwave radiation	RRTM/Dudhia scheme
Surface layer physics	Eta similarity
Land surface processes	Noah Land Surface Model
Planetary Boundary layer	Mellor-Yamada-Janjic scheme

Figure 6 shows the WRF prediction in the experimental period. It is clear that the WRF prediction can describe the overall variability of wind speed, even though the forecasting accuracy should be enhanced. The atmospheric dynamics information plays an important role in WRF prediction. In the next section, this result will be an exogenous input of the ARIMAX model; it is regarded as a reference value for the final prediction. Thus, the developed prediction procedure contains information not only from historical observations but also from physical-based WRF results.

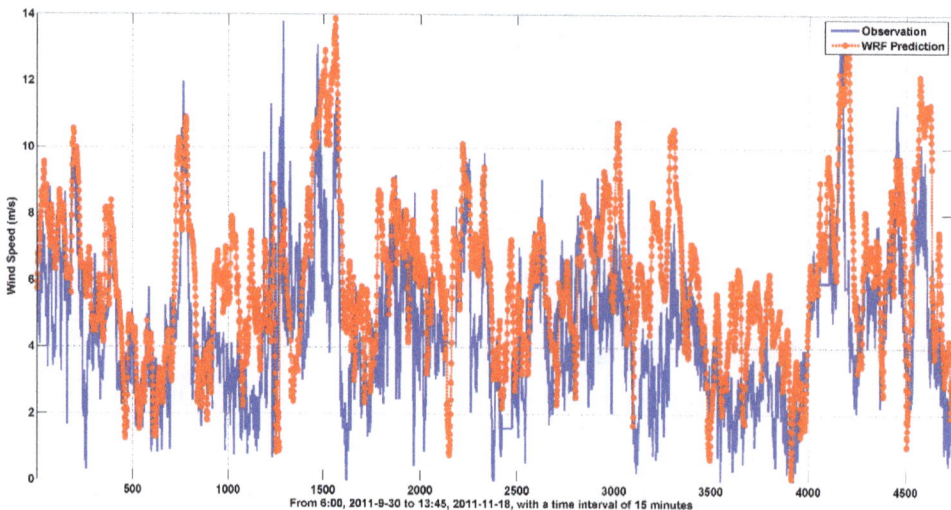

Figure 6. The WRF prediction in the study period.

5.2. Original ARMAX and ARIMAX Predictions

This section establishes the original ARIMAX model. The first 800 observations of the available data set are chosen as the initial training data set; the rest is used for rolling prediction and model testing. By using the BIC method, ARMAX (3,3,1) is established. Figure 7 shows the ARMAX results and absolute error. The absolute error of the ARMAX prediction is mainly distributed in a range of 0 to 4 m/s. The original ARMAX model has high MAPE in wind speed prediction. This may result from two aspects. One aspect is the strong fluctuation of the wind speed time series, which makes it difficult to capture the variation and randomness. The other aspect may be the fixed model parameters in the original ARMAX model.

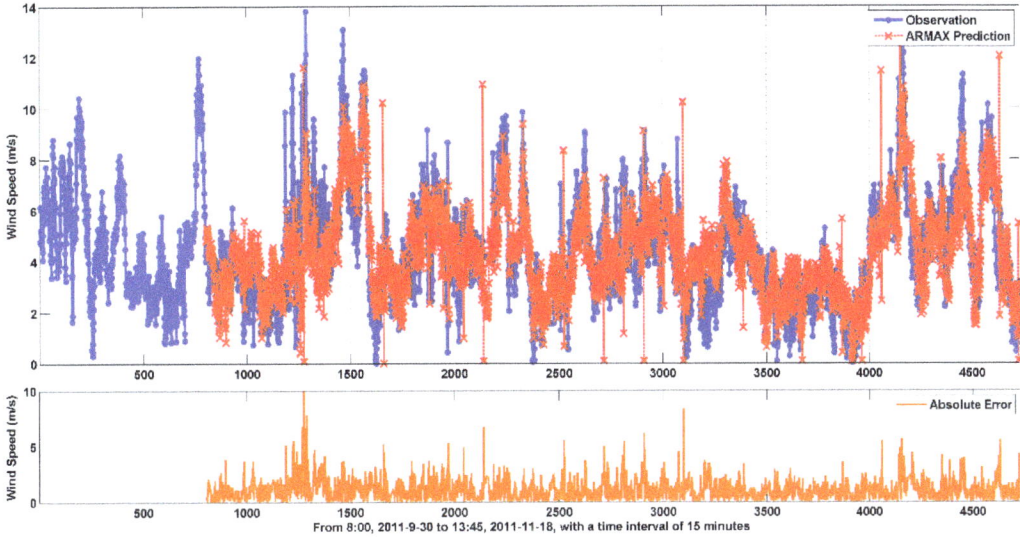

Figure 7. ARMAX prediction results.

To transform the original observation data into a stationary sequence, the first-order difference is adopted; therefore, the ARIMAX model is established. Figure 8 shows the transformed data series, and Figure 9 provides the ARIMAX predictions. Compared with Figure 7, the ARIMAX prediction has fewer statistical errors and this phenomenon also occurs in Table 3, which provides detailed error comparisons among the root mean absolute error (RMSE), Bias and correlation coefficient (R). Compared with ARMAX prediction, the ARIMAX model has a decline of 14.69% in RMSE.

At the same time, it can also be found from Figure 9 that the ARIMAX prediction has a strong fluctuation, and the predicted value is always considerably higher than the observation. Therefore, the model performance should be improved further. In the following sections, a method of adaptive model parameters is applied to both ARMAX and ARIMAX procedures. Simulation results show that predictions with adaptive parameters perform considerably better than the original ARMAX and ARIMAX predictions.

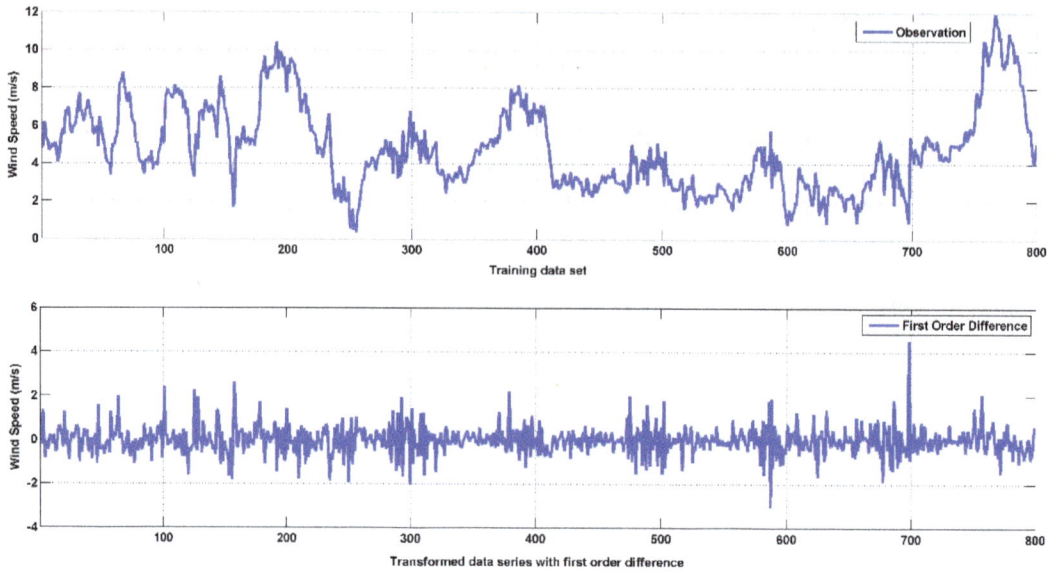

Figure 8. Transformed data set with first order difference.

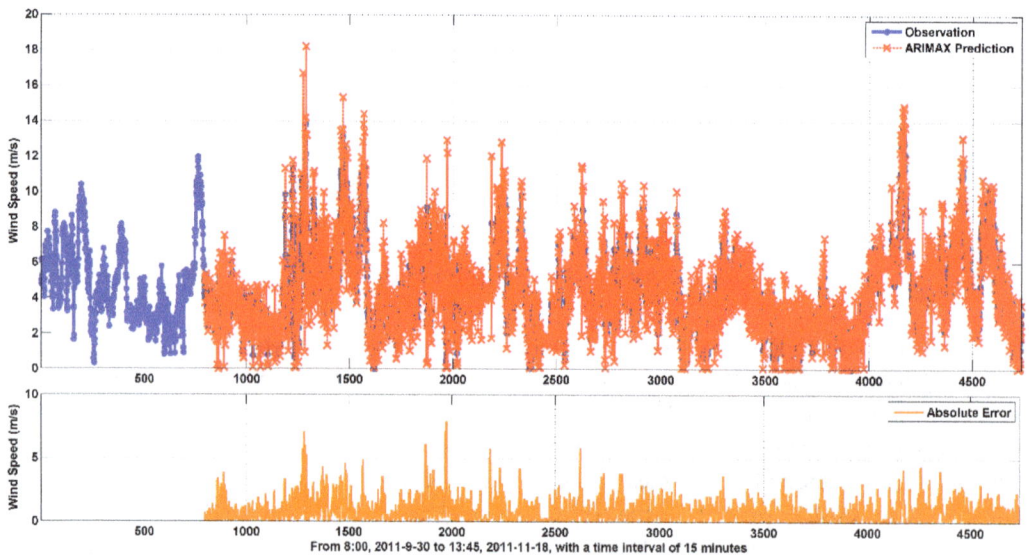

Figure 9. ARIMAX prediction results.

Table 3. Errors of ARMAX and ARIMAX models.

	RMSE (m/s)	Bias (m/s)	R
ARMAX	1.43	−0.10	0.76
ARIMAX	1.22	0.04	0.87

5.3. Developed SA-ARMAX-CPSO and SA-ARIMAX-CPSO Predictions

In this section, this paper develops a method of adaptive parameters. The developed method is applied to both SA-ARMAX and SA-ARIMAX procedures. The three weights, α, β, and γ, are obtained by the CPSO algorithm, as $\alpha = 0.96$, $\beta = 0.92$, $\gamma = 0.39$. Figures 10 and 11 show the prediction results.

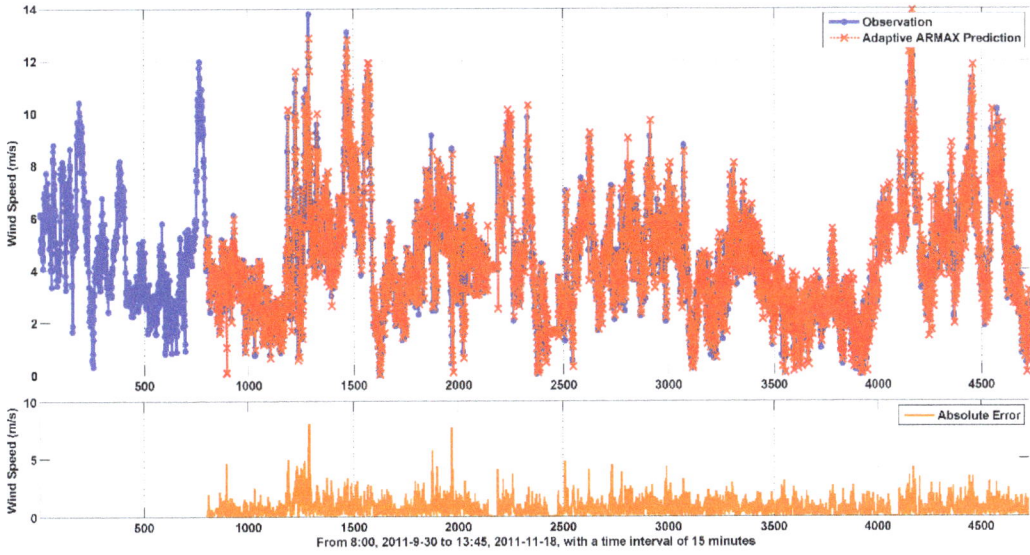

Figure 10. Adaptive ARMAX prediction results.

Compared to Figures 7 and 9 the adaptive models show significant improvements. Table 4 shows the detailed statistical errors of the adaptive ARMAX and adaptive ARIMAX models. The adaptive ARIMAX model performs slightly better than the adaptive ARMAX model, which can be found from all three criteria in Table 4.

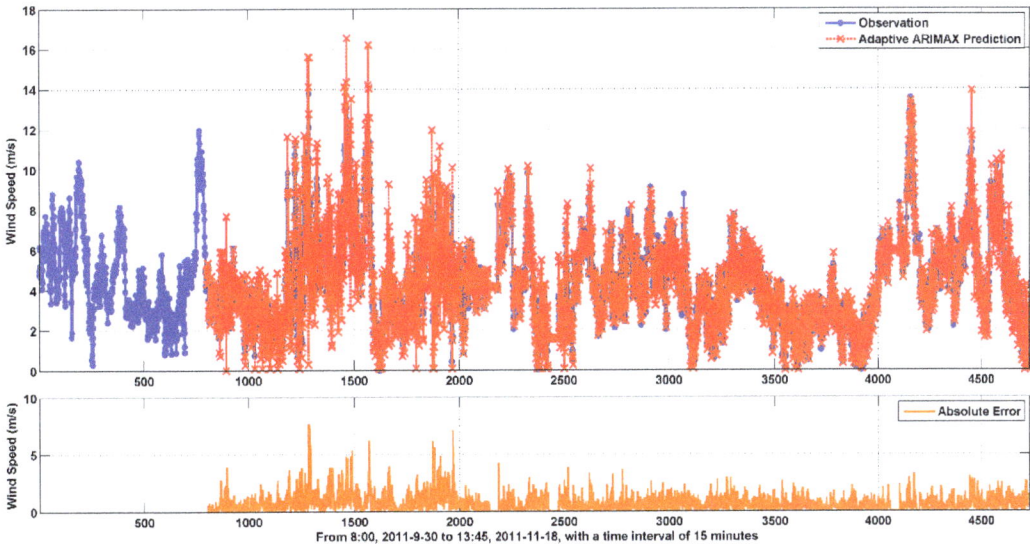

Figure 11. Adaptive ARIMAX prediction results.

However, compared with the original ARMAX and ARIMAX predictions, adaptive models work more efficiently. Specifically, when compared with the original ARMAX/ARIMAX models, the SA-ARMAX/ARIMAX-CPSO models show 23.78% and 12.30% lower RMSE, respectively. Bias of the adaptive methods declines and the correlation coefficients of the adaptive methods increase, compared with the original ARMAX/ARIMAX models. This is a benefit of the CPSO-driven SA strategy proposed in this paper. It indicates that the developed adaptive method effectively improves the original model performances.

Table 4. Errors of adaptive ARMAX and adaptive ARIMAX models.

	RMSE (m/s)	Bias (m/s)	R
Adaptive ARMAX	1.09	−0.05	0.87
Adaptive ARIMAX	1.07	0.03	0.89

5.4. Performance Comparison among Several Models

This paper employs the WRF model as an exogenous input to the ARMA/ARIMA methods, where this physical model runs once a day. Thus, it is significant to compare the model performance between the WRF and ARMAX/ARIMAX models at the start point of WRF prediction, which is 8:00 am CST in this paper. Figure 12 displays the comparison results among them.

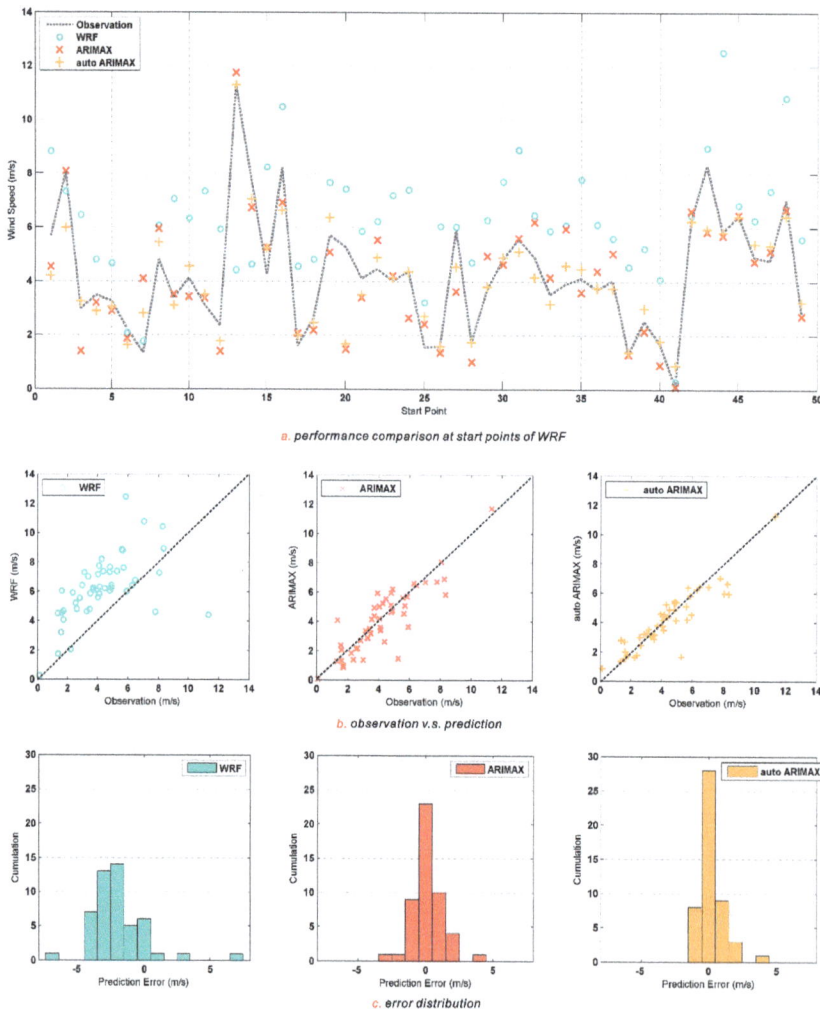

a. performance comparison at start points of WRF

b. observation v.s. prediction

c. error distribution

Figure 12. Performance comparison among WRF, ARIMAX and adaptive ARIMAX. (**a**) Performance comparison at start points of WRF; (**b**) Observation *vs*. prediction; (**c**) Error distribution.

Specifically, Figure 12a shows the observation and different predictions at the starting point of the WRF model. It can be found that no model always performs its best at each start point. Figure 12b contains three scatter plots between the observation and three different predictions. The smaller distance between the scatter points and the fixed line ($y = x$) refers to the better performance of the forecasting model. Obviously, the adaptive ARIMAX model shows the best performance among the three models. This can also be found in the Figure 12c, which displays the histogram for the forecasting error. The error distribution of WRF is wider than the other two, which indicates that the WRF simulation may have bad performances in some cases. This is disadvantageous when considering the very-short-term wind speed prediction. Different from it, the ARIMAX and adaptive ARIMAX models use the WRF prediction as an exogenous input; thus, the physical simulation can be a reference value to the final result. The error distribution of the adaptive ARIMAX model concentrates around zero value, which implies that the self-adaptive method is more stable and effective. Moreover, the comparison also indicates that although the physical prediction is regarded as a reference value, the final result of the ARIMAX prediction is not totally driven by the WRF simulation. The model performance benefits from both physical and statistical model processes. This can be also found in Table 5, which displays the comparison among WRF, ARIMAX and adaptive ARIMAX at the starting points of WRF simulation.

Table 5. Errors comparison at the starting points of WRF simulation.

	RMSE (m/s)	Bias (m/s)	R
WRF	2.54	1.61	0.57
ARIMAX	1.12	−0.14	0.87
Adaptive ARIMAX	0.91	−0.15	0.91

6. Further Discussions

The simulation in Section 5.2 shows that the proposed adaptive ARIMAX method performs better than the adaptive ARMAX and the original ARIMAX models, with a lower value of statistical errors. To provide a deeper understanding of how the forecasting errors can be reduced by the proposed method, this section has an additional discussion on this topic from the following aspects.

6.1. Contribution of the CPSO-Driven SA Strategy: Reduce the Forecasting Errors

As mentioned above, the principle of the proposed SA strategy is a combination of the historical model and the recently updated information, and the three parameters α, β, and γ determine the weight of this balance. The simulation result indicates that the CPSO-driven SA strategy contributes to the reduction of model errors.

- Contribution of the SA strategy. As well known, statistical models are established by finding the relationships inside the data records. An intuitionistic idea is that the newly fitted model contains the recently updated information and always leads to better results. However, series such as wind speed show continuous changes and strong variations, and the WRF simulation also contains unavoidable uncertainties itself. All of these factors may result in poor model performance. Thus, a combination of both the historical model and the recently updated information makes the forecasting process more stable.

- Contribution of the CPSO-driven optimization. Concerning the SA strategy, the most important task is to determine the balance between the historical model and the newly fitted model, which means the parameters α, β, and γ. Larger α, β, and γ prove that the model takes more information from the historical form, whereas the smaller values indicate the newly fitted model brings more influence. Under this circumstance, the CPSO algorithm is employed as a parameter searching tool to find the optimal value in the meaning of artificial intelligence.

6.2. Discussion on the Optimized Parameters

The optimal values are not identical for the three parameters. In this paper, the CPSO-driven optimal values of α, β, and γ are 0.96, 0.92, and 0.39, respectively. These three values are not identical, which means the three parts in an ARIMAX model, AR, MA, and X, perform in different ways.

- The CPSO-driven optimal values of α and β are similar. Taking $\alpha = 0.96$ as an example, it means that $A^{(t+1)}\left(z^{-1}\right)$ is nearly equal to $A^{(t)}\left(z^{-1}\right)$, and $\hat{A}^{(t+1)}\left(z^{-1}\right)$ only contributes to a very small percentage. A comparable discussion can be given for the parameter β. This indicates that the AR and MA parts heavily rely on the historical model parameters but not the recently updated model information.

- Different from α and β, the optimized value $\gamma = 0.39$ implies that parameter $C^{(t+1)}\left(z^{-1}\right)$ takes more information from the newly fitted $\hat{C}^{(t+1)}\left(z^{-1}\right)$ than $C^{(t)}\left(z^{-1}\right)$. These parameters correspond to the X part, which are related to the exogenous WRF input in this paper. The result shows that large weight should be assigned to recent WRF information. The reason may be that the WRF simulation describes the physical mechanism of atmosphere, and a recently updated simulation contains the approaching information of future atmospheric motion. Thus, parameters $C^{(t)}\left(z^{-1}\right)$ from the new information should be assigned to a larger weight compared with the previous one. This is helpful for short-term wind speed forecasting and is different from the principle of statistic parts.

7. Conclusions

As one of the most popular low-carbon resources, wind energy contributes not only to energy conservation but also to environmental protection. Wind speed prediction is a critical problem in wind power generation. In this paper, an adaptive ARIMAX model, which takes WRF results as an exogenous input and has adaptive model parameters, is developed for 15-min wind speed prediction.

The developed adaptive ARIMAX model performs better than the original ARMAX, ARIMAX, adaptive ARMAX and ANN models. It may result from several aspects. To begin with, considering both physical and statistical information, the proposed ARIMAX model in this paper chooses the WRF prediction as the exogenous input. The physics-based WRF describes the state of atmospheric motion and provides a believable prediction of wind speed with a forecast time horizon of three days. However, its forecasting accuracy should be improved when downscaled into a given area. Statistical predictions model the specific wind speed regulation by the historical information. Thus, taking both the atmospheric movement and the historical regulation into consideration is a good choice. In the developed ARIMAX model in this paper, the AR and MA parts model the statistical regulation among observations, while the X part imports the physical prediction result. Next, compared to the original ARIMAX model, the developed method contains adaptive model parameters. In the rolling forecasting procedure, the adaptive method can promptly bring new information into the prediction system. This method is a weighted average of the historical parameter and the parameter calculated from new information. The information from the historical parameter can maintain the stability of the forecasting system, while new information updates the system to obtain an accurate variation trend of the latest wind speed series. In addition, the ARIMAX model has not been applied in the area of wind speed prediction. Its usage in this paper is a new attempt for this topic to obtain better forecasting performance. Moreover, the developed method of adaptive model parameters can also be applied to other forecasting models.

Acknowledgments: This research was supported by the National Natural Science Foundation of China under Grant (71171102/G0107).

Author Contributions: Erdong Zhao planned the whole paper, designed the structure and suggested on the methodology. Jing Zhao contributed to the model selection and paper drafting. Liwei Liu contributed to the data collection and language editing. Zhongyue Su performed the experiments and simulations, Erdong Zhao and Ning An proofread the text.

Conflicts of Interest: The authors declare no conflict of interest.

Appendix A

Algorithm: CPSO

Input:
  $x(t) = x_1(t), x_2(t), \ldots, x_n(t)$
Output:
  $x(best)$, best values of input $x(t)$
Parameters:
  w, u, c_1, c_2

1 **INITIALIZATION**
   /* Initialize the position and velocity of each particle randomly in the n-dimensional problem space, using the uniform problem distribution function */
2 **FOR** each $(1 \leqslant i \leqslant n)$ **DO**
3    $x_i(t) = rand(x)$
4 **END FOR**
5 $iter = 1$
6 $x(iter) = x_1(iter), x_2(iter), \ldots, x_n(iter)$
7 **WHILE** $(iter \leqslant iter_{max})$ DO /* Find the best fitness value */
8   **FOR** EACH $(x_i^{iter} \in x)$ **DO**
     /* For each particle, evaluate the fitness value, set its pBest as the current position and value */
9   **IF** $(pBest_i > fitness(x_i^{iter}))$ THEN
10       $pBest_i = x_i^{iter}$
11   **END IF**
12   **END FOR**
13   /* Choose the particle with the best fitness value of all the particles */ **FOR** EACH $(x_i^{iter} \in x)$ DO
14   **IF** $(gBest > pBest_i)$ THEN
15        $gBest = pBest$
16        $xBest = x_i^{iter}$
17   **END IF**
18   **END FOR**
19   **FOR** EACH $(x_i^{iter} \in x)$ **DO**
20      $v_i(t+1) = w \cdot v_i(t) + c_1 \cdot ud_{i,j}(t) \cdot [pBest_i(t) - x_i(t)] + c_2 \cdot Ud_{i,j}(t) \cdot [gBest(t) - x_i(t)]$
21      $x_i(t+1) = x_i(t) + \Delta t \cdot v_i(t+1)$
22   **END FOR**
23    $iter = iter + 1$
24 **END WHILE**
25 **RETURN**
26 $x(best) = [x_1(best), x_2(best), \ldots, x_n(best)]$

Appendix B

Algorithm: SA-ARIMAX-CPSO

Input:
  WRF prediction X and historical records denoted as y
Output:
  $\hat{y}(T+1), \hat{y}(T+2) \ldots$
Parameters:
  α, β, γ-the weight of ARIMAX model
  q, p, r-the number of model order

1   *INITIALIZATION*
2   WRF prediction
3   Fit ARIMAX model. Parameters of ARIMAX prediction model (Equation (14)) are denoted as $A^{(t)}\left(z^{-1}\right)$, $B^{(t)}\left(z^{-1}\right)$ and $C^{(t)}\left(z^{-1}\right)$
4   *WHILE* ($t+1 < T$) DO
5   Adding new information at time (t+1)
6   Re-fit ARIMAX model as $\hat{A}^{(t)}\left(z^{-1}\right)$, $\hat{B}^{(t)}\left(z^{-1}\right)$ and $\hat{C}^{(t)}\left(z^{-1}\right)$
7   Calculate ARIMAX prediction model (Equation (14)) at time ($t+1$)
8   Calculate α, β, γ by using CPSO algorithm
9   Update the parameters of ARIMAX as $A^{(t+1)}\left(z^{-1}\right)$, $B^{(t+1)}\left(z^{-1}\right)$ and $C^{(t+1)}\left(z^{-1}\right)$
10   Obtained the prediction $\hat{y}\left(t+2\right)$
11 *END WHILE*
12 *RETURN*
13 $\hat{y}\left(T+1\right), \hat{y}\left(T+2\right) \ldots$

References

1. Mabel, M.C.; Fernandez, E. Analysis of wind power generation and prediction using ANN: A case study. *Renew. Energy* **2008**, *33*, 986–992. [CrossRef]
2. Lazic, L.; Pejanovic, G.; Zivkovic, M. Wind forecasts for wind power generation using the Eta model. *Renew. Energy* **2010**, *35*, 1236–1243. [CrossRef]
3. Monfared, M.; Rastegar, H.; Kojabadi, H.M. A new strategy for wind speed forecasting using artificial intelligent methods. *Renew. Energ.* **2009**, *34*, 845–848. [CrossRef]
4. De Giorgi, M.G.; Ficarella, A.; Tarantino, M. Assessment of the benefits of numerical weather predictions in wind power forecasting based on statistical methods. *Energy* **2011**, *36*, 3968–3978. [CrossRef]
5. Guo, Z.; Zhao, J.; Zhang, W.; Wang, J. A corrected hybrid approach for wind speed prediction in Hexi Corridor of China. *Energy* **2011**, *36*, 1668–1679. [CrossRef]
6. Ramirez-Rosado, I.; Fernandez-Jimenez, L. An advanced model for short term forecasting of mean wind speed and wind electric power. *Control Intell. Syst.* **2014**, *32*, 21–26. [CrossRef]
7. Sfetsos, A. A novel approach for the forecasting of mean hourly wind speed time series. *Renew. Energy* **2002**, *27*, 163–174. [CrossRef]
8. Liu, H.P.; Shi, J.; Erdem, E. Prediction of wind speed time series using modified Taylor Kriging method. *Energy* **2010**, *35*, 4870–4879. [CrossRef]
9. Skamarock, W.C.; Klemp, J.B.; Dudhia, J.; Gill, D.O.; Barker, D.M.; Duda, M.G.; Huang, X.-Y.; Wang, W.; Powers, J.G. A Description of the Advanced Research WRF Version 3. Available online: http://opensky.ucar.edu/islandora/object/technotes:500 (accessed on 10 December 2015).
10. Landberg, L. Short-term prediction of the power production from wind farms. *J. Wind Eng. Ind. Aerod.* **1999**, *80*, 207–220. [CrossRef]
11. Negnevitsky, M.; Potter, C.W. Innovative short-term wind generation prediction techniques. In proceedings of 2006 IEEE/PES Power Systems Conference and Exposition, Atlanta, GA, USA, 29 October–1 November 2006; pp. 60–65.
12. Carvalho, D.; Rocha, A.; Gomez-Gesteira, M.; Santos, C.S. Sensitivity of the WRF model wind simulation and wind energy production estimates to planetary boundary layer parameterizations for onshore and offshore areas in the Iberian Peninsula. *Appl. Energy* **2014**, *135*, 234–246. [CrossRef]
13. Carvalho, D.; Rocha, A.; Gomez-Gesteira, M.; Santos, C.S. Offshore wind energy resource simulation forced by different reanalyses: Comparison with observed data in the Iberian Peninsula. *Appl. Energy* **2014**, *134*, 57–64. [CrossRef]
14. Carvalho, D.; Rocha, A.; Silva Santos, C.; Pereira, R. Wind resource modelling in complex terrain using different mesoscale-microscale coupling techniques. *Appl. Energy* **2013**, *108*, 493–504. [CrossRef]
15. Liu, Y.; Guo, W.; Feng, J.; Zhang, K. A Summary of Methods for Statistical Downscaling of Meteorological Data. *Adv. Earth Sci.* **2011**, *26*, 837–847.

16. Ma, L.; Luan, S.Y.; Jiang, C.W.; Liu, H.L.; Zhang, Y. A review on the forecasting of wind speed and generated power. *Renew. Sustain. Energy Rev.* **2009**, *13*, 915–920.

17. Sanz, S.S.; Perez, A.B.; Ortiz, E.G.; Portilla-Figueras, A.; Prieto, L.; Paredes, D.; Correoso, F. Short-term Wind Speed Prediction by Hybridizing Global and Mesoscale Forecasting Models with Artificial Neural Networks. In Proceedings of the 8th International Conference on Hybrid Intelligent Systems, Barcelona, Spain, 10–12 September 2008; pp. 608–612.

18. Salcedo-Sanz, S.; Perez-Bellido, A.M.; Ortiz-Garcia, E.G.; Portilla-Figueras, A.; Prieto, L.; Correoso, F. Accurate short-term wind speed prediction by exploiting diversity in input data using banks of artificial neural networks. *Neurocomputing* **2009**, *72*, 1336–1341. [CrossRef]

19. Wu, S.J.; Lin, S.L. Intelligent Web-Based Fuzzy and Grey Models for Hourly Wind Speed Forecast. *Int. J. Comput.* **2010**, *4*, 235–242.

20. Li, J.; Zhang, B.; Xie, G.; Li, Y.; Mao, C. Grey predictor models for wind speed-wind power prediction. *Power Syst. Prot. Control* **2010**, *38*, 151–159.

21. Ren, C.; An, N.; Wang, J.; Li, L.; Hu, B.; Shang, D. Optimal parameters selection for BP neural network based on particle swarm optimization: A case study of wind speed forecasting. *Knowl.-Based Syst.* **2014**, *56*, 226–239. [CrossRef]

22. Guo, Z.; Zhao, W.; Lu, H.; Wang, J. Multi-step forecasting for wind speed using a modified EMD-based artificial neural network model. *Renew. Energy* **2012**, *37*, 241–249. [CrossRef]

23. Alexiadis, M.C.; Dikopoulos, P.S.; Sahsamanoglou, H.S.; Manousaridis, I.M. Short-term forecasting of wind speed and related electrical power. *Sol. Energy* **1998**, *63*, 61–68. [CrossRef]

24. Beyer, H.; Degner, T.; Haussmann, J.; Hoffman, M.; Rujan, P. Short term forecast of wind speed and power output of a wind turbine with neural networks. In Proceedings of the second European congress on intelligent techniques and soft computing, Aachen, Germany, 20–23 Septmber 1994; pp. 349–352.

25. Kariniotakis, G.N.; Stavrakakis, G.S.; Nogaret, E.F. Wind power forecasting using advanced neural networks models. *IEEE Trans. Energy Conver.* **1996**, *11*, 762–767. [CrossRef]

26. More, A.; Deo, M.C. Forecasting wind with neural networks. *Mar. Struct.* **2003**, *16*, 35–49. [CrossRef]

27. Wang, X.; Sideratos, G.; Hatziargyriou, N.; Tsoukalas, L.H. Wind speed forecasting for power system operational planning. In Proceedings of 2004 International Conference on Probabilistic Methods Applied to Power Systems, Ames, IA, USA, 16 Septmber 2004; pp. 470–474.

28. Wang, J.; Xiong, S. A hybrid forecasting model based on outlier detection and fuzzy time series—A case study on Hainan wind farm of China. *Energy* **2014**, *76*, 526–541. [CrossRef]

29. Lee, W.-J.; Hong, J. A hybrid dynamic and fuzzy time series model for mid-term power load forecasting. *Int. J. Electr. Power* **2015**, *64*, 1057–1062. [CrossRef]

30. Kamal, L.; Jafri, Y.Z. Time series models to simulate and forecast hourly averaged wind speed in Quetta, Pakistan. *Sol. Energy* **1997**, *61*, 23–32. [CrossRef]

31. Torres, J.L.; Garcia, A.; De Blas, M.; De Francisco, A. Forecast of hourly average wind speed with ARMA models in Navarre (Spain). *Sol. Energy* **2005**, *79*, 65–77. [CrossRef]

32. Erdem, E.; Shi, J. ARMA based approaches for forecasting the tuple of wind speed and direction. *Appl. Energy* **2011**, *88*, 1405–1414. [CrossRef]

33. Cadenas, E.; Rivera, W. Wind speed forecasting in the South Coast of Oaxaca, Mexico. *Renew. Energy* **2007**, *32*, 2116–2128. [CrossRef]

34. Kavasseri, R.G.; Seetharaman, K. Day-ahead wind speed forecasting using f-ARIMA models. *Renew. Energy* **2009**, *34*, 1388–1393. [CrossRef]

35. Zhang, G.P. Time series forecasting using a hybrid ARIMA and neural network model. *Neurocomputing* **2003**, *50*, 159–175. [CrossRef]

36. Cadenas, E.; Rivera, W. Wind speed forecasting in three different regions of Mexico, using a hybrid ARIMA-ANN model. *Renew. Energy* **2010**, *35*, 2732–2738. [CrossRef]

37. Giebel, G. The State-Of-The-Art in Short-Term Prediction of Wind Power A Literature Overview. Available online: http://ecolo.org/documents/documents_in_english/wind-predict-ANEMOS.pdf (accessed on 10 December 2015).

38. Schwarz, G. Estimating the Dimension of a Model. *Ann. Stat.* **1978**, *6*, 461–464. [CrossRef]

39. Xu, S.; Xie, W. An order estimation of ARMAX model. *Harbin Inst. Electr. Technol. J.* **1992**, *15*, 73–76.

40. Xu, S.; Xie, W. Several Problems of ARMAX model. *Harbin Inst. Electr. Technol. J.* **1989**, *12*, 391–395.

41. Krohling, R.A.; Hoffmann, F.; Coelho, L.D.S. Co-evolutionary Particle Swarm Optimization for min-max problems using Gaussian distribution. In Proceedings of Evolutionary Computation, 2004. CEC2004, Portland, OR, US, 19–23 June 2004; Volume 1, pp. 959–964.

42. Coelho, L.D.; Krohling, R.A. Predictive controller tuning using modified particle swarm optimization based on Cauchy and Gaussian distributions. *Adv. Soft Comput.* **2005**, 287–298.

43. Alatas, B.; Akin, E.; Ozer, A.B. Chaos embedded particle swarm optimization algorithms. *Chaos Soliton. Fract.* **2009**, *40*, 1715–1734. [CrossRef]

44. Weather Research and Forecasting Model 2.2 Documentation: A Step-by-step guide of a Model Run. Available online: http://citeseerx.ist.psu.edu/viewdoc/download?doi=10.1.1.447.3692&rep=rep1&type=pdf (accessed on 10 December 2015).

Vibration Durability Testing of Nickel Manganese Cobalt Oxide (NMC) Lithium-Ion 18,650 Battery Cells

James Michael Hooper [1,*], James Marco [1], Gael Henri Chouchelamane [2] and Christopher Lyness [2]

Academic Editor: K. T. Chau

[1] Warwick Manufacturing Group (WMG), University of Warwick, Coventry CV4 7AL, UK;
    james.marco@warwick.ac.uk

[2] Jaguar Land Rover, Banbury Road, Warwick, Coventry CV35 0XJ, UK;
    gchouch1@jaguarlandrover.com (G.H.C.); clyness@jaguarlandrover.com (C.L.)

*   Correspondence: j.m.hooper@warwick.ac.uk

Abstract: Electric vehicle (EV) manufacturers are employing cylindrical format cells in the construction of the vehicles' battery systems. There is evidence to suggest that both the academic and industrial communities have evaluated cell degradation due to vibration and other forms of mechanical loading. The primary motivation is often the need to satisfy the minimum requirements for safety certification. However, there is limited research that quantifies the durability of the battery and in particular, how the cells will be affected by vibration that is representative of a typical automotive service life (e.g., 100,000 miles). This paper presents a study to determine the durability of commercially available 18,650 cells and quantifies both the electrical and mechanical vibration-induced degradation through measuring changes in cell capacity, impedance and natural frequency. The impact of the cell state of charge (SOC) and in-pack orientation is also evaluated. Experimental results are presented which clearly show that the performance of 18,650 cells can be affected by vibration profiles which are representative of a typical vehicle life. Consequently, it is recommended that EV manufacturers undertake vibration testing, as part of their technology selection and development activities to enhance the quality of EVs and to minimize the risk of in-service warranty claims.

Keywords: vehicle vibration; electric vehicle (EV); Li-ion battery ageing; durability

1. Introduction

Within the automotive sector, the main driving force for technological innovation is the requirement to reduce fuel consumption and vehicle exhaust emissions. Legislative requirements are motivating original equipment manufacturers (OEM) and subsystem suppliers to develop and integrate new and innovative technologies into their fleet. Consequently, over the last few years, different types of electric vehicles (EV) have been built alongside conventional internal combustion engine (ICE) cars. Within the field of EV, a key technology enabling this reduction in fuel consumption and exhaust emissions is design and integration of rechargeable energy storage systems (RESS) [1,2].

Many OEMs are employing cylindrical format cells (e.g., 18,650) within the design and construction of RESS [3–6]. Cylindrical cells are often chosen in EV applications over their prismatic and pouch cell counterparts because of a combination of factors. For example, 18,650 cells are produced in very large quantity which makes them very cost effective [6–8]. Similarly, they have built-in safety systems such as a positive temperature coefficient (PTC) resistor that prevents high current surge and the use of a current interrupt device (CID) [6–8].

To ensure in-market reliability, OEMs perform a variety of life representative mechanical abuse and durability tests during the design and prototype stages of the development process. Firstly, these tests ensure that new vehicle sub-assemblies and components are fit-for-purpose. Secondly, it allows OEMs to obtain characterization data for simulations and computer aided engineering (CAE) activities. Thirdly, it ensures that the product meets requirements for vehicle homologation.

Vibration durability is one of these tests. The use of electromagnetic shaker (EMS) tables is often preferred to understand the behavior of a given vehicle system when subjected to mechanical induced vibration that is representative of the in-service environment and desired vehicle life (typically 100,000 to 150,000 miles of customer usage). Vibration durability tests play an important role in the selection of components. As discussed within [9–13], poorly integrated components, assemblies or structures subjected to vibration can result in a significantly reduced service life or the occurrence of catastrophic structural failure through fatigue cracking or work hardening of materials [11,14,15].

Within the context of individual Li-ion cells aggregated together to form larger vehicle battery RESS, a significant body of research exists that underpins the mechanical characterization of both pouch and cylindrical cell formats through static and dynamic test techniques. The motivation for such work is often to provide data to either parameterize or validate CAE models and simulations. For static test methods, there has been a clear focus on obtaining data from materials found within Li-ion cells such as mechanical strain and bending [16–19], force displacement [16–20], creep [19] and tolerance changes during charge and discharge [21]. Within the dynamic testing domain there has been a significant focus towards assessing the crashworthiness and robustness of Li-ion cells via mechanical crush [16,20,22], penetration [18,23], impact resistance [16,22], mechanical shock [16,24] and the effect of environmental changes such as temperature [25] and decompression [26]. Research in the static and dynamic domains is driven by a need to comply with whole vehicle crash homologation [27,28], to meet consumer focused accreditation requirements (e.g., Euro NCAP [29]) and mandatory transport legislation such as UN 38.3 [30].

Vibration-induced fatigue was reported for a lithium iron phosphate (LFP) battery [31,32]. However, the impact of vibration on cell performance was not presented. The vibration profile and subsequent test parameters were also not defined in either of these sources.

In a study conducted by [33,34], the authors examined the impact of vibration on different samples of 18,650 lithium cobalt oxide (LCO) 2.8 Ah Samsung Cells. Their study aimed to determine if the exposure to vibration would affect the charge/discharge behavior of the cells. The study concluded that the cells were not adversely affected by vibration and also reported "no discernible difference in the measured open-circuit voltage (OCV) before and after testing". Their conclusions were also supported by visual inspections that showed no significant damage to the cells.

A comprehensive test program in which numerous 18,650 Li-ion cells, of unknown chemistry, were subjected to a vibration profile along the Z-axis of the cells is presented in [24]. The cells were clamped to an EMS table and excited for 186 h with a swept-sine wave from 4 to 20 Hz and back to 20 Hz in 30 s. The authors reported that most of the cells exhibited an increase in resistance along with a reduction in their 1C discharge capacity. Additionally, they described that some cells underwent an internal short circuit. This performance degradation was attributed to the central mandrel becoming loose during the vibration test, which in turn damaged the upper and lower cell components, including the current collector and tabs. Within that study, legislative abuse tests were also performed in accordance with UN 38.8 Test 3—Swept Sine and UN 38.8 Test 4—Shock Testing. It was found that no significant degradation in electrical performance was observed. However, the central mandrel was displaced because of the mechanical loading along the Z-axis of the cells. Similarly, deformation of the CID, as well as damage to the current collector was also observed in a subset of 18,650 cells assessed in accordance with UN 38.3 Test 4 where the excitation was again applied along the Z-axis of the cell.

A doctoral thesis [35] assessed the robustness of 12 Ah Li-ion pouch cells of four different cell chemistries to vibration. The cells were exposed to vibration and temperature cycling according to ISO 16750-3 (Section 4.1.2.7). It was presented that none of the cells showed any electrochemical performance degradation as their 1C capacity after vibration was similar to that of the control samples.

As a result, this study concluded that vibration and temperature cycling did not significantly affect the electrochemical performance of the cells.

From this short review, it is clear that the academic literature within this field, presents conflicting evidence with regard to the susceptibility of Li-ion cells to vibration. In addition, the vibration profiles used within these studies were not representative of an EV application. For example, the UN 38.3 standard utilized in the study defined in [24] was derived for assessing the robustness of batteries for consumer electronics for air transit [11,36]. Similarly, the ISO 16750-3 (Section 4.1.2.7) utilized in [35] is designed to represent the vibration excitation experienced by a large commercial vehicle with chassis mounted component. Also the long term vibration testing discussed in [24] utilizes a swept sine profile which does not excite the cells within a road vehicle representative vibration spectra [11,37,38]. The vibration profiles also did not represent a known service life such as 100,000 miles of vehicle use. Equally, the swept-sine waves utilized within these durability assessments are an unrealistic representation of the vibration loading that occurs within road vehicles [11] and are suited to vibration characterization.

As a result, the authors are proposing to study the effect of mechanical induced vibration, using 2 random vibration cycles (Society of Automotive Engineers (SAE) J2380 and Warwick Manufacturing Group/Millbrook Proving Ground (WMG/MBK) profiles [36,37]), on a commercially available nickel manganese cobalt oxide (NMC) Samsung 2.2 Ah 18,650 cylindrical cell (model number ICR18650-22F). Both vibration test profiles were developed to underpin durability evaluation and to replicate a 100,000 miles of vehicle use. Both these specifications apply vibration loading through "random" excitation which is more representative of road-induced structural vibration [11]. Also unlike the previous studies discussed in [24,33–35], each cell evaluated experience vibration excitation in the X, Y and Z axis, as opposed to experiencing vibration in a single axis for the duration of the test, which is more representative of the vibration experienced by cells within an EV battery assembly. It is outside the scope of this study to discuss in detail the derivation of the vibration profiles. However, this information is available in [36,37]. These profiles were applied to cells with different states of charge (SOC) and a different in-pack orientation. The effect of mechanical induced vibration on the mechanical and electrochemical performances of the cells, before and after vibration, is also presented.

This paper is structured as follows: Section 2 of this paper provides a detailed overview of the experimental method employed, including the design of the test equipment and fixtures that are key to ensuring accuracy and repeatability of the measurements. Results are presented in Section 3 and include an assessment of the mechanical degradation (visual inspection and natural frequency) as well as changes in the electrochemical performance (capacity, impedance *etc.*) of the cell. Discussion, further work and conclusions are presented in Sections 4–6 respectively.

2. Experimental Method

The following section outlines the experimental method, including the test process, fixture design, rig assembly and cell characterization employed.

2.1. Test Fixture Design

Figure 1a presents the cell-mounting fixture that was designed and fabricated to support this study. Each fixture holds up to three cells and is intended to recreate a generic but representative 18,650 EV RESS mounting condition. 5 mm of each end of the cell are clamped within the cell test fixture.

Because a single axis EMS table was used, 3 cell-mounting fixtures were made, all based on the same design to allow the concurrent evaluation of multiple cells in different orientations during a single vibration test. The different cell orientations (X, Y and Z) were achieved by mounting the 3 fixtures onto different surfaces of the durability fixture. During the test programme, the cells were subjected to different axis of vibration by relocating the cell fixture onto different surfaces of the durability fixture. Installation of the durability fixture and the cell-mounting fixture onto the EMS table, complete with instrumentation, is presented in Figure 1b. The fixtures were constructed from aluminum due to the high specific stiffness [39] (also referred to as materials ratio within vibration testing [40,41] or specific modulus) associated with this material. Specific stiffness is defined in Equation (1) where the materials Young's modulus is defined as E (in GPa) and density as ρ (in g/cm^3) [39–41]. A high specific stiffness indicates a high material natural frequency [40,41] and results in a lower risk of undesirable fixture resonances impacting the accuracy of the experimentation.

$$Specific\ Stiffness\ =\ \frac{E}{\rho} \tag{1}$$

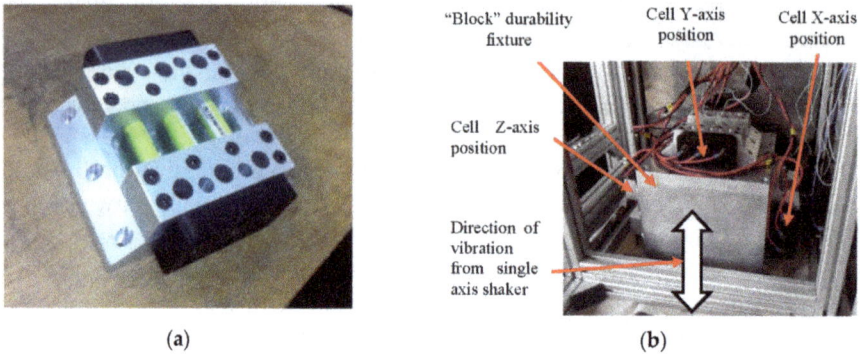

Figure 1. (**a**) Single test fixture (**b**) Assembled test fixture on shaker table with test positions.

2.2. Test Setup

The test environment used to mechanically induce vibration to the Li-ion cells is presented in Figure 2.

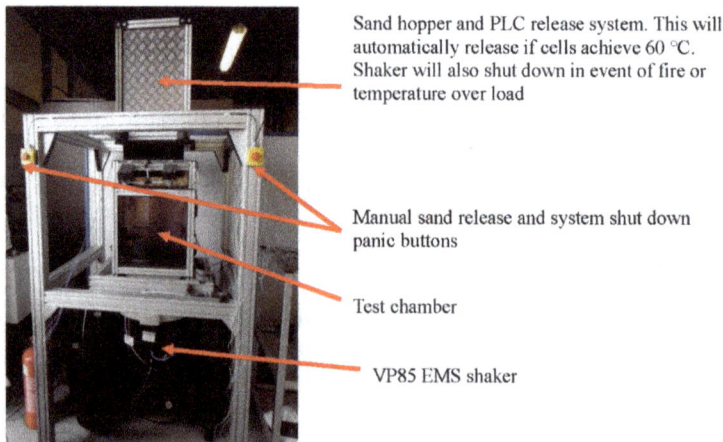

Figure 2. Test set up.

The test rig employs a 700 kgf, single axis, EMS table manufactured by Derritron (Hastings, UK; model number: VP85, serial number: 74). A LMS Scadas III (Leuven, Belgium; serial number: 23-4709-58) digital vibration controller was programmed with both the J2380 and WMG/MBK vibration profiles. To facilitate closed-loop vibration control, two single axis accelerometers (PCB 352C65, PCB Piezotronics Inc, Depew, NY, USA) were mounted at opposite sides of the durability fixture via HBM X60 adhesive (North Harrow, UK). A LabVIEW PXie-1075 chassis (National Instruments, Newbury, UK) was used with an integrated Ni-PXIe-8133 controller and input modules for 32 thermocouple sensors (Ni-PXIe-4353, National Instruments, Newbury, UK), 4 channels for accelerometer measurements (NI PXI-4462) and a multifunctional data module (NI PXIe-6363) is used for data acquisition. To mitigate against any potential risk of catastrophic cell failure during vibration testing, the EMS shaker table was installed within a blast proof enclosure. Integrated with the enclosure is a programmable logic controlled (PLC) fire extinguishing mechanism that would automatically activate if either a cell surface temperature greater than 60 °C or an increase in cell temperature of greater than 4 °C/s was observed. Within the test environment, K-type thermocouples and accelerometers are employed to provide suitable test accuracy and safety.

2.3. Rig and Fixture Pre-Testing Characterization

2.3.1. Response of EMS Shaker

As discussed within [42–44] prior to commencing any vibration study, a key requirement is to fully understand the frequency response of the EMS shaker to ensure that the armature assembly does not exhibit a resonance within the frequency range of interest. The vibration response of the EMS shaker was measured using a swept sine wave of amplitude 1 g_n (9.8 m/s^2) over a frequency range of 5–3700 Hz at 1 octave/minute prior to testing. Upon analyzing the response of the EMS shaker used in this study, no significant resonances were identified that would detrimentally impact the accuracy or reliability of the durability test programme.

2.3.2. Transmissibility of Fixtures

The primary requirement for durability testing is to ensure that the vibration profile demanded by the electronic controller is faithfully applied to the samples under test. This is achieved by designing the experimental fixture to maximize the transmissibility of the vibration energy from the EMS table to the sample and to concurrently minimize the cross-axis behavior of the durability fixture. Transmissibility is a comparison of the output signal to the input signal [45] and is determined by pre-test experimental evaluations of the fixture. At a transmissibility of unity, the output faithfully follows input [45]. To ensure a uniform transmission of acceleration from the vibration exciter, the fixture must carry the force to the test object with a minimum of loss and distortion. This is accomplished by ensuring the rigidity of the fixture so that the force is not deflected by the specimen load and that the fixture transfers motion with high fidelity [41,45,46]. Ideally, a dynamic test fixture couples the motion from the vibration shaker table to the specimen with zero distortion at all amplitudes and frequencies specified by the test procedure [45,47]. Practically, an ideal value of 1.0 over a wide test frequency cannot be met, therefore fixtures are characterized via swept sine resonance search evaluations prior to testing to ensure that no significant resonances occur in the three axis of the vibration fixture. The cross-axis behavior of the experimental set-up, (Figure 1b) was evaluated in accordance with BS EN 60,068 to ensure that "the maximum vibration amplitude in any axis perpendicular to the specified axis shall not exceed 50% of the specified amplitude up to 500 Hz or 100% for frequencies in excess of 500 Hz [44]".

To measure the vibration characteristics of the test fixture, accelerometers were placed in the X, Y and Z-axis of the assembled fixture, within or close to every cell mounting position. The fixture was excited in the Z-axis. The test samples were not installed into the fixture during the transmissibility investigation. Prior to conducting the transmissibility measurement activity several vibration test standards were consulted. It was noted that there is ambiguity within the regulatory and industrial guidelines with respect to assessing the suitability of fixtures for durability assessments. For example, standards such as the NASA GSFC-STD-7000A [48] and MIL-STD-810f [49] request that fixture transmissibility is assessed on the fixture in isolation. However BS 60,068 and DEF STAN 0035 (Part 3) [50] suggest testing both with and without the device under test installed.

Figure 3 shows the measured response of the fixture in the X, Y and Z-axis when excited in the Z-axis via a swept sine wave, of amplitude 1g_n over a frequency range from 5–800 Hz (800 Hz is peak frequency of durability profiles utilized within this study) at a rate of 1 octave per minute. The results show that the vibration responses measured, in all three axis, are within the limits specified by BS EN 60,068. Please note that the data presented in Figure 3 is a sample of the total data recorded for illustrative purposes.

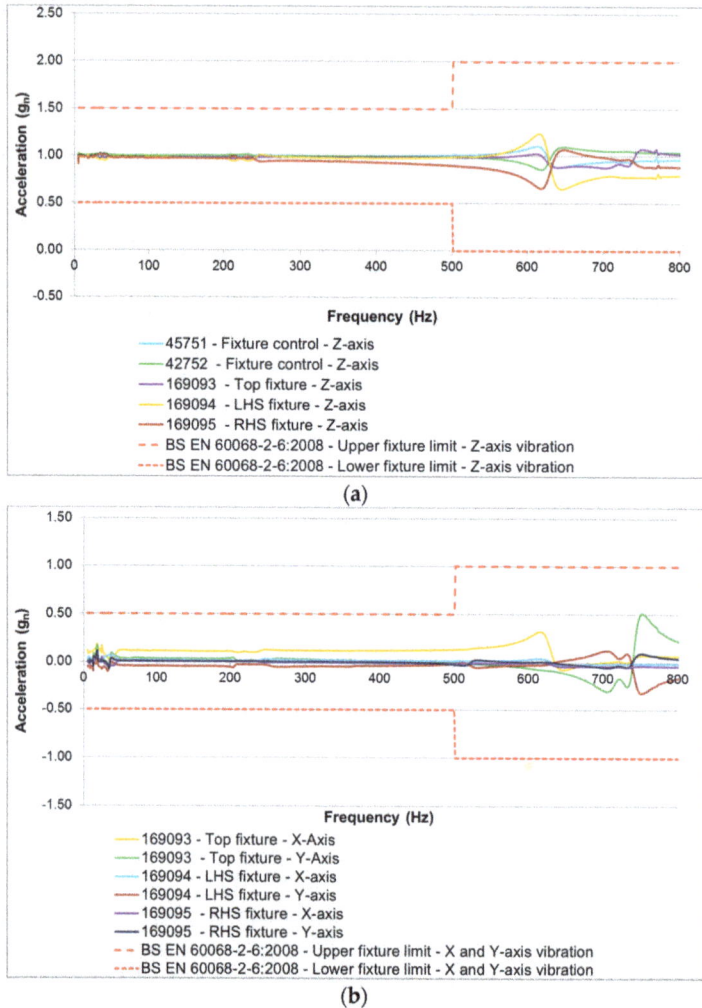

(a)

(b)

Figure 3. BS EN 60,068 resonance evaluation of 18,650 durability fixture: **(a)** Z-Axis of fixture **(b)** X and Y-Axis of fixture.

To mechanically characterize the cells at the start and end of test (discussed in Section 2.6.6), an additional "cell resonance search plate" fixture was fabricated to perform this assessment (shown in Figure 4a) which was designed to accept a three cell test fixture (as shown in Figure 1a). The assembled resonance search plate and three cell test fixture is shown in Figure 4b. This separate fixture plate was fabricated to allow for accurate natural frequency measurements outside the frequency range (>800 Hz) of the durability fixture assembly.

The resonance search plate fixture assembly as shown in Figure 4b was evaluated in accordance with BS EN 60,068 (without test samples installed) and was excited in the Z axis via a 1 g_n swept sine from 5 to 3700 Hz at a sweep rate of 1 octave/minute. To measure the vibration characteristics of the test fixture, accelerometers were placed in the X, Y and Z-axis of the assembled fixture, within or close to every cell mounting position. The resonance search plate with a single 18,650 three cell fixture installed met the requirements of BS EN 60,068 from 5 to 3700 Hz, however a 0.5 g_n resonance (which is within the limits of BS EN 60,068) was noted at 1500 Hz. The results from this assessment are shown in Figure 5.

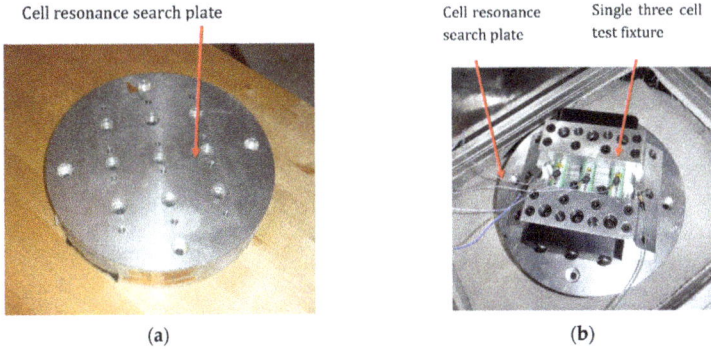

Figure 4. (**a**) Cell resonance search plate (**b**) Cell resonance search plate with single 18,650 three cell fixture installed on VP85 electromagnetic shaker (EMS).

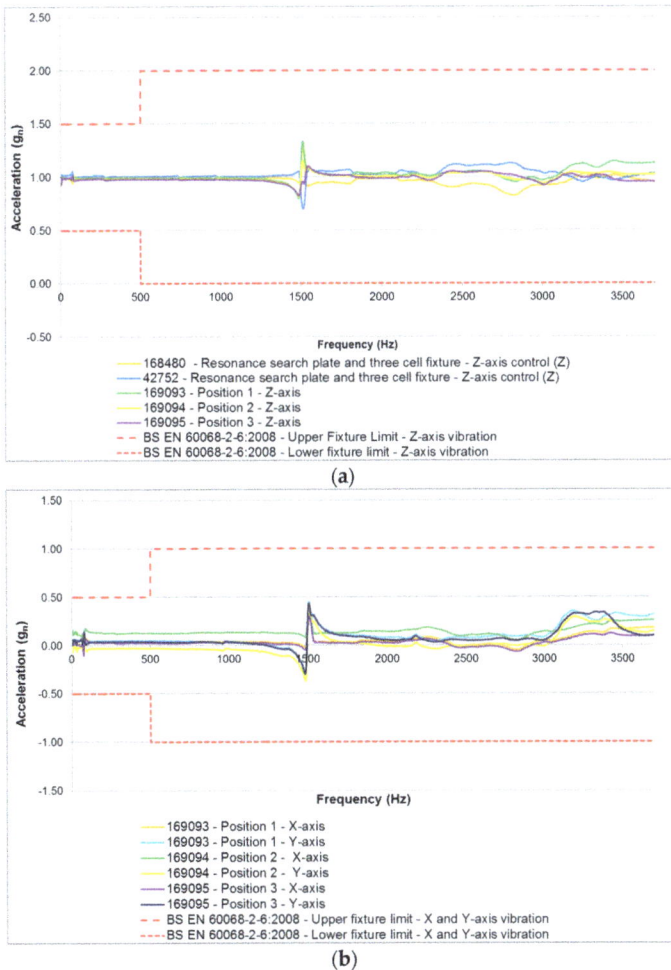

Figure 5. BS EN 60,068 Resonance evaluation of resonance search plate (**a**) Z-Axis of search plate (**b**) X and Y-Axis of search plate.

2.4. Test Samples

Twenty seven Samsung 2.2 Ah 18,650 cells (NMC) were evaluated. Each cell was pre-conditioned to a defined SOC prior to durability testing and allocated a test orientation with respect to the vehicle Z-axis. The details of sample preparation, cell SOC and cell orientation are defined in Table 1.

Table 1. Test sample information.

Sample No.	Test Profile	SOC (%)	Cell Orientation (Vehicle Axis: Cell Axis)
1	Control sample-In permanent storage	25%	Control
2	Control sample-Followed J2380 test samples	25%	Control
3	Control sample-Followed WMG/MBK profile test samples	25%	Control
4	Control sample-In permanent storage	50%	Control
5	Control sample-Followed J2380 test samples	50%	Control
6	Control sample-Followed WMG/MBK profile test samples	50%	Control
7	Control sample-In permanent storage	75%	Control
8	Control sample-Followed J2380 test samples	75%	Control
9	Control sample-Followed WMG/MBK profile test samples	75%	Control
10	J2380	25%	Z:Z
11	J2380	25%	Z:X
12	J2380	25%	Z:Y
13	J2380	50%	Z:Z
14	J2380	50%	Z:X
15	J2380	50%	Z:Y
16	J2380	75%	Z:Z
17	J2380	75%	Z:X
18	J2380	75%	Z:Y
19	WMG/MBK	25%	Z:Z
20	WMG/MBK	25%	Z:X
21	WMG/MBK	25%	Z:Y
22	WMG/MBK	50%	Z:Z
23	WMG/MBK	50%	Z:X
24	WMG/MBK	50%	Z:Y
25	WMG/MBK	75%	Z:Z
26	WMG/MBK	75%	Z:X
27	WMG/MBK	75%	Z:Y

A detailed explanation of the two test profiles defined in Table 1 are discussed in Section 2.5, whilst the test orientation is discussed in greater detail in Section 2.7.

2.5. Vibration Cycles

Nine cells were subjected to the vibration profile defined in the SAE J2380 standard and nine were subjected to WMG/MBK vibration profile. Both test specifications utilized random vibration profiles. A full explanation of the derivation of both profiles can be found in [11,36–38,51]. For completeness, these profiles are presented in Figure 6 and the vibration profile used for each cell is presented in Table 1.

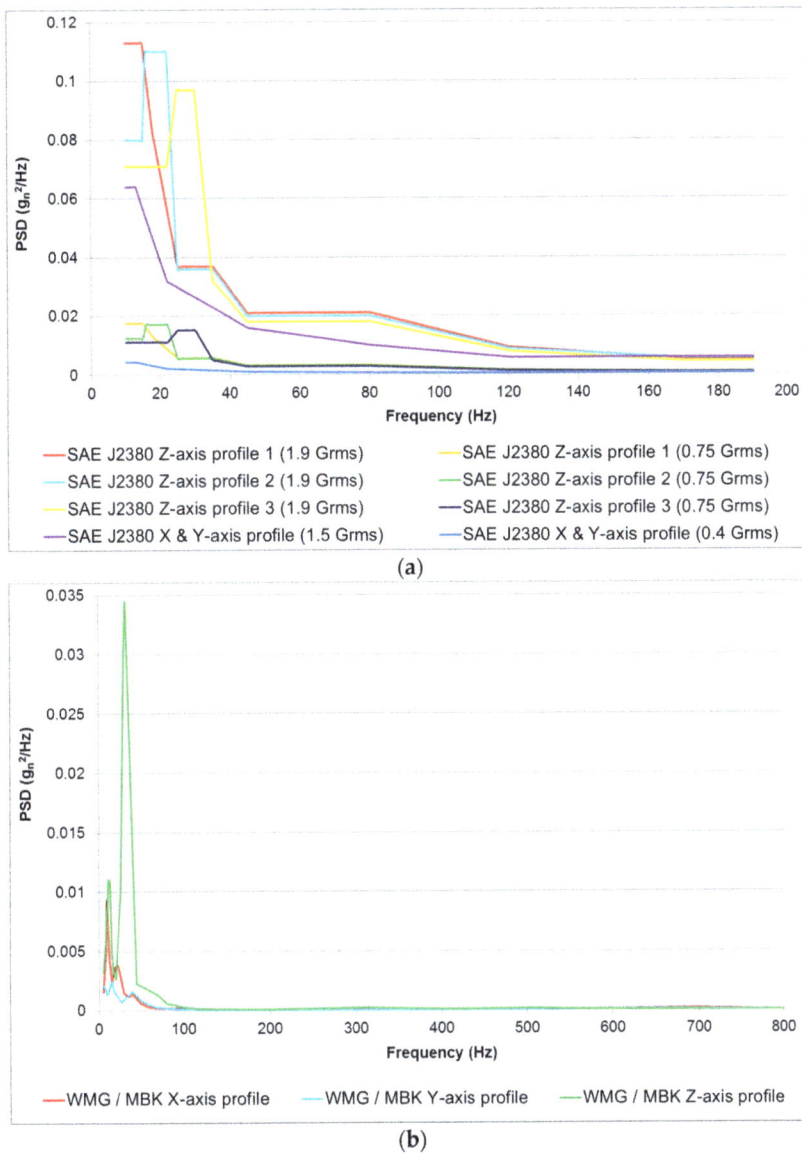

Figure 6. (a) SAE J2380 vibration power spectral density (PSD) profiles for testing samples 10 to 18; (b) Warwick Manufacturing Group/Millbrook Proving Ground WMG/MBK vibration PSD profiles for testing samples 19 to 27.

Whilst the WMG/MBK profile was developed through a previous study undertaken by the authors, SAE J2380 was selected as it is currently the only internationally recognized vibration test standard that has been correlated to 100,000 miles of road vehicle durability.

2.6. Mechanical and Electrochemical Testing

The following tests were performed on the cells at SOT and EOT (after vibration).

2.6.1. SOC Adjustment

The cell SOC was adjusted by fully charging the cells with a constant current of 1.1 A (C/3) to 4.2 V followed by a constant voltage phase at 4.2 V until the current fell to 0.05 A (C/65). At the end of charge, the cells were allowed to rest for 4 h prior to being discharged at 1C for 45, 30 and 15 min, to achieve a cell SOC of 25%, 50% or 75%, respectively. The cells were allowed to reach equilibrium for 4 h before the application of vibration energy.

2.6.2. 1C Capacity

The cells were fully charged using a constant current phase of 1.1 A (C/3) to 4.2 V followed by a constant voltage phase at 4.2 V until the current reduced to 0.05 A (C/65). The cells were allowed to rest for 4 h prior to being fully discharged at 1C to 2.75 V that represents the lower voltage threshold defined by the manufacturer. The energy extracted from the cells during the discharge was recorded as a measure of the 1C capacity.

2.6.3. Pulse Power

To determine the DC resistance of the cells (R_{DC}), a series of pulses was applied to the cells when conditioned to 50% SOC. Each current pulse was of 10 s in duration, with a magnitude of 20%, 40%, 60%, 80% and 100% of the cell's rated maximum discharge current. The maximum discharge current is defined by the manufacturer In the case of the Samsung 18,650, the maximum discharge current is specified as 4400 mA. A rest interval of 30 min was employed between consecutive pulses. DC resistance was calculated as described in Equation 2. V_{OCV} is the voltage prior the application of the current pulse (I_{max}), V_{10s} is the cell voltage at the end of the 10 s current pulse at I_{max}:

$$R_{DC} = \frac{(V_{OCV} - V_{10\,s})}{I_{max}} \tag{2}$$

2.6.4. Open Circuit Voltage (OCV)

The OCV of the cells under evaluation was measured with the cell isolated from any electrical load using a standard laboratory voltmeter. The OCV was recorded at the start and end of test. It was also recorded prior to moving the samples on the durability fixture.

2.6.5. Electrochemical Impedance Spectroscopy (EIS)

EIS data was recorded 4 h after the last pulse of the pulse power tests, as suggested by Barai et al. [52] and was performed at 50% SOC. The EIS measurement was carried out in a galvanostatic mode using a ModuLab® (Solartron, Leicester, UK) electrochemical system model 2100 A fitted with a 2 A booster and driven by Modulab® ECS software. The EIS spectra were collected within the frequency range of 10 mHz to 10 kHz using 10 frequency points per decade. The amplitude of the applied current was 200 mA (RMS). No DC current was superimposed on the RMS value. The commercially available Z view® software was employed to fit the EIS spectra to an equivalent circuit model (ECM) of the cell, thereby facilitating the quantification of key cell parameters such as: DC and charge transfer resistance.

2.6.6. Natural Frequency

The cells natural frequency was recorded at SOT and EOT to quantify the mechanical characteristics of each cell. Changes in natural frequency can indicate a change in material properties (such as stiffness) through mechanisms such as cracking or work hardening. The natural frequency of each cell was measured by fastening the respective cell to the EMS table (as illustrated in Figure 4) and applying a swept sine wave from 5 to 3700 Hz, of amplitude 1 g_n at a rate of 1 octave/min.

The response of the cell in relation to this 1 g_n excitation, was recoded via a lightweight, single axis, accelerometer (PCB 352A24, PCB Piezotronics Inc, Depew, NY, USA) mounted as shown in Figure 7. These were secured to the center of the cell using a petro wax adhesive. With a weight of only 0.8 g (1.9% additional mass for each cell), their inclusion within the experimental set-up was not deemed to have any significant impact on test accuracy through the addition of extra mass.

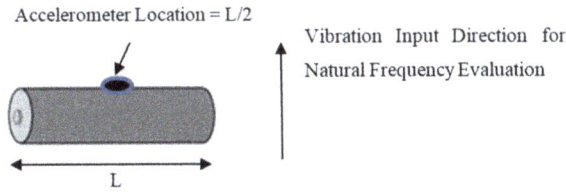

Figure 7. Location of cell accelerometer for natural frequency measurement via swept sine frequency sweep.

Two control accelerometers were secured at opposite ends at the top of the test fixture. Each control accelerometer was mounted close to the specimens. An averaging control strategy was employed during the natural frequency measurement. Data was recorded at 2.5 times the desired peak frequency in accordance with Nyquist rate guidelines [53]. With this test program, the peak desired frequency was the maximum achievable frequency of the VP85 EMS, defined by the manufacturer as 3.7 kHz. Therefore, accelerometer data was measured at a frequency greater than 9.25 kHz. Post natural frequency measurement each cell was allowed to rest for a minimum of 3 h before commencing the vibration durability test profiles. It noteworthy that no control samples were measured during the natural frequency assessment, this was to ensure that they were not subjected to any mechanical loading.

2.7. Test Procedure

The 27 cells presented in Table 1 were characterized as described in Section 2.6 (SOT characterization). They were then divided into three equal batches, comprising nine cells each. Batch 1 was subjected to the vibration profile defined in the SAE J2380 standard and batch 2 was subjected to WMG/MBK vibration profile. The remaining nine cells (batch 3) were defined as control samples. These were co-located within the same environmental conditions, but not subjected to any vibration loading. Both vibration tests comprise a vertical (Z-axis) profile in addition to vibration profiles defined for the horizontal plane (X-axis and Y-axis). As part of the experimental procedure, each profile is sequentially applied to the cells to achieve the desired 100,000 miles of representative EV life. For a complete execution of either J2380 or WMG/MBK, the three different combinations of vibration loads with respect to each cell orientation are defined below:

- Z:Z to X:X to Y:Y
- Z:X to X:Y to Y:Z
- Z:Y to X:Z to Y:X

Using the above notation, for each pair of letters, the first letter refers to the vehicle axis, whilst the second refers to cell orientation. For simplicity this paper identifies the cell orientation in relationship to the vertical (Z axis) of the vehicle. For example a cell that was subjected to the vibration sequence of Z:X to X:Y to Y:Z, is referred to as being evaluated in the Z:X orientation. Figure 8a illustrates the axis convention for the vehicle axis, whilst Figure 8b illustrates the axis convention for the 18,650 cell.

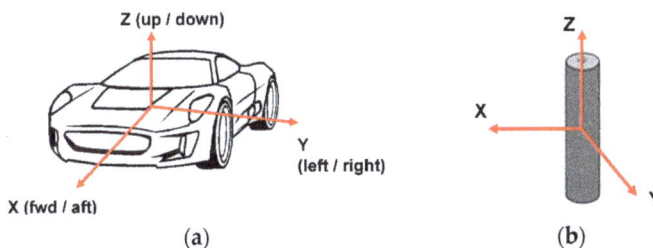

Figure 8. (a) Axis convention of vehicle vibration durability profiles, (b) Axis convention of cells.

Figure 9 presents the orientation of the nine cells mounted onto the durability fixture for the three orientation test conditions.

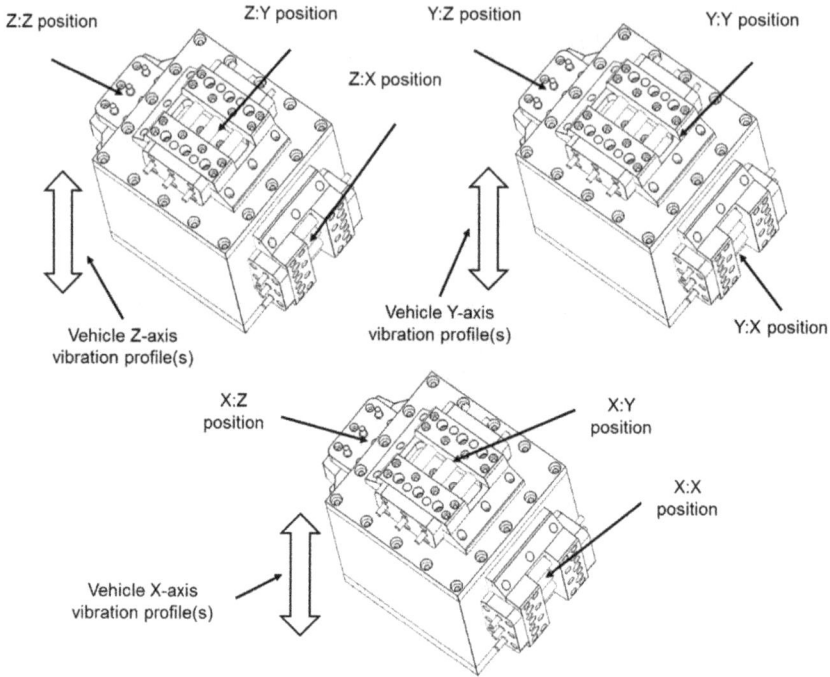

Figure 9. Experimental orientations and test positions on durability fixture.

Due to limited equipment availability, a single axis shaker was employed for the durability testing. Because the orientation of the EMS could not be changed, the cells had to be rotated on the durability fixture between X, Y and Z axis profile changes to achieve the correct loading. This test methodology is termed as not testing "with respect to gravity" and does not allow for changes in sample mass during the re-orientation of cells with respect to the input axis of vibration. While the authors believe that this limitation will not significantly impact the results, this limitation is discussed further in Section 5, where the authors propose to repeat the study using a multi-axis shaker to identify if any correlation does exists within the experimental data that can be attributed to experimental approach employed.

All testing was conducted within an air-conditioned room at a temperature of $21 \pm 5\,°C$. The closed loop application of the vibration profile was achieved by using an averaging control strategy, as defined within [54] which included ± 3 dB alarm limits and ± 6 dB abort limits. Once the cells were installed to the durability fixture and mounted onto the EMS table, the Z-axis vibration profile of either J2380 or WMG/MBK was applied first (Table 2). The calculation of Grms levels (defined in Tables 2–4) is discussed in [51]. On completion, the cells were left to stabilize for a minimum of 4 h.

The cells were then moved on the durability fixture to the corresponding vehicle X-axis and subjected to the X-axis vibration profile (Table 3).

Finally, the cells were repositioned on the durability fixture to facilitate the application of the vehicle Y-axis vibration profile (Table 4). At the end of the vibration profile, the cells were left to stabilize for 4 h prior to visual inspection.

The cells were then characterized at EOT as described in 2.5. The complete experimental procedure followed during this test programme is summerized in Figure 10.

Table 2. Z axis vibration profiles schedule.

SAE J2380 (Samples 10 to 18)		WMG/Millbrook (Samples 19 to 27)	
Profile Description and Grms Level	**Duration (HH:MM)**	**Profile Description and Grms Level**	**Duration (HH:MM)**
Subject cells to 9 min of Z-axis profile 1 at 1.9 Grms in the vertical orientation of the cells under assessment.	00:09	Subject cells to 150 h of Z-axis profile in the vertical axis orientation of the cells under assessment. Grms 0.67 for Z-axis.	150:00
Subject cells to 5 h and 15 min of Z-axis profile 1 at 0.75 Grms in the vertical axis orientation of the cells under assessment.	05:15		
Subject cells to 9 min of Z profile 2 at 1.9 Grms in the vertical axis orientation of the cells under assessment.	00:09		
Subject cells to 5 h and 15 min of Z-axis profile 2 at 0.75 Grms in the vertical axis orientation of the cells under assessment.	05:15		
Subject cells to 9 min of Z-axis profile 3 at 1.9 Grms in the vertical axis orientation of the cells under assessment.	00:09		
Subject cells to 5 h and 15 min of Z-axis profile 3 at 0.75 Grms in the vertical axis orientation of the cells under assessment.	05:15		
Total for Z Axis	16:12	Total for Z Axis	150:00

Table 3. X axis vibration profiles schedule.

SAE J2380 (Samples 10 to 18)		WMG/Millbrook (Samples 19 to 27)	
Profile Description and Grms Level	**Duration (HH:MM)**	**Profile Description and Grms Level**	**Duration (HH:MM)**
Subject cells to 5 min of longitudinal X & Y-axis profile at 1.5 Grms in the X axis orientation of the cells under assessment.	00:05	Subject cells to 150 h of X-axis profile in the X-axis orientation of the cells under assessment. Grms 0.384 for X-axis.	150:00
Subject cells to 19 h of longitudinal X & Y-axis profile at 0.4 Grms in the X axis orientation of the cells under assessment.	19:00		
Subject cells to 5 min of longitudinal X & Y-axis profile at 1.5 Grms in the X axis orientation of the cells under assessment.	00:05		
Subject cells to 19 h of longitudinal X & Y-axis profile at 0.4 Grms in the X axis orientation of the cells under assessment.	19:00		
Total for X Axis	38:10	Total for X Axis	150:00

Table 4. Y axis vibration profiles schedule.

SAE J2380 (Samples 10 to 18)		WMG/Millbrook (Samples 19 to 27)	
Profile Description and Grms Level	**Duration (HH:MM)**	**Profile Description and Grms Level**	**Duration (HH:MM)**
Subject cells to 5 min of longitudinal X & Y-profile at 1.5 Grms in the Y axis orientation of the cells under assessment.	00:05	Subject cells to 150 h of Y-axis profile in the Y-axis orientation of the cells under assessment. Grms 0.306 for Y-axis.	150:00
Subject cells to 19 h of longitudinal X & Y-profile at 0.4 Grms in the Y axis orientation of the cells under assessment.	19:00		
Subject cells to 5 min of longitudinal X & Y-profile at 1.5 Grms in the Y axis orientation of the cells under assessment.	00:05		
Subject cells to 19 h of longitudinal X & Y-profile at 0.4 Grms in the Y axis orientation of the cells under assessment.	19:00		
Total	38:10	Total	150:00

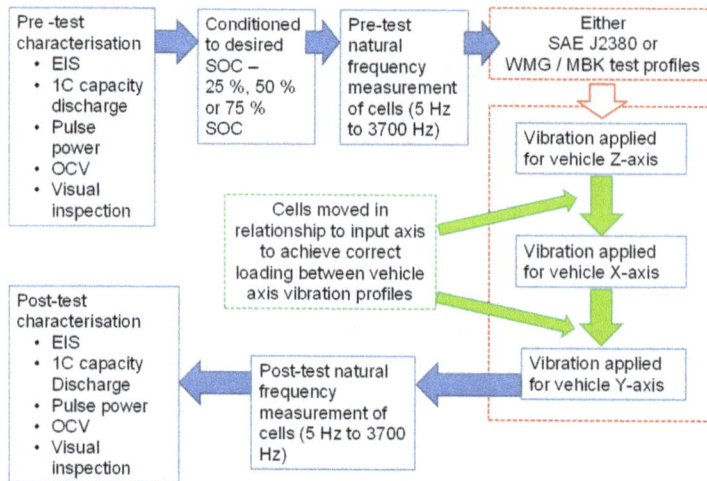

Figure 10. Schematic of test process.

3. Results

3.1. *Mechanical Characterization* via *Natural Frequency Measurements*

Tables 5 and 6 present the natural frequency and amplitude of the first resonant mode of each cell, respectively. The results show that every cell subjected to a vibration ageing profile, irrespective of the type of profile used exhibited a reduction in its natural frequency. Similarly, the amplitude of the first resonant mode is also affected by the vibration ageing cycle.

Table 5. Summary of change in frequency of observed first cell resonance for all test samples.

Sample No.	Test Profile	SOC%	Orientation	First Natural Frequency Greater Than 2 g_n (Hz)		
				SOT (Hz)	EOT (Hz)	Percentage Change (%)
10	J2380	25%	Z:Z	2579	1877	27.2
16	J2380	75%	Z:Z	3400	2834	16.6
11	J2380	25%	Z:X	2335	2039	12.7
15	J2380	50%	Z:Y	3200	2913	9.0
14	J2380	50%	Z:X	3189	2914	8.6
18	J2380	75%	Z:Y	2937	3165	7.8
23	WMG/MBK	50%	Z:X	1962	2097	6.9
17	J2380	75%	Z:X	1940	1816	6.4
13	J2380	50%	Z:Z	3209	3011	6.2
26	WMG/MBK	75%	Z:X	3354	3467	3.4
27	WMG/MBK	75%	Z:Y	3700	3584	3.1
21	WMG/MBK	25%	Z:Y	3061	2970	3.0
25	WMG/MBK	75%	Z:Z	3182	3156	0.8
19	WMG/MBK	25%	Z:Z	3641	3614	0.7
22	WMG/MBK	50%	Z:Z	3700	3674	0.7
24	WMG/MBK	50%	Z:Y	3061	3070	0.3
12	J2380	25%	Z:Y	2572	2579	0.3
20	WMG/MBK	25%	Z:X	3694	3700	0.2

	SOT	EOT
Standard deviation for J2380 samples (Hz)	488.77	522.61
Mean for J2380 samples (Hz)	2817.89	2572.00
Standard deviation for WMG/MBK samples (Hz)	558.97	514.99
Mean for WMG/MBK samples (Hz)	3261.67	3259.11

Table 6. Summary of change in amplitude of observed first cell resonance for all test samples.

Sample No.	Test Profile	SOC (%)	Orientation	Amplitude of First Natural Frequency Greater Than 2 g_n (g_n)		
				SOT (g_n)	EOT (g_n)	Percentage Change (%)
25	WMG/MBK	75%	Z:Z	1.62	2.91	79.6
18	J2380	75%	Z:Y	2.95	4.01	35.9
14	J2380	50%	Z:X	2.89	1.90	34.3
19	WMG/MBK	25%	Z:Z	5.03	3.46	31.2
26	WMG/MBK	75%	Z:X	2.96	2.21	25.3
27	WMG/MBK	75%	Z:Y	4.09	3.08	24.7
10	J2380	25%	Z:Z	3.32	2.51	24.4
20	WMG/MBK	25%	Z:X	4.52	3.61	20.1
23	WMG/MBK	50%	Z:X	2.61	2.30	11.9
15	J2380	50%	Z:Y	2.85	3.17	11.2
12	J2380	25%	Z:Y	1.99	1.82	8.5
22	WMG/MBK	50%	Z:Z	3.63	3.89	7.2
16	J2380	75%	Z:Z	3.31	3.51	6.0
11	J2380	25%	Z:X	2.02	1.91	5.4
24	WMG/MBK	50%	Z:Y	2.80	2.70	3.6
17	J2380	75%	Z:X	1.97	1.90	3.6
13	J2380	50%	Z:Z	2.34	2.30	1.7
21	WMG/MBK	25%	Z:Y	3.01	2.96	1.7
				SOT	EOT	
Standard deviation for J2380 samples (g_n)				0.55	0.81	
Mean for J2380 samples (g_n)				2.63	2.56	
Standard deviation for WMG/MBK samples (g_n)				1.06	0.57	
Mean for WMG/MBK samples (g_n)				3.36	3.01	

The change in natural frequency was previously attributed to a change in the material properties caused for example by internal cracking, delamination or fracture. The change in the amplitude of the acceleration at the natural frequency was found to indicate a change in the level component damping [12]. As none of the cells experienced any reduction in mass after either of the vibration ageing cycles, these results may indicate a reduction in the internal stiffness of each cell. More specifically, for the cells which underwent the J2380 vibration profiles, the cell orientated along the Z:Z axis and pre-conditioned at 25% SOC exhibited the greatest change in natural frequency, 27.2%. Likewise, the cell orientated along the Z:Y axis and pre-conditioned at 25% SOC exhibited the least change in natural frequency, 0.3%. In addition, it was observed that the cell positioned along the Z:Y axis and pre-conditioned at 75% SOC and the cell positioned along the Z:Z axis and pre-conditioned to 50% SOC showed the greatest and least change in amplitude, 35.9% and 1.7%, respectively. For the cells that underwent the WMG/MBK vibration profile, the cell placed along the Z:Z axis and pre-conditioned at 50% SOC exhibited the greatest change in natural frequency, 6.9%. Similarly, the cell orientated along the Z:X axis and pre-conditioned at 25% SOC exhibited the least change in natural frequency, 0.2%. Table 5 also shows that the shift in resonance frequency for the cells which underwent the WMG/MBK vibration profile is less pronounced than that observed for cells subjected to the J2380 vibration profile. Furthermore, the cell positioned along the Z:Z axis and pre-conditioned at 75% SOC and the cell positioned along the Z:Y axis and pre-conditioned to 25% SOC showed the greatest and least change in amplitude, 79.6% and 1.7%, respectively. The effect of cell orientation and cell SOC on the resonance frequency and the amplitude of the cells that underwent the J2380 and WMG/MBK vibration ageing cycles are summarized in Tables 7 and 8 respectively.

The data shows that the overall rankings for cell orientation and SOC are in reverse order when comparing both parameters. However, a larger spread of measurements for the change in amplitude of the first resonance than for the change in natural frequency can be observed. This may indicate that there are limitations in the measurement method employed to record these parameters. These may include the accuracy of the attachment of accelerometers to the cells or the amount of Petro wax used to attach the accelerometers to the cells. Equally, it may indicate an actual change in material properties. A more conclusive analysis that aims to determine the root cause of the results is discussed in Section 5.

Table 7. Assessment ranking of orientation by test.

Assessment	Test	Orientation Ranking by Assessment and Test Profile		
		Least Change	$\Longrightarrow$	Greatest Change
Electrical characterization				
Pulse power	J2380	Z:Z	Z:X	Z:Y
	WMG/MBK	Z:Y	Z:X	Z:Z
EIS (R_o)	J2380	Z:Z	Z:X	Z:Y
	WMG/MBK	Z:Y	Z:X	Z:Z
EIS (R_{CT})	J2380	Z:Z	Z:Y	Z:X
	WMG/MBK	Z:Z	Z:X	Z:Y
OCV	J2380	Z:Z	Z:X	Z:Y
	WMG/MBK	Z:X	Z:Y	Z:Z
Capacity—1C discharge	J2380	Z:X	Z:Y	Z:Z
	WMG/MBK	Z:Y	Z:X	Z:Z
Mechanical characterization				
Resonance (change in frequency)	J2380	Z:Y	Z:X	Z:Z
	WMG/MBK	Z:Z	Z:X	Z:Y
Resonance (change in amplitude)	J2380	Z:Z	Z:X	Z:Y
	WMG/MBK	Z:Y	Z:X	Z:Z

Table 8. Assessment ranking of state of charge (SOC) by test.

Assessment	Test	SOC Ranking by Assessment and Test Profile		
		Least Change	$\Longrightarrow$	Greatest Change
Electrical characterization				
Pulse power	J2380	50%	25%	75%
	WMG/MBK	75%	50%	25%
EIS (R_o)	J2380	25%	50%	75%
	WMG/MBK	50%	75%	25%
EIS (R_{CT})	J2380	25%	50%	75%
	WMG/MBK	75%	25%	50%
OCV	J2380	50% = 25%		75%
	WMG/MBK	25%	75%	50%
Capacity—1C discharge	J2380	25%	50%	75%
	WMG/MBK	50%	75%	25%
Mechanical characterization				
Resonance (change in frequency)	J2380	50%	75%	25%
	WMG/MBK	25%	75%	50%
Resonance (change in amplitude)	J2380	25%	75%	50%
	WMG/MBK	50%	25%	75%

3.2. Visual Inspection

At EOT, irrespective of the vibration profile, cell orientation or cell SOC, no mechanical damage was observed for any of the samples tested. None of the cells showed any change in dimensionality, leaked electrolyte or showed any sign of external fatigue (e.g., cracking). The one exception to this was a small external irregularity, approximately 3 mm in length, noted on the surface of cell sample 16, outside of the clamped area. It is believed that this is a manufacturing defect in the cell casing as the defect is not conducive to damage caused by excessive clamping loading.

3.3. Electrical Characterization

3.3.1. 1C Discharge Capacity

Table 9 presents the 1C discharge capacity for each cell at SOT and EOT. The results show a tendency for samples orientated in the Z:Z axis and pre-conditioned to 50% and 75% SOC to exhibit a higher capacity fade than other samples. The effect of cell orientation and cell SOC on the cell capacity for the cells which underwent the J2380 and WMG/MBK vibration ageing cycles are summarized in Tables 7 and 8 respectively.

Table 9. Summary of change in 1C discharge capacity performance of all test cells.

Sample No.	Test Profile	SOC (%)	Orientation	Cell Capacity at SOT (Ah)	Cell Capacity at EOT (Ah)	Percentage Change in Ah (%)
16	J2380	75%	Z:Z	2.21	1.94	−12.22
18	J2380	75%	Z:Y	2.21	2.06	−6.79
22	WMG/MBK	50%	Z:Z	2.18	2.08	−4.59
11	J2380	25%	Z:X	2.15	2.10	−2.33
15	J2380	50%	Z:Y	2.18	2.14	−1.83
13	J2380	50%	Z:Z	2.23	2.19	−1.79
26	WMG/MBK	75%	Z:X	2.10	2.07	−1.43
17	J2380	75%	Z:X	2.15	2.13	−0.93
20	WMG/MBK	25%	Z:X	2.12	2.12	0.00
19	WMG/MBK	25%	Z:Z	2.11	2.11	0.00
21	WMG/MBK	25%	Z:Y	2.09	2.09	0.00
10	J2380	25%	Z:Z	2.19	2.19	0.00
27	WMG/MBK	75%	Z:Y	2.09	2.09	0.00
12	J2380	25%	Z:Y	2.18	2.19	0.46
14	J2380	50%	Z:X	2.15	2.17	0.93
25	WMG/MBK	75%	Z:Z	2.07	2.10	1.45
23	WMG/MBK	50%	Z:X	2.05	2.13	3.90
24	WMG/MBK	50%	Z:Y	1.92	2.12	10.42
8	J2380	75%	Control	2.16	2.06	−4.63
2	J2380	25%	Control	2.20	2.13	−3.18
9	WMG/MBK	75%	Control	2.13	2.12	−0.47
3	WMG/MBK	25%	Control	2.14	2.14	0.00
5	J2380	50%	Control	2.18	2.19	0.46
6	WMG/MBK	50%	Control	1.92	2.09	8.85
				SOT	**EOT**	
Standard deviation for J2380 samples (Ah)				0.03	0.08	
Mean for J2380 samples (Ah)				2.18	2.12	
Standard deviation for WMG/MBK samples (Ah)				0.07	0.02	
Mean for WMG/MBK samples (Ah)				2.08	2.10	

3.3.2. Pulse Power Performance

Table 10 presents the DC resistance of the cells at SOT and EOT. It can be seen that irrespective of the vibration profile employed (including variations in cell SOC and orientation) each cell exhibits an increase in internal resistance. As discussed within [24], this potentially indicates a reduction in contact area within the current collectors or possible internal fatigue of components within the cell due to the mechanical load.

Sample 16 displays the greatest variation in resistance with a 128.1% increase from SOT to EOT. Conversely, sample 27 shows the least amount of increase (17.4%). Interestingly, both cells were pre-conditioned to 75% SOC. However; the vibration profiles employed were different, as was their respective orientation when mechanically loaded. Similarly, nine out of 10 of the cells that exhibit the greatest rise in resistance were pre-conditioned to either 25% or 75% SOC. This result correlates well with Section 3.3.1, that identified the same trend between cell SOC.

Table 10. Summary of change in pulse power performance resistance.

Sample No.	Test Profile	SOC (%)	Orientation	DC Resistance (SOT) (mΩ)	DC Resistance (EOT) (mΩ)	Percentage Change in DC Resistance between SOT and EOT (%)
16	J2380	75%	Z:Z	74.32	169.55	128.1
17	J2380	75%	Z:X	73.41	162.50	121.4
12	J2380	25%	Z:Y	72.73	151.59	108.4
20	WMG/MBK	25%	Z:X	72.27	142.95	97.8
25	WMG/MBK	75%	Z:Z	73.18	142.95	95.3
15	J2380	50%	Z:Y	71.82	138.18	92.4
11	J2380	25%	Z:X	72.27	131.82	82.4
18	J2380	75%	Z:Y	73.18	131.36	79.5
19	WMG/MBK	25%	Z:Z	72.27	126.59	75.2
21	WMG/MBK	25%	Z:Y	72.05	121.36	68.5
10	J2380	25%	Z:Z	71.14	117.05	64.5
14	J2380	50%	Z:X	72.73	116.36	60.0
22	WMG/MBK	50%	Z:Z	72.73	115.00	58.1
24	WMG/MBK	50%	Z:Y	72.05	109.32	51.7
13	J2380	50%	Z:Z	70.45	102.27	45.2
23	WMG/MBK	50%	Z:X	71.82	93.64	30.4
26	WMG/MBK	75%	Z:X	73.64	90.00	22.2
27	WMG/MBK	75%	Z:Y	73.18	85.91	17.4
5	J2380	50%	Control	75.23	86.14	14.5
2	J2380	25%	Control	72.27	73.64	1.9
3	WMG/MBK	25%	Control	71.14	71.82	1.0
6	WMG/MBK	50%	Control	72.05	72.05	0.0
9	WMG/MBK	75%	Control	73.18	72.73	−0.6
8	J2380	75%	Control	74.09	73.41	−0.9

		SOT	EOT
Standard deviation for J2380 samples (mΩ)		1.19	22.35
Mean for J2380 samples (mΩ)		72.45	135.63
Standard deviation for WMG/MBK samples (mΩ)		0.63	21.48
Mean for WMG/MBK samples (mΩ)		72.58	114.19

For the cells that underwent the J2380 vibration profile, the cell orientated along the Z:Z axis and pre-conditioned at 75% SOC exhibited the greatest change in DC resistance, 128.1%. Likewise, the cell orientated along the Z:Z axis and pre-conditioned at 50% SOC exhibited the least change in DC resistance, 17.4%. Moreover, for the cells which underwent the WMG/MBK vibration profile, the cell placed along the Z:X axis and pre-conditioned at 25% SOC exhibited the greatest change in DC resistance, 97.8%. Similarly, the cell orientated along the Z:Y axis and pre-conditioned at 75% SOC exhibited the least change in DC resistance, 45.2%. The effect of cell orientation and cell SOC on the cell DC resistance for the cells which underwent the J2380 and WMG/MBK vibration ageing cycles are summarized in Tables 7 and 8.

3.3.3. OCV

Table 11 presents the measured OCV for each cell both at SOT and EOT. None of the tested cells displayed any significant change in OCV, irrespective of the vibration profile, the cell SOC or cell orientation employed. The voltage difference recorded is within the tolerance of the error of the equipment. This supports the results presented in [33,34] that also noted that OCV is not adversely affected by vibration loading.

Table 11. Start and end of test open-circuit voltage (OCV) measurements of all cells evaluated.

Sample No.	Test Profile	SOC	Orientation	Voltage (V)		
				SOT	EOT	Percentage Change (%)
27	WMG/MBK	75%	Z:Y	3.882	3.891	0.23
24	WMG/MBK	50%	Z:Y	3.658	3.666	0.22
25	WMG/MBK	75%	Z:Z	3.878	3.885	0.18
22	WMG/MBK	50%	Z:Z	3.663	3.670	0.19
23	WMG/MBK	50%	Z:X	3.661	3.668	0.19
3	WMG/MBK	25%	Control	3.588	3.595	0.2
19	WMG/MBK	25%	Z:Z	3.584	3.590	0.17
20	WMG/MBK	25%	Z:X	3.587	3.590	0.08
26	WMG/MBK	75%	Z:X	3.883	3.886	0.08
18	J2380	75%	Z:Y	3.897	3.899	0.05
5	J2380	50%	Control	3.678	3.676	0.05
21	WMG/MBK	25%	Z:Y	3.581	3.580	0.03
16	J2380	75%	Z:Z	3.894	3.895	0.03
17	J2380	75%	Z:X	3.895	3.896	0.03
2	J2380	25%	Control	3.599	3.598	0.03
15	J2380	50%	Z:Y	3.674	3.675	0.03
6	WMG/MBK	50%	Control	3.668	3.669	0.03
11	J2380	25%	Z:X	3.591	3.590	0.03
8	J2380	75%	Control	3.897	3.897	0
9	WMG/MBK	75%	Control	3.889	3.889	0
13	J2380	50%	Z:Z	3.674	3.674	0
14	J2380	50%	Z:X	3.674	3.674	0
10	J2380	25%	Z:Z	3.599	3.599	0
12	J2380	25%	Z:Y	3.598	3.598	0
				SOT	EOT	
Standard deviation for J2380 samples (V)				0.13	0.14	
Mean for J2380 samples (V)				3.72	3.72	
Standard deviation for WMG/MBK samples (V)				0.13	0.13	
Mean for WMG/MBK samples (V)				3.71	3.71	

3.3.4. EIS

Tables 12 and 13 show the ohmic resistance (R_o) and the charge transfer resistance (R_{CT}) of the cells at SOT and EOT as measured by EIS. In addition, comparing Tables 10 and 12 it can be seen that the increase in R_{DC} of the cell is accompanied by an increase in R_o. Though the magnitude may differ from one technique to another, this concurs with the origin of these parameters. A complete explanation of EIS results is beyond the scope of this study and is already well documented in a number of academic and educational texts [55,56]. Figure 11 presents typical Nyquist plot of the cells pre and post vibration test.

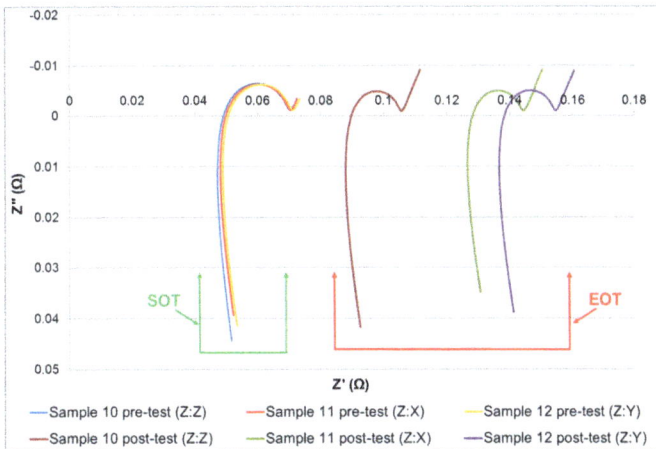

Figure 11. Typical electrochemical impedance spectroscopy (EIS) pre and post test results.

Table 12. electrochemical impedance spectroscopy (EIS) ohmic resistance (R_o) results for all tested samples.

Sample No.	Test Profile	SOC	Orientation	SOT (mΩ)	EOT (mΩ)	Percentage Change (%)
16	J2380	75%	Z:Z	*	*	*
19	WMG/MBK	25%	Z:Z	46.7	167.1	257.82
15	J2380	50%	Z:Y	46.4	164.5	254.53
25	WMG/MBK	75%	Z:Z	46.8	155.0	231.20
20	WMG/MBK	25%	Z:X	46.9	153.5	227.29
17	J2380	75%	Z:X	46.2	143.1	209.74
18	J2380	75%	Z:Y	46.5	141.2	203.66
12	J2380	25%	Z:Y	47.7	135.5	184.07
11	J2380	25%	Z:X	46.9	126.0	168.66
26	WMG/MBK	75%	Z:X	47.0	114.3	143.19
14	J2380	50%	Z:X	47.3	114.2	141.44
21	WMG/MBK	25%	Z:Y	46.1	102.2	121.69
23	WMG/MBK	50%	Z:X	46.4	90.0	93.97
10	J2380	25%	Z:Z	45.9	86.9	89.00
13	J2380	50%	Z:Z	46.0	84.0	82.61
22	WMG/MBK	50%	Z:Z	50.0	84.9	69.80
24	WMG/MBK	50%	Z:Y	46.8	66.8	42.74
27	WMG/MBK	75%	Z:Y	47.0	64.9	38.09
5	J2380	50%	Control	49.6	60.8	22.58
2	J2380	25%	Control	46.2	49.3	6.71
9	WMG/MBK	75%	Control	46.3	47.5	2.60
3	WMG/MBK	25%	Control	46.7	47.3	1.40
6	WMG/MBK	50%	Control	46.2	46.8	1.30
8	J2380	75%	Control	47.5	47.5	0.00

	SOT	EOT
Standard deviation for J2380 samples (mΩ)	0.64	28.05
Mean for J2380 samples (mΩ)	46.61	124.43
Standard deviation for WMG/MBK samples (mΩ)	1.13	39.02
Mean for WMG/MBK samples (mΩ)	47.08	110.97

* = No data available due to cell issue.

Table 13. EIS charge transfer resistance (R_{CT}) results for all tested samples.

Sample No.	Test Profile	SOC	Orientation	SOT (mΩ)	EOT (mΩ)	Percentage Change (%)
16	J2380	75%	Z:Z	*	*	*
18	J2380	75%	Z:Y	23.90	15.50	35.15
21	WMG/MBK	25%	Z:Y	24.08	15.75	34.59
22	WMG/MBK	50%	Z:Z	24.50	16.10	34.29
23	WMG/MBK	50%	Z:X	23.73	15.73	33.71
24	WMG/MBK	50%	Z:Y	23.90	16.06	32.80
19	WMG/MBK	25%	Z:Z	22.43	15.67	30.14
17	J2380	75%	Z:X	24.50	17.50	28.57
20	WMG/MBK	25%	Z:X	21.64	15.47	28.51
26	WMG/MBK	75%	Z:X	22.90	16.43	28.25
27	WMG/MBK	75%	Z:Y	22.56	16.20	28.19
14	J2380	50%	Z:X	24.70	18.20	26.32
13	J2380	50%	Z:Z	23.10	17.40	24.68
15	J2380	50%	Z:Y	23.70	18.00	24.05
25	WMG/MBK	75%	Z:Z	21.50	16.51	23.21
10	J2380	25%	Z:Z	23.72	18.70	21.16
11	J2380	25%	Z:X	23.00	18.23	20.74
12	J2380	25%	Z:Y	22.80	19.10	16.23
9	WMG/MBK	75%	Control	28.89	16.38	43.30
2	J2380	25%	Control	28.80	16.70	42.01
6	WMG/MBK	50%	Control	25.44	15.73	38.17
8	J2380	75%	Control	24.60	15.70	36.18
3	WMG/MBK	25%	Control	22.51	15.25	32.25
5	J2380	50%	Control	24.40	17.50	28.28

	SOT	EOT
Standard deviation for J2380 samples (mΩ)	1.15	0.91
Mean for J2380 samples (mΩ)	23.04	16.40
Standard deviation for WMG/MBK samples (mΩ)	0.70	1.36
Mean for WMG/MBK samples (mΩ)	23.59	17.27

* = No data available due to cell issue.

Table 12 highlights that irrespective of the vibration profile, SOC or cell orientation, all the cells exhibit a significant increase in R_o at EOT. For the cells that underwent the J2380 vibration profile, the cell orientated along the Z:Y axis and pre-conditioned at 50% SOC exhibited the greatest change in R_o, 254.53%. Moreover, the cell orientated along the Z:Z axis and pre-conditioned at 50% SOC exhibited the least change in R_o, 82.61%. Similarly, for the cells that underwent the WMG/MBK vibration profile, the cell positioned along the Z:Z axis and pre-conditioned at 25% SOC exhibited the greatest change in R_o, 257.82%. Likewise, the cell placed along the Z:Y axis and pre-conditioned at 75% SOC exhibited the least change in R_o, 38.09%. The increase in R_o was found to originate from an increase in cell contact resistance or delamination of the material layers [55,56]. Post vibration, it was observed that cell sample 16 could not undergo an EIS procedure. This is assumed to relate to internal fatigue damage within the cell. Further investigation into the exact nature of the failure mode, is beyond the scope of this study, but is discussed further in Section 5.

Table 13 shows that all cells that underwent vibration testing show a similar decrease in R_{ct} as the control samples. Consequently, it suggests that this parameter is unaffected by vibration. The effect of cell orientation and SOC on the measured values of R_o and R_{ct} is summarized in Tables 7 and 8 respectively.

4. Discussion

The primary conclusion from this study is that both the electrical performance and the mechanical properties of the Li-ion cells are affected by exposure to vibration energy that is commensurate with a typical vehicle life. Experimental data suggests that the rate of degradation is not uniform and varies considerably with respect to cell SOC and orientation to the applied axis of vibration. Further, this investigation highlights that even cells that have comparable characteristics at SOT, key measure of performance, such as impedance, diverge considerably after the application of vibration energy. However, the results do not show a consistent trend. Consequently at this stage of the research, the magnitude and spread of that performance change is unpredictable. This is highlighted by Table 14.

Table 14. Comparison of cell performance ranking by post-test assessment.

Sample No.	Test Profile	SOC	Orien-tation	Electrical Characterization					Mechanical Characterization	
				Pulse Power	EIS: R_o	EIS: R_{CT}	OCV	Capacity	Resonance Frequency	Resonance Amplitude
10	J2380	25%	Z:Z	11	14	16	15	12	1	7
11	J2380	25%	Z:X	7	9	17	13	4	3	14
12	J2380	25%	Z:Y	3	8	18	15	14	17	11
13	J2380	50%	Z:Z	15	15	13	15	6	9	17
14	J2380	50%	Z:X	12	11	12	15	15	5	3
15	J2380	50%	Z:Y	6	3	14	13	5	4	10
16	J2380	75%	Z:Z	1	1	1	10	1	2	13
17	J2380	75%	Z:X	2	6	8	10	8	8	16
18	J2380	75%	Z:Y	8	7	2	9	2	6	2
19	WMG/MBK	25%	Z:Z	9	2	7	6	10	14	4
20	WMG/MBK	25%	Z:X	4	5	9	7	9	18	8
21	WMG/MBK	25%	Z:Y	10	12	3	10	11	12	18
22	WMG/MBK	50%	Z:Z	13	16	4	3	3	15	12
23	WMG/MBK	50%	Z:X	16	13	5	3	17	7	9
24	WMG/MBK	50%	Z:Y	14	17	6	2	18	16	15
25	WMG/MBK	75%	Z:Z	5	4	15	3	16	13	1
26	WMG/MBK	75%	Z:X	17	10	10	7	7	10	5
27	WMG/MBK	75%	Z:Y	18	18	11	1	13	11	6

Ranking Key:

Greatest reduction in performance												Least reduction in performance		
1	2	3 4	5	6	7	8	9	10	11	12	13	14	15 16 17	18

The table shows that a series of complex interactions are potentially triggered whilst the cells are undergoing a vibration load that activates several failure modes and/or degradation mechanisms. The implications of these observations are further explored in Section 5. The remainder of this section discusses further the results obtained and highlights the implications of this research for the design of a RESS for future EV applications.

4.1. Impact of Cell Orientation

Table 7 collates the electromechanical EOT results and highlights, for each test-type, the individual cell ranking with respect to effect of vibration cycle on cell orientation. For cells subjected to the J2380 vibration profiles, the greatest overall change in electrical performance was experienced by the cell placed in the Z:Y orientation. Similarly, for the WMG/MBK dataset, the results suggest that cells orientated in the Z:Z axis experienced the greatest overall amount of electrical performance degradation. This cell behavior is also shown in the amplitude of the first resonance frequency. However, it is not present in the shift of the natural frequency of the cells. The differences in cell behavior, when exposed to the different vibration profiles, may be attributed to the difference in acceleration levels within the profiles. This in turn is related to the amount of time compression applied to synthesis the vibration standard from measured vehicle data [36]. The evidence tends to suggest that a correlation may exist between electrical performance degradation and a change in the mechanical properties of the cell post vibration.

In relation to the study discussed in [24] the WMG/MBK tested samples correlated well with the orientation conclusion from this study in that samples oriented in the Z:Z axis displayed a greater amount of performance decrease. However it must be noted, that the samples within [24] only accumulated vibration in one axis throughout the whole test, when this study has applied vibration in all three axis of the cell in a sequential fashion and therefore is more representative of the accumulation of damage that an automotive battery cell would achieve. This sequential axis vibration damage accumulation may also explain why the horizontally oriented samples in [24] were relatively unaffected by the application of vibration.

4.2. Impact of Cell State of Charge

Table 8 collates the electromechanical EOT results and presents the ranking with respect to the different values of SOC used to pre-condition each cell prior to subjecting them to the different vibration excitations. The results for the J2380 vibration profile show that the cells pre-conditioned to 75% SOC experienced the greatest level electrical performance degradation. Similarly, the results obtained from the WMG/MBK profile highlight that the cells pre-conditioned to 25% SOC displayed the greatest electrical performance degradation. Conversely, the cells pre-conditioned to 50% SOC exhibited the lowest levels of electrical performance degradation. It has previously been shown that a potential reason for this difference can be attributed to the changes that occur within the mechanical structure of the cell at the different levels of SOC [57,58]. However Table 14 indicates that no correlation can be established between the measured data for changes in mechanical properties and the degradation in electrical performance post vibration testing. This initial conclusion is supported by related research that reported the difficulty in correlating electromechanical ageing mechanisms [24,58–62].

4.3. Implications for Vehicle Design

The results from this study show that both the electrical performance and the mechanical properties of Li-ion cells can be affected by exposing the cell to vibration energy that is representative of a typical vehicle life. Whilst this is evident from the data presented, the underlying causality is not yet clear. As a result, it is not possible to quantify the relationship that defines cell ageing caused by vibration excitation. Irrespective of this limitation, both the electrical and mechanical data show that cells subject to vibration have a much greater spread in the internal resistance, energy capacity and natural frequency. Managing this diversity may potentially drive further complexity in the systems engineering functions required to scale individual cells into a complete RESS. A number of articles discuss the need to minimize cell-to-cell variations within the system as a mean to reduce the differential current flows and heat generation with the pack. This research highlights that even for a RESS that is initially well designed; the impact of vibration-induced ageing may require greater levels of cell balancing and thermal management.

The results summarized in Tables 7, 8 and 14 highlight that both the SOC and orientation are as important parameters to consider when designing a RESS as the contribution of the vibration induced profile. It is expected that variations in SOC within the RESS will be observed, especially for an EV, where a large depth of discharge (DOD) is required to maximize vehicle range. Consequently, SOC may be a parameter that engineers consider more greatly than orientation. However, to maximize the volumetric energy density and minimize the footprint of the RESS, engineers may need to account for the impact of cell orientation on the performance of the RESS. Consequently, the authors suggest that as part of the technology selection process, OEMs should study the susceptibility of the chosen cells to mechanically induced vibration profiles at different SOC and cell orientation to mitigate their effects through improved system design.

5. Further Work

One of the limitations of the methodology employed within this study is that electrical and mechanical characterization data was only measured at SOT and EOT. As a result, no discussion or conclusions can be made about the rate of degradation throughout the vehicle's life. It is recommended that a future study should characterize the cells at intermediate points during the test programme, e.g., intervals representative of 10,000 miles of vehicle use. This would facilitate further investigation into both the absolute value of degradation, but also the expected in-service rate of capacity and power fade over the life of the vehicle.

The experimental approach may also be improved, by revising the derivation of the vibration profile employed to exercise the cells. Both the J2380 standard and the WMG/MBK profiles are derived from real-world automotive data. Derivation of the WMG/MBK profile, as discussed within [36,37], used vibration data recorded directly from the battery packs of commercially available EVs. This vibration may not however directly correlate to that observed within the battery assembly, since cell restraints and packaging may induce further resonant modes and damping. Further research should measure directly the in-pack vibration. If significant differences exist between this data and that recorded externally to the vehicle RESS, then a new durability profile should be synthesized and the research repeated.

The results collected from this study imply that the rate of cell degradation is not uniform and varies considerably with cell orientation (relative to the axis of vibration) and SOC. However, given the limited dataset employed for this initial study, definitive conclusions regarding the underlying causality between the different ageing mechanisms cannot be made. The authors believe that these initial results warrant further research. Firstly, using novel cell imaging and autopsy methods, as discussed within [24], to better quantify the changes that occur within the material composition and structure of the cell post vibration. Secondly, to reduce the potential impact of cell-to-cell variations, by expanding the scope of the experimental study to encompass a greater number of cells of a given type. Expanding the experimental programme should also include using cells from a broader cross-section of manufacturers and chemistries. This will identify if the experimental results presented here are transferable to other cell technologies.

Due to the equipment availability at the time of testing, the experiment was conducted using a single axis EMS. As a result, the cells were not tested with respect to gravity. This could have caused unrepresentative loading due to the effects of mass loading associated with rotating the samples on the durability fixture. It is therefore recommended that a future experiment is conducted using either a multi-axis shaker table or single axis EMS with slip table capability so that the samples are evaluated with respect to gravity.

6. Conclusions

Both vibration profiles synthesized to represent 100,000 miles of vehicle operation resulted in a performance decrease within the tested Samsung 2.2 Ah 18,650 cells. However the two different vibration profiles of SAE J2380 and WMG/MBK resulted in two different results with respect to the effect of SOC and cell orientation. Of the samples evaluated to SAE J2380, cells in the Z:Z orientation displayed the least amount of degradation, whilst cells in the Z:Y orientation displayed the greatest. Whilst samples evaluated to the Z:X and Z:Z orientation displayed the least and greatest amount of degradation when exposed to the WMG/MBK profile, respectively. Of the samples evaluated to SAE J2380, items conditioned to 75% SOC displayed the greatest degradation, whilst WMG/MBK, items conditioned to 25% SOC displayed the greatest degradation. Samples conditioned to 50% SOC typically displayed the least degradation regardless of the test profile.

In conclusion, the experimental results presented highlight the potential for key electrical and mechanical properties within the cell to diverge, over time, due to the application of vibration energy that is consummate with a typical road vehicle life. Unless this phenomenon is well understood at the design stage of the vehicle, it may drive further complexity into design of the RESS in addition to causing in-service warranty claims. At this stage, the underlying causality between the application of vibration energy and cell SOC and orientation are not fully understood. Defining these relationships is the focus of on-going research within the University. For example; by using novel cell imaging and autopsy methods to quantify changes in material composition and structure. Expanding the experimental programme to also include cells of different form-factor and chemistry will identify if the experimental results presented here are transferable to other cell technologies.

Acknowledgments: The research presented within this paper is supported by the Engineering and Physical Science Research Council (EPSRC—EP/I01585X/1) through the Engineering Doctoral Centre in High Value, Low Environmental Impact Manufacturing. The research was undertaken in collaboration with the WMG Centre High Value Manufacturing Catapult (funded by Innovate UK) and Jaguar Land Rover. The authors would also like to express their gratitude to Millbrook Proving Ground Ltd. (Component Test Laboratory) fortheir support and advice throughout the test program.

Author Contributions: James Michael Hooper—Primary researcher and lead author. James Marco—Academic research supervision and co-author. Gael Henri Chouchelamane—Experimental researcher (electrical characterisation) and co-author. Christopher Lyness—Industrial research support and peer-review.

Conflicts of Interest: The authors declare no conflict of interest.

References

1. Jackson, N. Technology road map, R & D agenda and UK capabilities. In *Cenex Low Carbon Vehicle Show 2010*; Automotive Council UK: Bedford, UK, 2010; pp. 1–16.
2. Parry-Jones, R. *Driving Success—A Strategy for Growth and Sustainability in the UK Automotive Sector*; Automotive Council UK: London, UK, 2013; pp. 1–87.
3. Day, J. Johnson Controls' Lithium-Ion Batteries Power Jaguar Land Rover's 2014 Hybrid Range Rover. Available online: http://johndayautomotiveelectronics.com/johnson-controls-lithium-ion-batteries-power -2014-hybrid-range-rover/ (accessed on 17 February 2015).
4. Rawlinson, P.D. Integration System for a Vehicle Battery Pack. U.S. Patent 20120160583 A1, 28 June 2012.
5. Berdichevsky, G.; Kelty, K.; Straubel, J.; Toomre, E. *The Tesla Roadster Battery System*; Tesla Motors: Palo Alto, CA, USA, 2007; pp. 1–5.
6. Kelty, K. *Tesla—The Battery Technology behind the Wheel*; Tesla Motors: Palo Alto, CA, USA, 2008; pp. 1–41.
7. Paterson, A. *Our Guide to Batteries*; Axeon: Aberdeen, UK, 2012; pp. 1–22.
8. Anderman, M. *Tesla Motors: Battery Technology, Analysis of the Gigafactory, and the Automakers' Perspectives*; The Tesla Battery Report; Advanced Automotive Batteries: Oregon House, CA, USA, 2014; pp. 1–39.
9. Karbassian, A.; Bonathan, D.P. *Accelerated Vibration Durability Testing of a Pickup Truck Rear Bed*; 2009-01-1406; SAE International: Warrendale, PA, USA, 2009; pp. 1–5.

10. Risam, G.S.; Balakrishnan, S.; Patil, M.G.; Kharul, R.; Antonio, S. *Methodology for Accelerated Vibration Durability Test on Electrodynamic Shaker*; 2006-32-0081; SAE International: Warrendale, PA, USA, 2006; Volume 1, pp. 1–9.

11. Harrison, T. *An Introduction to Vibration Testing*; Bruel and Kjaer Sound and Vibration Measurement: Naerum, Denmark, 2014; p. 11.

12. Hooper, J.; Marco, J. Experimental modal analysis of lithium-ion pouch cells. *J. Power Sources* **2015**, *285*, 247–259. [CrossRef]

13. Moon, S.I.; Cho, I.J.; Yoon, D. Fatigue life evaluation of mechanical components using vibration fatigue analysis technique. *J. Mech. Sci. Technol.* **2011**, *25*, 611–637. [CrossRef]

14. Halfpenny, A.; Hayes, D. *Fatigue Analysis of Seam Welded Structures Using Ncode Designlife*; nCode: Ahmedabad, India, 2010; pp. 1–21.

15. Halfpenny, A. Methods for accelerating dynamic durability tests. In Proceedings of the 9th International Conference on Recent Advances in Structural Dynamics, Southampton, UK, 17–19 July 2006; pp. 1–19.

16. Avdeev, I.; Gilaki, M. Structural analysis and experimental characterization of cylindrical lithium-ion battery cells subject to lateral impact. *J. Power Sources* **2014**, *271*, 382–391. [CrossRef]

17. Zhang, X.; Wierzbicki, T. Characterization of plasticity and fracture of shell casing of lithium-ion cylindrical battery. *J. Power Sources* **2015**, *280*, 47–56. [CrossRef]

18. Choi, H.Y.; Lee, J.S.; Kim, Y.M.; Kim, H. *A Study on Mechanical Characteristics of Lithium-Polymer Pouch Cell Battery for Electrci Vehicle*; 13-0115; Hongik University: Seoul, Korea, 2013; pp. 1–10.

19. Berla, L.; Lee, S.W.; Cui, Y.; Nix, W. Mechanical behavior of electrochemically lithiated silicon. *J. Power Sources* **2015**, *273*, 41–51. [CrossRef]

20. Greve, L.; Fehrenbach, C. Mechanical testing and macro-mechanical finite element simulation of the deformation, fracture, and short circuit initiation of cylindrical lithium ion battery cells. *J. Power Sources* **2012**, *214*, 377–385. [CrossRef]

21. Oh, K.-Y.; Siegel, J.; Secondo, L.; Kim, S.U.; Samad, N.; Qin, J.; Anderson, D.; Garikipati, K.; Knobloch, A.; Epureanu, B.; *et al.* Rate dependence of swelling in lithium-ion cells. *J. Power Sources* **2014**, *267*, 197–202. [CrossRef]

22. Sahraei, E.; Meiera, J.; Wierzbicki, T. Characterizing and modeling mechanical properties and onset of short circuit for three types of lithium-ion pouch cells. *J. Power Sources* **2014**, *247*, 503–516. [CrossRef]

23. Feng, X.; Sun, J.; Ouyang, M.; Wang, F.; He, X.; Lu, L.; Peng, H. Characterization of penetration induced thermal runaway propagation process within a large format lithium ion battery module. *J. Power Sources* **2015**, *275*, 261–273. [CrossRef]

24. Brand, M.; Schuster, S.; Bach, T.; Fleder, E.; Stelz, M.; Glaser, S.; Muller, J.; Sextl, G.; Jossen, A. Effects of vibrations and shocks on lithium-ion cells. *J. Power Sources* **2015**, *288*, 62–69. [CrossRef]

25. Liu, X.; Stoliarov, S.; Denlinger, M.; Masias, A.; Snyder, K. Comprehensive calorimetry of the thermally-induced failure of a lithium ion battery. *J. Power Sources* **2015**, *280*, 516–525. [CrossRef]

26. Spinner, N.; Field, C.; Hammond, M.; Williams, B.; Myers, K.; Lubrano, A.; Rose-Pehrsson, S.; Tuttle, S. Physical and chemical analysis of lithium-ion battery cell-to-cell failure events inside custom fire chamber. *J. Power Sources* **2015**, *279*, 713–721. [CrossRef]

27. Nations, U. *ECE R100—Battery Electric Vehicles with Regard to Specific Requirements for the Construction, Functional Safety and Hydrogen*; United Nations: Lake Success, NY, USA, 2002.

28. Economic Commission for Europe (ECE). *Proposal for the 02 Series of Amendments to Regulation*; No. 100 (Battery Electric Vehicle Safety), ECE/TRANS/WP.29/2012/102; United Nations Economic and Social Council: New York, NY, USA, 2013; pp. 1–54.

29. The Tests Explained. Available online: http://www.euroncap.com/testprocedures.aspx (accessed on 9 February 2015).

30. Nations, U. *Transport of Dangerous Goods—Manual of Tests and Criteria*, 5th ed.; Amendment 1; United Nations: Lake Success, NY, USA, 2011; p. 62.

31. Pohl, D. Lithium iron phospahate: What factors influence the durability of storage systems. In *Solar Energy Storage*; Levran, A., Ed.; EES International: Pforzheim, Germany, 2014; Volume 2015, pp. 1–3.

32. Suttman, A. *Lithium Ion Battery Aging Experiments and Algorithm Development for Life Estimation*; The Ohio State University: Columbus, OH, USA, 2011.

33. Chapin, J.T.; Alvin, W.; Carl, W. *Study of Aging Effects on Safety of 18650-Type Licoox Cells*; Underwriters Laboratory Inc.: Northbrook, IL, USA, 2011.

34. Wu, A. *Study on Aging Effects on Safety of 18650 Type Licoox Cells*; Product Safety Engineering Society: Austin, TX, USA, 2012.

35. Svens, P. *Methods for Testing and Analyzing Lithium-Ion Battery Cells Intended for Heavy-Duty Hybrid Electric Vehicles*; KTH Royal Institute of Technology: Stockholm, Sweden, 2014.

36. Hooper, J. *Study into the Vibration Inputs of Electric Vehicle Batteries*; Cranfield University: Cranfield, UK, 2012.

37. Hooper, J.; Marco, J. Characterising the in-vehicle vibration inputs to the high voltage battery of an electric vehicle. *J. Power Sources* **2014**, *245*, 510–519. [CrossRef]

38. Hooper, J.; Marco, J. Understanding vibration frequencies experienced by electric vehicle batteries. In Proceedings of the 4th Hybrid and Electric Vehicles Conference (HEVC 2013), London, UK, 6–7 November 2013; IET: London, UK, 2013; pp. 1–6.

39. Kaw, A. *Mechanics of Composite Materials*, 2nd ed.; CRC Press: Boca Raton, FL, USA, 2006; p. 475.

40. Buckley, K.; Chiang, L. *Design Principles for Vibration Test Fixtures*; MIT Lincoln Laboratory: Lexington, MA, USA, 2011; pp. 1–12.

41. Coe, S. *Fixtures for Vibration Testing*; Data Physics: Hailsham, UK, 2013; pp. 1–47.

42. Harrison, T. *Resonance*; Bruel and Kjaer: Naerum, Denmark, 2014; Volume 5, p. 11.

43. Harrison, T. *A practical Guide to Vibration Testing*; Brüel & Kjær Royston: Hertfordshire, UK, 2014; p. 11.

44. Standards, B. *BS EN 60068 Environmental Testing*; British Standards; BSI Group: London, UK, 2008.

45. Reddy, T.S.; Reddy, K.V.K. Design and analysis of vibration test bed fixtures for space launch vehicles. *Indian J. Sci. Technol.* **2010**, *3*, 592–595.

46. Harrison, T. *Basic Fixture Design*; Bruel and Kjaer: Naerum, Denmark, 2014; Volume 12, p. 15.

47. Avitabile, P. Why you can't ignore those vibration fixture resonances. *Sound Vib.* **1999**, *3*, 20–26.

48. National Aeronautics and Space Administration (NASA). *General Environmental Verfication Standard (GEVS)—For GSFC Flight Programs and Projects*; NASA: NASA Goddard Space Flight Centre: Greenbelt, MD, USA, 2013.

49. United States Department of Defence. *MIL-STD-810f*; United States Department of Defence: Fort Belvoir, VA, USA, 2000; pp. 1–539.

50. Ministry of Defence (MoD). *Ministary of Defence Standard 00-35, Environmental Handbook for Defence Materiel, Part 3*; Ministry of Defence: Glasgow, UK, 2006.

51. Harrison, T. *Random Vibration Theory*; Bruel and Kjaer Sound and Vibration Measurement: Naerum, Denmark, 2014.

52. Barai, A.; Chouchelamane, G.H.; Guo, Y.; McGordon, A.; Jennings, P. A study on the impact of lithium-ion cell relaxation on electrochemical impedance spectroscopy. *J. Power Sources* **2015**, *280*, 74–80. [CrossRef]

53. Coe, S. *Vibration and Signal Processing*; Data Physics: Hailsham, UK, 2006; pp. 1–33.

54. Harrison, T. *The Vibration System*; Bruel and Kjaer Sound and Vibration Measurement: Naerum, Denmark, 2014; p. 15.

55. Chouchelamane, G. *Electrochemical Impedance Spectroscopy*; University of Warwick: Coventry, UK, 2013; pp. 1–24.

56. Birkl, C.; Howey, D. Model identification and parameter estimation for lifepo4 batteries. In Proceedings of the 4th Hybrid and Electric Vehicles Conference (HEVC 2013), London, UK, 6–7 November 2013; IET: London, UK, 2013; pp. 1–6.

57. Wanga, J.; Liua, P.; Hicks-Garnera, J.; Shermana, E.; Soukiaziana, S.; Verbruggeb, M.; Tatariab, H.; Musserc, J.; Finamorec, P. Cycle-life model for graphite-LiFePO$_4$ cells. *J. Power Sources* **2010**, *196*, 3942–3948. [CrossRef]

58. Bourlot, S.; Blanchard, P.; Robert, S. Investigation of aging mechanisms of high power Li-ion cells used for hybrid electric vehicles. *J. Power Sources* **2011**, *196*, 6841–6846. [CrossRef]

59. Bono, R. Transducer mounting and test setup configurations. In Proceedings of the IMAC XXVI: Conference & Exposition on Structural Dynamics—Technologies for Civil Structures, Orlando, FL, USA, 4–7 February 2008; The Modal Shop: Sharonville, OH, USA, 2011.

60. Sujatha, C. *Vibration and Acoustics*, 1st ed.; Tata McGraw Hill Education Private Ltd.: Dehli, India, 2010; p. 513.

61. Jeevarajan, J.; Duffield, B.; Orieukwu, J. Safety of lithium at different states of charge. In *Space Safety is No Accident*, Proceedings of the 7th IAASS Conference, Friedrichshafen, Germany, 20–22 October 2014; Sgobba, T., Rongier, I., Eds.; Springer: Friedrichshafen, Germany, 2015; pp. 131–134.

62. Ramadass, P.; Haran, B.; White, R.; Popov, B.N. Mathematical modeling of the capacity fade of Li-ion cells. *J. Power Sources* **2003**, *123*, 230–240. [CrossRef]

3

A Lithium-Ion Battery Simulator Based on a Diffusion and Switching Overpotential Hybrid Model for Dynamic Discharging Behavior and Runtime Predictions

Lan-Rong Dung [1,*], Hsiang-Fu Yuan [2,*], Jieh-Hwang Yen [2], Chien-Hua She [3] and Ming-Han Lee [1]

Academic Editor: K. T. Chau

[1] Department of Electrical and Computer Engineering, National Chiao-Tung University, 1001 Ta-Hsueh Rd., Hsinchu 30010, Taiwan; may168889.eed02g@nctu.edu.tw
[2] Institute of Electrical Control Engineering, National Chiao-Tung University, 1001 Ta-Hsueh Rd., Hsinchu 30010, Taiwan; jiehyen@gmail.com
[3] MiTAC International Corp., No. 1, R & D 2nd Road, Hsinchu Science-Based Industrial Park, Hsinchu 30010, Taiwan; eric.she@mic.com.tw
* Correspondence: lennon@faculty.nctu.edu.tw (L.-R.D.); kane1984.ece97g@nctu.edu.tw (H.-F.Y.)

Abstract: A new battery simulator based on a hybrid model is proposed in this paper for dynamic discharging behavior and runtime predictions in existing electronic simulation environments, e.g., PSIM, so it can help power circuit designers to develop and optimize their battery-powered electronic systems. The hybrid battery model combines a diffusion model and a switching overpotential model, which automatically switches overpotential resistance mode or overpotential voltage mode to accurately describe the voltage difference between battery electro-motive force (EMF) and terminal voltage. Therefore, this simulator can simply run in an electronic simulation software with less computational efforts and estimate battery performances by further considering nonlinear capacity effects. A linear extrapolation technique is adopted for extracting model parameters from constant current discharging tests, so the EMF hysteresis problem is avoided. For model validation, experiments and simulations in MATLAB and PSIM environments are conducted with six different profiles, including constant loads, an interrupted load, increasing and decreasing loads and a varying load. The results confirm the usefulness and accuracy of the proposed simulator. The behavior and runtime prediction errors can be as low as 3.1% and 1.2%, respectively.

Keywords: battery simulator; overpotential; linear extrapolation; diffusion model; equivalent circuit model (ECM); rate capacity effect; recovery effect

1. Introduction

With a high energy density, a high voltage level and a compact volume, lithium-ion batteries have become one of the most attractive power sources for prevalent portable electronic devices, hybrid electric vehicles (HEVs), electric vehicles (EVs), *etc.* For these applications, the ability of the power source to maintain good operational functionalities is very important to users. For example, long runtime is a key feature in the user's perception for purchasing the electronic products that use Li-ion batteries as power sources. To achieve this goal, a useful battery simulator is an essential tool for tracking battery state-of-charge (SOC), simulating dynamic behavior and predicting runtime [1–5]. The battery simulator should possess not only good accuracy, but also flexibility and feasibility, so that power circuit designers can co-simulate batteries and electrical circuits in existing electronic simulation environments for optimizing system performances.

All battery simulators rely on a precise battery model. However, it is difficult to model a Li-ion battery, because these complex and nonlinear electro-chemical reactions that take place in batteries have many time-varying parameters. In the past, a variety of battery models has been introduced and can be classified into three categories: electrochemical models [6–10], mathematical models [11–17] and electrical equivalent circuit models (ECMs) [18–24]. The electrochemical models, such as dual-foil, are the most accurate models, but also have very high complexity. This kind of model is established in the electrochemical point of view, so users should require professional knowledge in this field to manipulate these models, which makes them hard to be used by most of the power circuit designers. In addition, the electrochemical models normally contain many nonlinear differential equations and numerous parameters, so the simulations require heavy computational efforts.

The mathematical models, such as Peukert's law [13], the kinetic battery model (KiBaM) [14] and the diffusion model [15–17], estimate the battery runtime or the remaining capacity with specific math equations that are derived from the empirical law or the kinetic reactions and diffusion processes inside a battery. This type of model is simpler than electrochemical models, so it has less computational efforts in simulation. Besides, KiBaM and diffusion models are able to describe nonlinear capacity effects, such as the rate capacity effect and the recovery effect, so both of them are popularly mathematical models. However, the mathematical models cannot provide battery dynamic behavior, like voltage transient responses of current changes, so they are still not suitable for circuit designers to develop battery-powered electronic systems.

The ECMs are adopted by many researchers in the studies of Li-ion battery monitoring and management systems. The electrical circuit components, like resistance, capacitance and voltage source, are basic elements to build the electrical models. As a result, these models are perfect for circuit designers to simulate battery charging/discharging processes in electronic simulation tools and to co-design with electrical circuits and systems. Despite the usefulness, the ECMs can only describe linear capacity calculation. These nonlinear capacity effects are not considered in these models. This is prone to cause some simulation errors on battery performance predictions and makes the simulation results inaccurate. Recently, several hybrid models that combine a mathematical and an electrical model have been proposed in [25–27] to integrate the ability of nonlinear capacity estimation with the second-order electrical battery model [19]. However, these hybrid models increase the computational efforts again due to the complexity of mathematical models and nonlinear relations between battery SOC and electrical model parameters, so some of the hybrid models are not really feasible to run in electronic simulation environments.

In this paper, a new battery simulator is proposed for battery dynamic behavior and runtime predictions in power electronic simulation environments. The proposed simulator imitates a real battery with a hybrid battery model that combines the diffusion model to enhance the ability of capturing the nonlinear capacity effects and a switching overpotential model to simulate discharging voltage responses. The switching overpotential model only uses an overpotential resistance or an overpotential voltage, which is dependent on the SOC region, to describe the voltage difference between battery electro-motive force (EMF) and terminal voltage. Thus, this simulator has low computational efforts and makes itself more feasible to run in an electronic simulation software. A linear extrapolation technique is applied in this paper to extract model parameters and avoid the EMF hysteresis problem. The simulator is validated by experiments and simulations in the MATLAB and PSIM environments. The validation results of both simulation tools show that the proposed simulator has high accuracy. The behavior and runtime prediction errors can be as low as 3.1% and 1.2% in six predefined test profiles.

This paper is organized as follows: the related works about the well-known second-order electrical model and the diffusion model are introduced in Section 2. The proposed battery simulator and linear extrapolation technique to extraction model parameters are addressed in Section 3.

Section 4 illustrates the model validation plan, the experimental setup and processes. Section 5 gives the experimental results of parameter extraction and dynamic load tests. Section 6 discusses the model comparison between the second-order ECM and the proposed battery simulator. Final conclusions are made in Section 7.

2. Related Work

2.1. Second-Order ECM

Currently, a famous second-order ECM for circuit simulation is proposed in [19]. The electrical model shown in Figure 1 is composed of two subcircuits. The first subcircuit includes an RC pair ($R_{self\_dis}//C_{FCC}$), and a current-controlled current source (CCCS) is for SOC tracking and runtime prediction. The C_{FCC} represents a full charge capacity (FCC) capacitor, which is used to store charge. When the capacitor is full, the voltage of C_{FCC} or $V_{SOC}(t)$ goes to 1 V, and this case represents 100% SOC. Otherwise, the voltage drops down to 0 V or 0% SOC if the C_{FCC} is empty. The R_{self_dis}, which is connected to the C_{FCC}, models the self-discharge phenomenon, and its value is dependent on the charge retention rate of a real battery. The CCCS senses the battery current flowing into or out of Subcircuit 2 and reproduces it in Subcircuit 1 for charge accumulation.

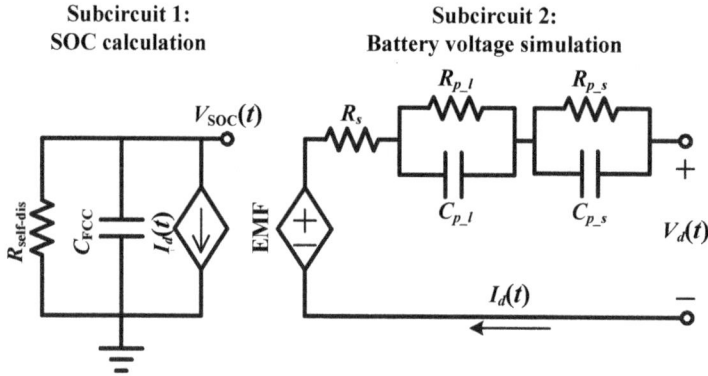

Figure 1. Second-order equivalent circuit model (ECM).

Subcircuit 2 consists of an RC network and a voltage-controlled voltage source (VCVS). The VCVS characterizes the battery EMF, and the voltage value is controlled by the $V_{SOC}(t)$ in Subcircuit 1. The RC network is made up of a series resistance R_s and two RC parallel circuits ($R_{p\_s}//C_{p\_s}$ and $R_{p\_l}//C_{p\_l}$). The series resistance R_s is an equivalent resistance that comes from the pure resistances of two electrodes, the electrolyte and the separator in a battery. The series resistance is responsible for the instantaneous voltage drop when batteries have a step current change. Both RC parallel circuits are responsible for the transient voltage change, but have different orders of the time constant. The $R_{p\_s}//C_{p\_s}$ is a short transient RC pair, while the $R_{p\_l}//C_{p\_l}$ is a long transient RC pair.

Subcircuit 2 is able to simulate the battery I-V characteristics and transient responses. The output voltage is approximated to Equation (1).

$$V_d(t) = \text{EMF} - I_d(t)R_s - I_d(t)R_{p\_s}\left(1 - e^{-\frac{t}{R_{p\_s}C_{p\_s}}}\right) - I_d(t)R_{p\_l}\left(1 - e^{-\frac{t}{R_{p\_l}C_{p\_l}}}\right) \tag{1}$$

$$\{\text{EMF}, R_s, R_{p\_s}, R_{p\_l}, C_{p\_s}, C_{p\_l}\} = f(\text{SOC(t)}) \tag{2}$$

where $V_d(t)$ stands for discharging battery voltages and $I_d(t)$ is the discharging current. The $\{EMF, R_s, R_{p\_s}, R_{p\_l}, C_{p\_s}, C_{p\_l}\}$ parameters are all functions of SOC. However, in this standard ECM, the battery SOC is calculated by the Coulomb counter, which can only describe linear capacity calculation.

$$SOC(t) = (1 - \frac{\int I_d(t)dt}{Q_{nom}}) \times 100\% \tag{3}$$

where Q_{nom} is the nominal capacity claimed by manufacturers. Hence, battery nonlinear capacity effects, such as rate capacity effect and recovery effect, cannot be captured. This introduces some SOC errors between actual batteries and this model and reduces the accuracy of simulation results.

2.2. Rakhmatov and Vrudhula Diffusion Model

The battery rate capacity effect and the recovery effect describe that the usable charge capacity inside a battery normally reduces when the discharging current increases. However, the unavailable charge will be useful after the discharging current changes to a light load or no load condition. In 2001 and 2003, Rakhmatov and Vrudhula developed a diffusion model [15] to capture these nonlinear capacity effects for battery time-to-failure prediction. The diffusion model uses Fick's laws to model the one-dimensional diffusion mechanism of the concentration of electroactive species in the electrolyte. The derivation result is given in Equation (4).

$$\alpha = \int_0^L I_d(\tau)d\tau + 2\sum_{m=1}^{\infty} \int_0^L I_d(\tau)e^{-\beta^2 m^2(L-\tau)}d\tau \tag{4}$$

where α is the total charge capacity of a battery, L is the time-to-failure and β is a constant related to the diffusion rate. In this equation, battery charge capacity is not simply calculated by a Coulomb counter. The total charge capacity α is the sum of a usable capacity, which is current integration in the first term, and an unavailable capacity in the second term. The unavailable capacity is a summation of infinite integrations from $m = 1$ to $m = \infty$, but the first 10 terms mainly determine the final result. The diffusion model estimates the battery remaining capacity by taking this unavailable capacity into account and provides a more accurate time-to-failure prediction than the Coulomb counter in the case of a constant current load or a piece-wise constant current load. However, this model is unable to simulate the battery dynamic voltage responses of these various loads, so it is difficult to apply in electronic simulation environments.

3. Proposed Battery Simulator

In this paper, the proposed battery simulator in Figure 2 is developed based on a hybrid model that combines the diffusion model and a switching overpotential model. The diffusion model replaces the Coulomb counter in the second-order ECM to evaluate the battery SOC, so that the nonlinear capacity effects are able to reflect in the proposed simulator. The switching overpotential model generates the overpotential based on the SOC reported by the diffusion model to simulate the battery voltage behavior. The battery overpotential is the difference between the battery EMF, which is equal to the sum of the equilibrium potentials of two electrodes inside a cell, and the actual terminal voltage during current-flowing conditions. When a battery is charging or discharging, the overpotential is produced by ohmic losses in the two electrodes and the electrolyte, as well as the overpotential of the charge-transfer reaction, which includes two reactions: kinetic aspects and mass transport phenomena.

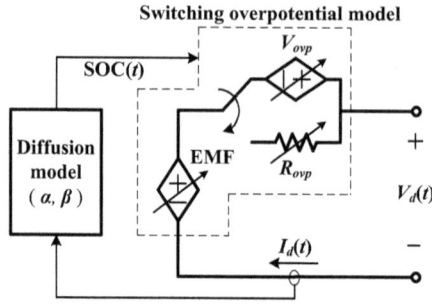

Figure 2. Hybrid battery model that combines a diffusion model and a switching overpotential model.

The switching overpotential model generates the overpotential using a resistance mode or a voltage mode. While simulating the voltage behavior, the proposed battery simulator applies different modes in different SOC regions. In the past research, the measured data of the internal resistance of a Li-ion battery cell show that the internal resistance in the flat region that is normally 20% to 80% SOC [19,28,29] is less affected by the current rate effect, so ECMs usually model the internal resistance in this SOC region as a constant resistance. However, when the battery SOC is close to the empty or full state, the SOC effect and current rate effect cannot be ignored. The internal resistance in the two regions cannot just be considered as a simple constant resistance. The factors of the SOC effect and current rate effect should all be taken into account.

To clearly define the SOC regions, the variance of the overpotential resistance, or Var(R_{ovp}), is used. The SOC region with a small resistance variance is defined when the Var(R_{ovp}) of an SOC point is less than two-times the average value of the Var(R_{ovp}) curve. Thus, two SOC levels (SOC$_H$ and SOC$_L$) can be defined when the Var(R_{ovp}) of an SOC point is larger than this threshold. The SOC$_H$ is the upper limit of the flat region, and the SOC$_L$ is the lower limit of the flat region. In the flat region, the proposed simulator adopts the resistance mode. Instead of using a constant resistance, the proposed battery simulator adopts the average resistance of different current rates at each SOC point to determine a battery overpotential voltage. The reason is that the average resistance can generate a more realistic and smoother voltage response than a constant resistance. Thus, the overpotential resistance is a function of SOC. The overpotential voltage in this region is expressed in Equation (5).

$$V_{ovp}(t) = R_{ovp}(\mathrm{SOC}(t)) \cdot I_d(t) \tag{5}$$

Unlike the voltage mode, the resistance mode only requires one lookup table for the R_{ovp} parameter.

In high and low SOC regions (SOC > SOC$_H$ and SOC < SOC$_L$), the proposed simulator adopts the voltage mode. In these two regions, the overpotential resistance or overpotential voltage is a function of the SOC and current rate. To thoroughly model the voltage, the voltage mode directly models the overpotential voltage as expressed in Equation (6).

$$V_{ovp}(t) = A(\mathrm{SOC}(t)) \cdot I_d(t) + B(\mathrm{SOC}(t)), \text{if SOC} > \mathrm{SOC}_H \text{ or SOC} < \mathrm{SOC}_L \tag{6}$$

where the A and B parameters are functions of SOC; the unit of A is Ω, and the unit of B is volts. With this method, two parameter lookup tables (A and B) at every SOC sampling point are required. To sum up, the overpotential voltages in the three SOC regions are summarized as a single mathematical form in Equation (7).

$$V_{ovp}(t) = \begin{cases} A(\mathrm{SOC}(t)) \cdot I_d(t) + B(\mathrm{SOC}(t)), & \text{if SOC}(t) > \mathrm{SOC}_H \\ R_{ovp}(\mathrm{SOC}(t)) \cdot I_d(t), & \text{if SOC}_L < \mathrm{SOC}(t) < \mathrm{SOC}_H \\ A(\mathrm{SOC}(t)) \cdot I_d(t) + B(\mathrm{SOC}(t)), & \text{if SOC}(t) < \mathrm{SOC}_L \end{cases} \tag{7}$$

The flowchart of the proposed simulator is presented in Figure 3. After the simulator measures the discharging current, the diffusion model calculates the change of nonlinear capacity and reports the SOC to the switching overpotential model. The switching overpotential model finds the EMF value from a lookup table stored in the EMF-SOC curve. The simulator then calculates the overpotential voltage by the resistance mode or voltage mode, which is dependent on the SOC region. Finally, the EMF and the overpotential voltage is added up and outputted to the terminal voltage for discharging simulations.

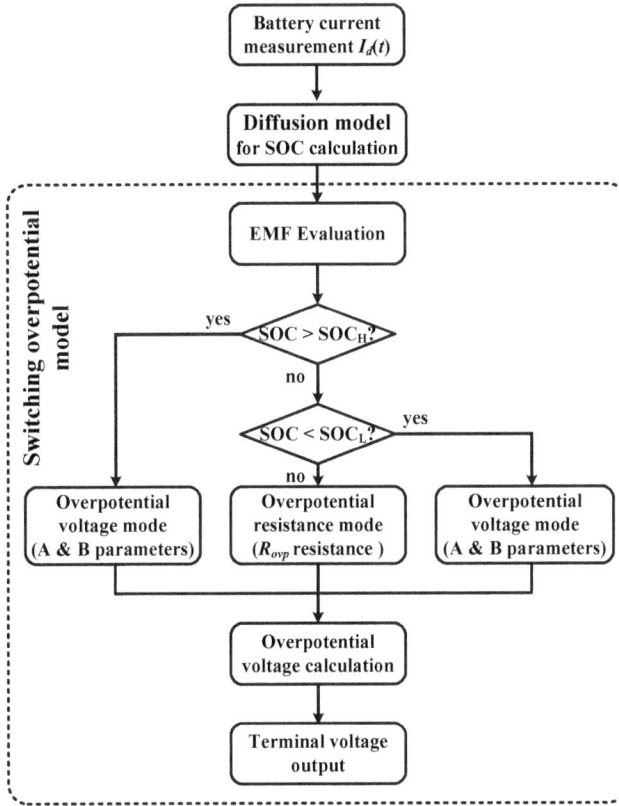

Figure 3. Flowchart of the proposed battery simulator for voltage simulation.

The switching overpotential model in the proposed simulator is modeled using the linear extrapolation technique. The detailed modeling processes for EMF, overpotential resistance and overpotential voltage are illustrated as follows. In addition, the method of nonlinear capacity estimation is also presented.

3.1. Nonlinear Capacity Estimation

By sensing the discharging current, the accumulated and unavailable capacities are calculated using Equation (8).

$$Q_d(t) = \int_0^t I_d(\tau)d\tau + 2\sum_{m=1}^{10} \int_0^t I_d(\tau)e^{-\beta^2 m^2(t-\tau)}d\tau \tag{8}$$

where $Q_d(t)$ is the nonlinear capacity, and its maximum value is α during discharging periods. For the consideration of computational efforts, the suggestion in [15] is adopted to compute only the first 10-term summation of the unavailable capacity. Battery SOC is then estimated by Equation (9).

$$\text{SOC}(t) = \left(1 - \frac{Q_d(t)}{\alpha}\right) \times 100\% \tag{9}$$

To find the α and β parameters of a battery cell, several constant discharging current loads are tested. Constant current is a special case for the diffusion model, and Equation (4) can be simplified to Equation (10).

$$\alpha = I_d L \left[1 + 2 \sum_{m=1}^{10} \frac{1 - e^{-\beta^2 m^2 L}}{\beta^2 m^2 L}\right] \tag{10}$$

where I_d is the magnitude of a constant discharging current. An optimized set of the α and β is acquired by minimizing the following equation.

$$\sum_{n=1}^{N} |I_{d(n)} - \hat{I}_{d(n)}|^2 \tag{11}$$

where N is the total number of constant current loads. The equation in Equation (11) can be expanded as Equation (12).

$$\left| I_{d(1)} - \frac{\alpha}{L_{(1)} + \sum_{m=1}^{10} \frac{1 - e^{-\beta^2 m^2 L_{(1)}}}{\beta^2 m^2}} \right|^2 + \cdots + \left| I_{d(N)} - \frac{\alpha}{L_{(N)} + \sum_{m=1}^{10} \frac{1 - e^{-\beta^2 m^2 L_{(N)}}}{\beta^2 m^2}} \right|^2 \tag{12}$$

where $\{I_{d(1)}, \ldots, I_{d(N)}\}$ is the set of test current and $\{L_{(1)}, \ldots, L_{(N)}\}$ is the set of runtime measured from experiments. The least squares method is the tool used in this optimization analysis for finding the best α and β values.

3.2. Linear Extrapolation for EMF Extraction

Several EMF extraction methods have been introduced in the past, like voltage relaxation, linear interpolation and linear extrapolation. The voltage relaxation is a popular method to extract the battery EMF-SOC curve by repeated charge-relaxation or discharge-relaxation tests. However, the required relaxation time is usually long because it is hard to make sure whether or not the battery voltage is completely relaxed. To shorten the measurement time, [30] gives the test cell a short relaxation time and finds the average EMF-SOC curve from both of the charge-relaxation and discharge-relaxation test results. This method speeds up the test time, but the effect of EMF hysteresis [31,32] will lead to some errors.

Linear interpolation uses a similar concept to measure an average EMF-SOC curve from test results of a fully charging and discharging cycle. The charging and discharging currents should be the same and small enough to reduce measurement errors. Nevertheless, the occurrence of the EMF hysteresis effect still causes errors in the extracted EMF-SOC curve. In this paper, a linear extrapolation method is chosen, because it can avoid the hysteresis effect and acquire an accurate EMF-SOC curve.

The linear extrapolation method applies different discharging currents to the test cell and then linearly extrapolates battery voltages to the case of zero current for obtaining EMF values at various SOC points. A demonstration of linear extrapolation for EMF extraction is illustrated in Figure 4. There are two discharging currents, $I_{d(1)}$ and $I_{d(\text{ref})}$, for example. The $I_{d(\text{ref})}$ is a reference current and is smaller than $I_{d(1)}$. At the same SOC point, a high discharging current leads to a small terminal voltage. As a result, $V_d(\text{SOC}_{(2)}, I_{d(1)})$ is smaller than $V_d(\text{SOC}_{(2)}, I_{d(\text{ref})})$, and $V_d(\text{SOC}_{(1)}, I_{d(1)})$ is smaller than $V_d(\text{SOC}_{(1)}, I_{d(\text{ref})})$. With the information about test current rates and measured voltages, the EMF values at $\text{SOC}_{(1)}$ and $\text{SOC}_{(2)}$ can be easily extrapolated by Equations (13) and (14).

$$\text{EMF}(\text{SOC}_{(1)}) = \frac{V_d(\text{SOC}_{(1)}, I_{d(\text{ref})}) - V_d(\text{SOC}_{(1)}, I_{d(1)})}{\left| I_{d(\text{ref})} - I_{d(1)} \right|} \times I_{d(1)} + V_d(\text{SOC}_{(1)}, I_{d(1)}) \tag{13}$$

$$\text{EMF}(\text{SOC}_{(2)}) = \frac{V_d(\text{SOC}_{(2)}, I_{d(\text{ref})}) - V_d(\text{SOC}_{(2)}, I_{d(1)})}{\left| I_{d(\text{ref})} - I_{d(1)} \right|} \times I_{d(1)} + V_d(\text{SOC}_{(2)}, I_{d(1)}) \tag{14}$$

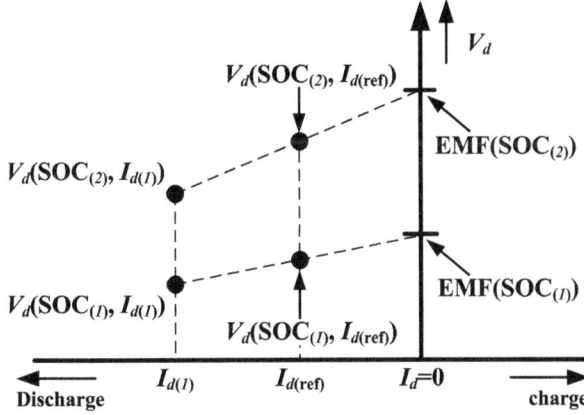

Figure 4. Demonstration of linear extrapolation for EMF extraction using $I_{d(1)}$ and $I_{d(\text{ref})}$.

Based on the same concept, a general equation to extract EMF values at other SOC points is formulated with Equation (15).

$$\text{EMF}(\text{SOC}_{(j)}) = \frac{V_d(\text{SOC}_{(j)}, I_{d(\text{ref})}) - V_d(\text{SOC}_{(j)}, I_{d(1)})}{\left| I_{d(\text{ref})} - I_{d(1)} \right|} \times I_{d(1)} + V_d(\text{SOC}_{(j)}, I_{d(1)}) \tag{15}$$

where $\text{SOC}_{(j)}$ is in the range from 0 to 100%, and $j = 1, 2, \ldots, J$.

To improve the accuracy, EMF at a particular SOC point can be obtained by more discharging currents, such as the current set $\{I_{d(1)}, \ldots, I_{d(N)}\}$ tested in the α and β estimation. Equation (15) is then modified to Equation (16).

$$\text{EMF}(\text{SOC}_{(j)}) = \frac{1}{N} \sum_{n=1}^{N} \left[\frac{V_d(\text{SOC}_{(j)}, I_{d(\text{ref})}) - V_d(\text{SOC}_{(j)}, I_{d(n)})}{\left| I_{d(\text{ref})} - I_{d(n)} \right|} \times I_{d(n)} + V_d(\text{SOC}_{(j)}, I_{d(n)}) \right] \tag{16}$$

However, due to the voltage measurement errors, which may be contributed by the quantization error of analog-to-digital converters (ADCs) or slight temperature variations, a small voltage error (e_v) can result in a significant EMF error (e_f). The location of the reference current is very important for improving the accuracy of the measured EMF-SOC curve. If the reference current is close enough to the zero current, the impacts of voltage errors on EMF values decrease. To search a proper location of the reference current, the $I_{d(\text{ref})}$ is defined as $I_{d(n)}/X$ in Figure 5, and the value of X is determined by $e_{v(n)}$ and $e_{f(n)}$. For a small $e_{v(n)}$, the relation between $e_{v(n)}$ and $e_{f(n)}$ is shown in Equation (17).

$$e_{f(n)} = \frac{e_{v(n)}}{\left| I_{d(\text{ref})} - I_{d(n)} \right|} \times I_{d(n)} - e_{v(n)} \tag{17}$$

Substituting $I_{d(n)} = X I_{d(\text{ref})}$ into Equation (17), the relation is simplified to Equation (18).

$$e_{f(n)} = \frac{X}{X-1} e_{v(n)} - e_{v(n)} \tag{18}$$

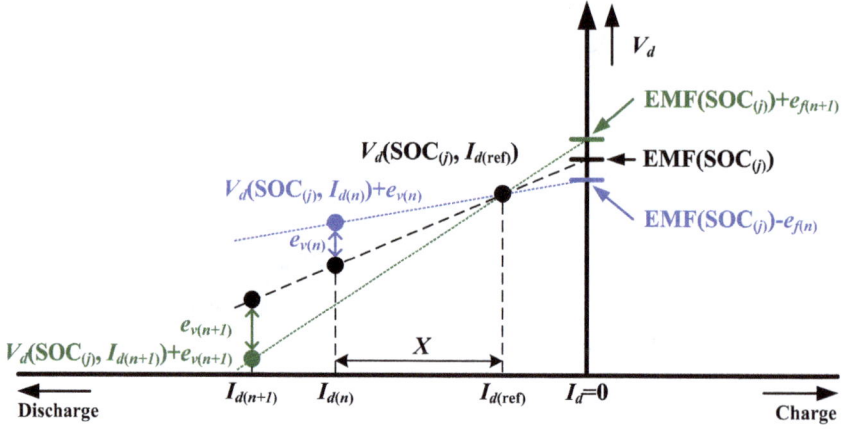

Figure 5. The $I_{d(\text{ref})}$ should be close enough to the zero current, so the impacts of voltage errors on EMF values can be reduced.

In this paper, the ratio of e_f to e_v is further defined as an error ratio in Equation (19).

$$\text{error ratio} = \frac{e_{f(n)}}{e_{v(n)}} = \frac{1}{X-1}, \; X > 1 \tag{19}$$

Thus, given a specification of the error ratio, the location of the reference current can be quickly determined.

3.3. Overpotential Resistance Modeling

In the SOC region with small resistance variations, the overpotential resistance models the voltage difference between the EMF value and the terminal voltage. According to the proposed model in Figure 2, the expression of battery voltage is given by Equation (20).

$$V_d(\text{SOC}_{(j)}, I_{d(n)}) = \text{EMF}(\text{SOC}_{(j)}) - I_{d(n)} \cdot R_{ovp}(\text{SOC}_{(j)}, I_{d(n)}) \tag{20}$$

where R_{ovp} is a function of SOC and discharging current. To get R_{ovp}, Equation (20) is rewritten and becomes Equation (21).

$$R_{ovp}(\text{SOC}_{(j)}, I_{d(n)}) = \frac{\text{EMF}(\text{SOC}_{(j)}) - V_d(\text{SOC}_{(j)}, I_{d(n)})}{I_{d(n)}} \tag{21}$$

Therefore, the $\text{EMF}(\text{SOC}_{(j)})$ calculated in Equation (16) is needed before the overpotential resistance modeling.

Compared to Equation (15), another approach to acquire the R_{ovp} is presented in the following equation.

$$R_{ovp}(\text{SOC}_{(j)}, I_{d(n)}) = \frac{V_d(\text{SOC}_{(j)}, I_{d(\text{ref})}) - V_d(\text{SOC}_{(j)}, I_{d(n)})}{\left| I_{d(\text{ref})} - I_{d(n)} \right|} \tag{22}$$

This approach only requires test currents and measured voltages, but is validated when the reference current is close enough to the zero current. The average overpotential resistance is then obtained by averaging the resistances of different current rates at each SOC point.

$$R_{ovp}(SOC_{(j)}) = \frac{1}{N} \sum_{n=1}^{N} R_{ovp}(SOC_{(j)}, I_{d(n)}) \tag{23}$$

3.4. Overpotential Voltage Modeling

In the SOC region with high resistance variations, the simulator runs in the overpotential voltage mode. The overpotential voltage at a particular SOC point and a discharging current rate is the difference between the extracted EMF value and the measured battery voltage as Equation (24).

$$V_{ovp}(SOC_{(j)}, I_{d(n)}) = EMF(SOC_{(j)}) - V_d(SOC_{(j)}, I_{d(n)}) \tag{24}$$

When the discharging current increases, the decreasing $V_d(SOC_{(j)}, I_{d(n)})$ usually leads to a high overpotential voltage. Figure 6 is an example of relations between the discharging current and overpotential voltages at various SOC points. The overpotential voltage can be modeled to a linear equation of a discharging current as Equation (25).

$$V_{ovp}(SOC_{(j)}, I_{d(n)}) = A(SOC_{(j)}) \cdot I_{d(n)} + B(SOC_{(j)}) \tag{25}$$

where $A(SOC_{(j)})$ and $B(SOC_{(j)})$ are parameters related to the battery SOC. When the simulator runs in the voltage mode, the two parameters are found firstly according to the SOC value. Then, the overpotential is generated by Equation (25). To obtain these two parameters, linear regression analyses are utilized for this curve fitting problem.

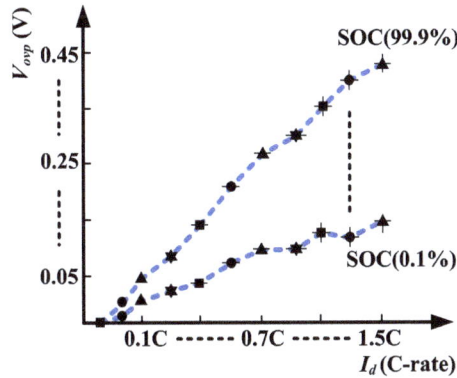

Figure 6. Example of the relations between the battery discharging current and overpotential voltages at various SOC points.

4. Model Validation

To validate the proposed battery simulator, experiments of model parameter extraction and dynamic load prediction are carried out on a single cell. The experimental setup is shown in Figure 7. The test cell is a Nokia BL-5C polymer Li-ion battery, which has a 1020-mAh nominal capacity and a 3.7-V nominal voltage. A GW Instek GBT-2211 battery tester (Good Will Instrument Co., Ltd., New Taipei City, Taiwan) charges and discharges the Li-ion cell and acquires measurement data every second. The GBT-2211 battery tester has maximum 250-W charging and discharging powers for testing different types of battery, such as nickel-cadmium, nickel-metal hydride, Li-ion or polymer

Li-ion batteries. A GMB GOV-103 constant climate chamber adjusts the ambient temperature within 26 to 28 °C.

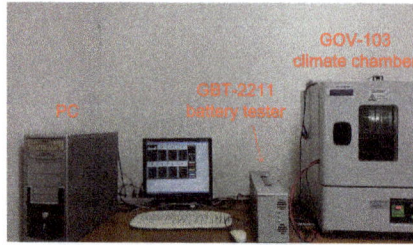

Figure 7. Experimental setup of the model validation.

4.1. Model Parameter Extraction

The parameters of the switching overpotential model and the diffusion model are all extracted from 11 constant discharging current tests. The experimental procedure is illustrated as follows.

- Charging process:
  The test cell is charged with a constant current-constant voltage (CC-CV) profile. The charging current in the CC regime is 0.8 C. The charging voltage in the CV regime is 4.2 V. The cutoff current is 0.05 mA. When the charging current drops down to this level, the test cell is fully charged or 100% SOC.
- Discharging process:
  In total, 10 constant currents are applied firstly. The current range is from 0.1 C to 1.0 C, and the increment is 0.1 C. In addition, a small current rate, 0.02 C, is tested, and the result is the reference current for EMF measurement. The cutoff voltage in discharging processes is 3.0 V. When the cell voltage is less than 3.0 V, the battery is fully discharged or 0% SOC.
- Relaxation time:
  One hour of relaxation time is required after charging or discharging processes and makes sure that the cell is fully recovered for the next test.

4.2. Dynamic Load Prediction

A series of discharging profiles is defined to check the feasibility and accuracy of the proposed battery simulator. There are six profiles in total, including two constant current loads, an interrupted current load, an increasing current load, a decreasing current load and a varying current load. The detailed configurations regarding the six profiles are listed in Table 1. The cutoff condition for each test is either the voltage drops below 3 V or the estimated capacity is empty. The same profiles are also simulated in the MATLAB and PSIM environments for comparisons. Figure 8 shows the simulation environment in PSIM. The results of experiments and simulations are all provided to validate the usefulness of the proposed battery simulator.

Table 1. Predefined dynamic discharging profiles.

Profile	Description	Configuration: Current (Time)
P_1	heavy load	0.9 C (100 min)
P_2	light load	0.5 C (150 min)
P_3	interrupted load	1.1 C (20 min)-rest (20 min)-1.1 C (100 min)
P_4	increasing load	0.4 C (20 min)-rest (10 min)-0.6 C (20 min)-rest (10min) 0.8 C (20 min)-rest (10 min)-1.0 C (100 min)
P_5	decreasing load	1.0 C (20 min)-rest (10 min)-0.8 C (20 min)-rest (10 min) 0.6 C (20 min)-rest (10 min)-0.4 C (100 min)
P_6	varying load	0.2 C (20 min)-0.5 C (10 min)-0.9 C (10 min)-0.6 C (10 min) 0.3 C (60 min)-0.2 C (20 min)-0.5 C (10 min)-0.9 C (60 min)

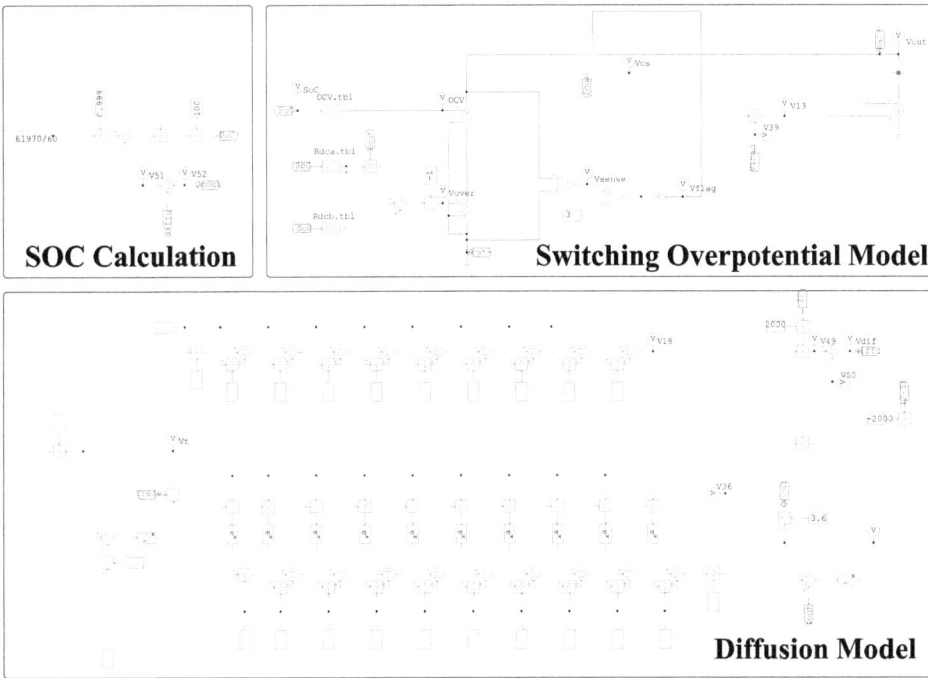

Figure 8. PSIM environment for battery simulation.

5. Experimental Results

5.1. Results of Parameter Extraction

Experimental results of the 11 constant discharging currents are shown in Figure 9. The discharging time is summarized in Table 2. The discharging time of the 1 C test is 54 min, while the 0.02 C test spends almost 46 h. The diffusion model parameters are estimated by the experimental results of 0.1 C to 1.0 C tests. The discharging time of the 0.1 C to 1.0 C tests is $\{L_{(1)}, \ldots, L_{(10)}\}$, and the discharging currents are $\{I_{d(1)}, \ldots, I_{d(10)}\}$. By taking the measurement results into Equation (12), an error surface in Figure 10 is plotted in the range of α = 58,000 mA·min to 66,000 mA·min and β = 0.5 to 2. An optimization point is found at the location of α = 61,970 mA·min (or 1032.83 mAh) and β = 1.28.

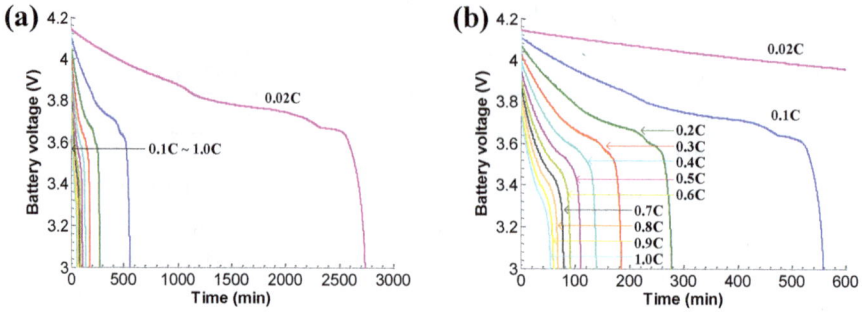

Figure 9. Battery voltage curves of the constant discharging current tests.

Table 2. Measured runtime of the constant discharging current tests.

C-rate (C)	1.0	0.9	0.8	0.7	0.6	0.5
Discharge time (min)	54.37	60.67	68.73	75.75	91.95	110.683

C-rate (C)	0.4	0.3	0.2	0.1	0.02
Discharge time (min)	139.03	185.15	278.38	558.08	2740.93

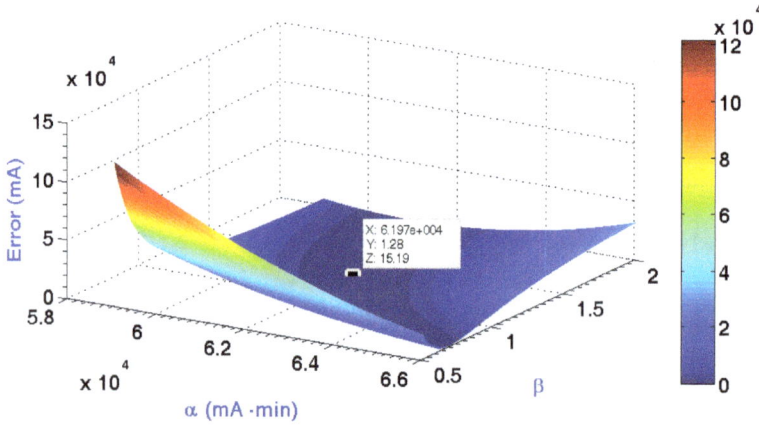

Figure 10. Error surface to find the optimization set of the α and β parameters.

Taking the measured α and β into Equations (8) and (9) obtains SOCs of the 11 discharging tests for extracting parameters of the switching overpotential model, e.g., the EMF-SOC curve, R_{ovp}-SOC curve and V_{ovp}-I_d curves. The battery voltage-SOC curves are shown in Figure 11a. The result of the 0.02 C current rate is a reference voltage curve for linearly extrapolating EMF-SOC curves. Figure 11b presents the results of EMF-SOC curves extracted from 0.1 C to 1.0 C current rates. Only the EMF-SOC curve of the 0.1 C test has a high error ratio in that the X is five in this case. Other EMF-SOC curves all present good consistency because the error ratios are less than 1%. To improve the accuracy of EMF extraction, the final EMF-SOC curve is an average result of these extracted curves. The value at each SOC point is computed by Equation (16). Figure 11c compares the voltage curves of a discharge-relaxation method, the linear extrapolation method and the 0.02 C discharging results. The discharge-relaxation method uses a 0.1 C discharging rate to measure the EMF-SOC curve. The EMF voltage is measured every 4.5% SOC. The relaxation time is one hour at each test SOC point to ensure that the battery voltage is fully recovered. The EMF errors between the discharge-relaxation curve and the extrapolation curve and between the discharge-relaxation curve and the 0.02 C discharging curve are all shown in Figure 11d. The EMF error is calculated based on Equation (26).

$$\text{EMF error} = \frac{\text{EMF}_{\text{dis-rel}} - V}{\text{EMF}_{\text{dis-rel}}} \times 100\% \qquad (26)$$

where $\text{EMF}_{\text{dis-rel}}$ is the EMF value measured by the discharge-relaxation method and V can be the voltage of the linear extrapolation method or the 0.02 C discharging result. From Figure 11d, it is obvious to observe that the error curve of the 0.02 C discharging result is higher than the error curve of the extrapolation method. The average error of the 0.02 C discharging result is about 0.5%, but the average error of the extrapolation method reduces to 0.15%. Thus, the extrapolation method is necessary for obtaining an accurate EMF-SOC curve.

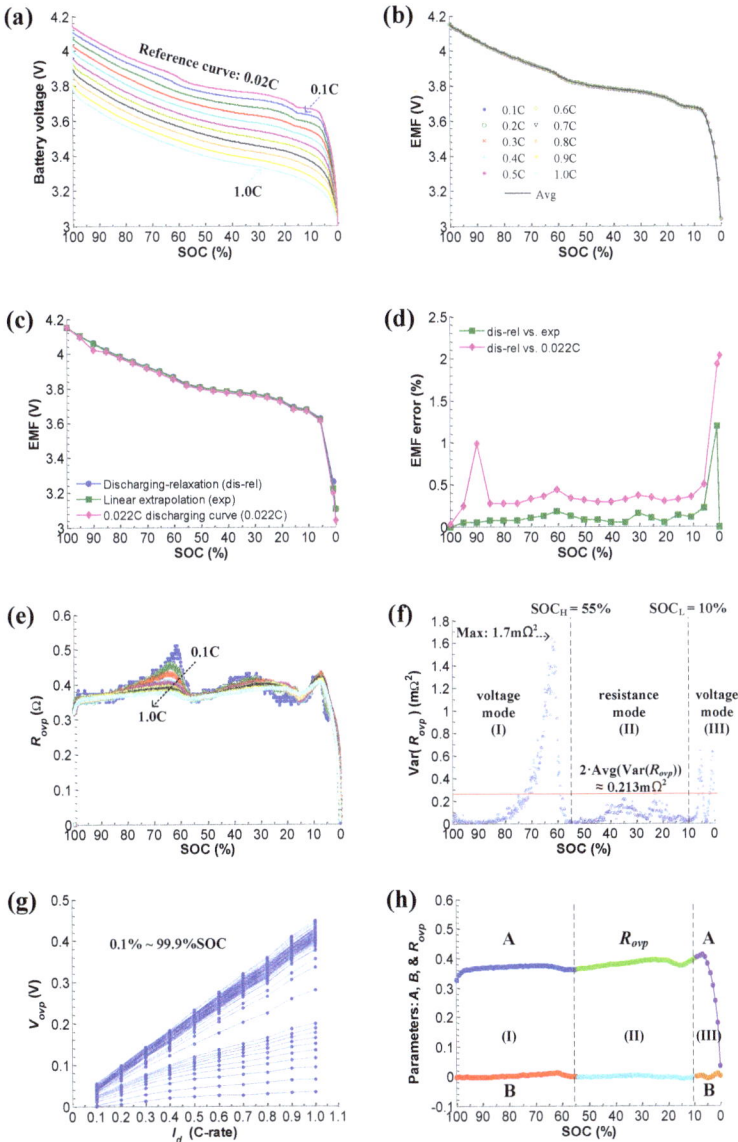

Figure 11. Experimental results of parameter extraction: (a) battery voltage-SOC curves; (b) EMF-SOC curves; (c) EMF comparisons; (d) EMF errors; (e) R_{ovp}-SOC curves; (f) variance of R_{ovp}; (g) V_{ovp}-I_d curves; (h) parameters: A, B and R_{ovp}.

With battery voltage-SOC and EMF-SOC curves in hand, R_{ovp}-SOC curves are then extracted from Equation (21). The results are shown in Figure 11e. The overpotential resistances in 10% to 90% SOC vary in 0.35 Ω to 0.52 Ω, but have large variations near 60% SOC. The variance of R_{ovp}, or Var(R_{ovp}), is presented in Figure 11f. The maximum variance is 1.7 mΩ^2 and happened in the 0.1 C test result due to the low current rate. These variations are also observed in the measured voltage curves and are probably caused by the chemical reactions, such as phase transformations in lithium electrode materials [33]. The overpotential resistances have sharp changes above 90% SOC and below 10% SOC. Consequently, most of the battery models have low accuracy when they operate in the two SOC regions.

Figure 11g analyzes the overpotential voltages in the SOC range from 0.1% to 99.9%. The linear regression analysis outputs the curve-fitting results of A, B and R_{ovp} parameters in Figure 11h. According to the average variance of R_{ovp}, the SOC is divided into three regions. The first SOC region, or Region I, begins from 100% to 55%. This region covers the maximum variance, so the battery simulator runs in the voltage mode. The second SOC region, or Region II, begins from 54.9% to 10.1%. The second region has a stable R_{ovp}. The average variance is less than 0.1 mΩ^2, and the average R_{ovp} is 0.38 Ω. In the second region, the battery simulator operates in the resistance mode. The third SOC region, or Region III, is close to empty capacity. The SOC is in the range from 10% to 0.1%. The R_{ovp} variance increases again, so the voltage mode is a better choice. In the third region, parameter A has a sharp decrement, so the overpotential voltage would still have small prediction errors.

5.2. Results of Dynamic Load Prediction

Figure 12 makes comparisons between experimental and MATLAB simulation results. For the constant current loads, the proposed simulator shows good matches in Figure 12a,b. In both cases, the rate capacity effect is reflected at the beginning of the simulations, so the terminal voltages in experiments and in simulations match well when the test cell is close to empty. The experimental runtimes of the heavy load and the light load tests are 63.13 and 116.87 min, respectively. The simulated runtime of both cases has less than 0.2% prediction errors. The normalized root mean square deviation (NRMSD), which is defined as Equation (27), recognizes the differences between simulated curves and experimental curves.

$$\text{NRMSD} = \frac{1}{V_{b,\max} - V_{b,\min}} \sqrt{\frac{\sum\limits_{t=1}^{K} (\hat{V}_b(t) - V_b(t))^2}{K}} \times 100\% \tag{27}$$

where K is the total measured points, $\hat{V}_b(t)$ is the simulated voltage value at time t and $V_b(t)$ is the experimental voltage value. The NRMSDs of the two cases are 2.2% and 1.16%. The small errors confirm that the proposed simulator is able to predict dynamic discharging behavior.

For the interrupted load, the simulation result in Figure 12c also shows a good agreement with the experimental result. The capacity recovery effect occurs during the rest period from 20 min to 40 min. After the rest period, the rate capacity effect takes place again, so the simulator successfully predicts the final runtime with a 0.92% error and dynamic behavior with 2.33% NRMSD.

For the increasing and decreasing loads, the rate capacity effect and recovery effect happen alternatively. The simulation result of the increasing load in Figure 12d predicts the runtime with a 0.54% error, while the prediction error of the decreasing load in Figure 12e slightly increases. In the overpotential voltage mode, the prediction error only grows to 1.2%. The error is mainly generated near the fully-discharged condition. A possible reason for this error would be that the low current in the finalstep lasts a long time in the low SOC region. As a result, prediction error is accumulated and increases the voltage differences between the simulation and experimental values. The NRMSD of the increasing load is 1.36%, so the proposed simulator has a good behavior prediction on the increasing load. By comparison, the NRMSD of the decreasing load is 3.14%.

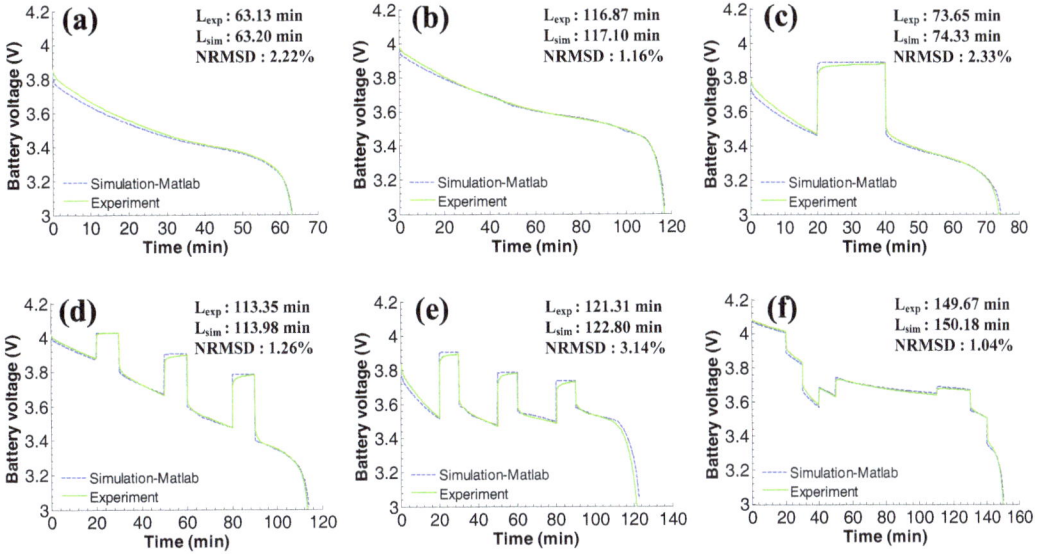

Figure 12. Model validation by MATLAB simulations: (**a**) heavy load (P_1); (**b**) light load (P_2); (**c**) interrupted load (P_3); (**d**) increasing load (P_4); (**e**) decreasing load (P_5); (**f**) varying load (P_6).

For the varying load, there is no rest time in this profile. Besides, the current changes without any rule. The simulation result in Figure 12f validates again that the proposed simulator can accurately predict the dynamic discharging behavior and further evaluate the runtime under various discharging profiles.

Figure 13 checks the battery voltage-SOC curves of the P_3 to P_5 profiles. In these plots, the SOCs estimated by the proposed simulator and the Coulomb counter are compared. The Coulomb counter estimates SOCs using the nominal capacity as the maximum available capacity. During the rest time, the recovering capacities are observed by the proposed battery simulator. However, the SOCs reported by the Coulomb counter do not have any change in this period of time, so this causes some runtime prediction errors.

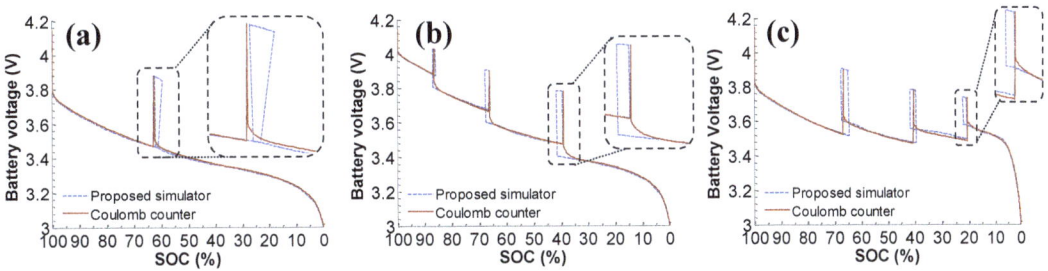

Figure 13. Capacity recovery effects in the P_3 to P_5 profiles.

Figure 14 presents the PSIM simulations for the same profiles. The simulation comparisons are made in Table 3. The runtime and behavior errors show that there are very small differences between PSIM simulation results and MATLAB simulation results. This means that the proposed battery simulator is successfully implemented in the PSIM simulation environment for dynamic discharging behavior and runtime predictions. For both simulation tools, the simulation time of the P_1 to P_6 profiles is smaller than 5 s.

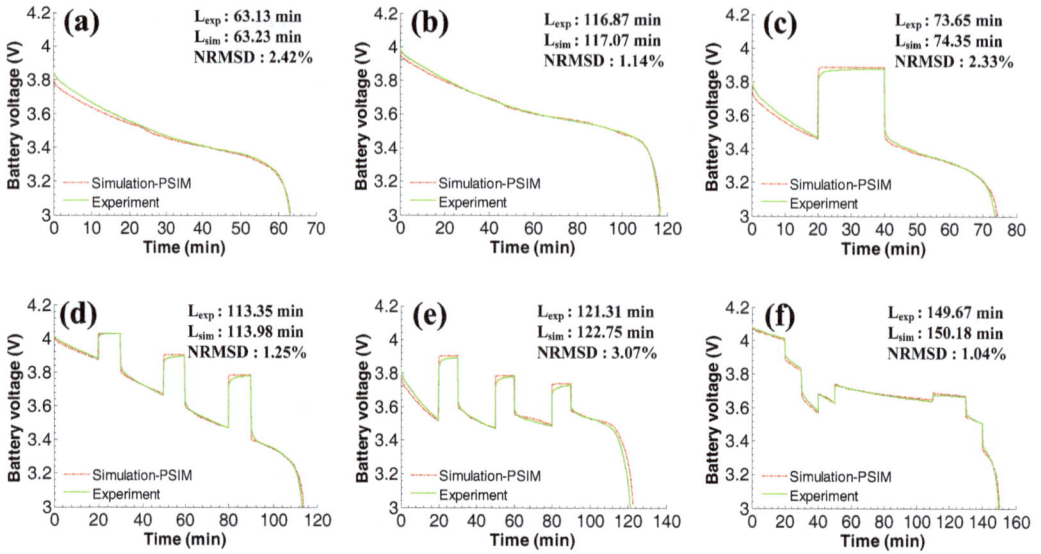

Figure 14. Model validation by PSIM simulations: (**a**) heavy load (P_1); (**b**) light load (P_2); (**c**) interrupted load (P_3); (**d**) increasing load (P_4); (**e**) decreasing load (P_5); (**f**) varying load (P_6).

Table 3. Simulation result comparisons. NRMSD, normalized root mean square deviation.

		P_1	P_2	P_3	P_4	P_5	P_6
MATLAB	runtime error	0.11%	0.20%	0.92%	0.56%	1.19%	0.34%
	behavior error (NRMSD)	2.22%	1.16%	2.33%	1.26%	3.14%	1.04%
PSIM	runtime error	0.15%	0.17%	0.95%	0.56%	1.15%	0.34%
	behavior error (NRMSD)	2.42%	1.14%	2.33%	1.25%	3.07%	1.04%

A more aggressive discharge profile to prove the predictive capability of the battery simulator is the discharge-relaxation experiment that has been applied to find the EMF-SOC curve in Figure 11c. The experiment and PSIM simulation results are shown in Figure 15. The experiment time is 1753.4 min, or 29.2 h. The runtime predicted by the proposed simulator is 1760.4 min, or 29.3 h. The runtime error is 0.4%. The behavior prediction error shows that the NRMSD is 0.96%. The PSIM simulation time of the 0.1C discharge-relaxation test is 44 s.

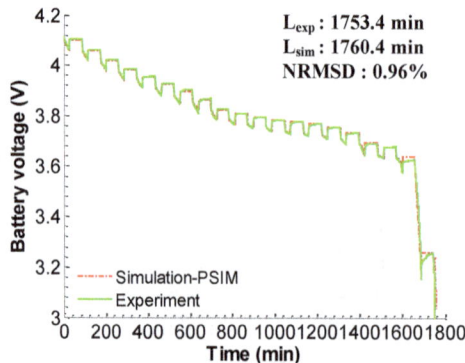

Figure 15. Validation of the 0.1 C discharge-relaxation test.

6. Discussion

A model comparison between the standard second-order ECM in [19] and the proposed battery simulator was conducted. To show the improvement, a dynamic current load and a constant current load were tested in simulations and experiments. Both current profiles have the same average current. The profile configuration of the dynamic current load is listed in Table 4. Although the time of 0.909 C is 100 min, the actual time is determined by the battery voltage. When the battery voltage drops to the 3.0-V cutoff voltage, simulations and experiments are terminated. The dynamic current load was tested firstly to realize in real time 0.909 C for calculating the average current rate. The constant load was then performed with this average current rate.

Table 4. Profile configuration of dynamic current load.

Profile	Configuration: Current (Time)
Dynamic current load	0.182 C (20 min)-0.364 C (10 min)-0.545 C (10 min)-0.727 C (10 min)-0.909 C (100 min)

The simulation results of the standard second-order ECM are shown in Figure 16. Table 5 summarizes the runtime and discharging capacity of the two current profiles. For the second-order ECM, the time of 0.909 C in the dynamic current load is 42.4 min, so the total discharging time is 92.4 min. The average current rate is 0.63 C. The discharging capacity is 830.022 mAh. The simulation results of the constant current load with 0.63 C show that the runtime and discharging capacity are almost the same. The runtime of the constant current load slightly increases 0.33% when it is compared to the result of the dynamic current load. The discharging capacity of the constant current load slightly increases 0.29%. The test results show that basically the two current profiles are the same for the standard ECM. However, this is not true for a real battery. In reality, the increasing current of the dynamic current profile will cause the available discharging capacity to reduce due to the rated capacity effect. The runtime will also reduce. Thus, the discharging capacity and the runtime of the dynamic current load should be lower than the results of the constant current load.

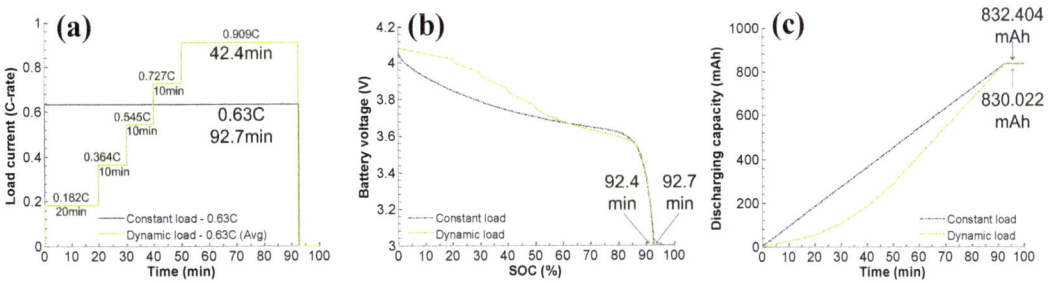

Figure 16. ECM simulation results: (a) load current; (b) battery voltage; and (c) discharging capacity.

Table 5. ECM simulation results of two different profiles with a 0.63 C average current.

Profile	Runtime	Discharging capacity
Dynamic current load	92.4 min	830.022 mAh
Constant current load	92.7 min	832.404 mAh
Growth rate	(+0.33%)	(+0.29%)

The same tests were conducted in experiments with the Nokia BL-5C polymer Li-ion battery cell and simulations with the proposed battery simulator. The experimental data and simulation results are shown in Figure 17. Two current profiles for testing in experiments are shown in Figure 17a. Figure 17b shows not only experimental battery voltages of the two current profiles, but also the simulated battery voltages of the proposed battery simulator for validations. Figure 17c shows the discharging capacities of the two current loads. Table 6 summarizes the runtime and discharging capacity. Experimental data show that the time of 0.909 C in the dynamic current load lasts only 37.3 min. The total experimental time of the dynamic current load reduces to 87.3 min, and the average current is 0.61 C. Comparing to the 0.6 C constant current, which was performed in model parameter extraction tests to obtain the α and β parameters in Section 4, the runtime is 91.95 min. This result is much higher than the runtime of the dynamic current load.

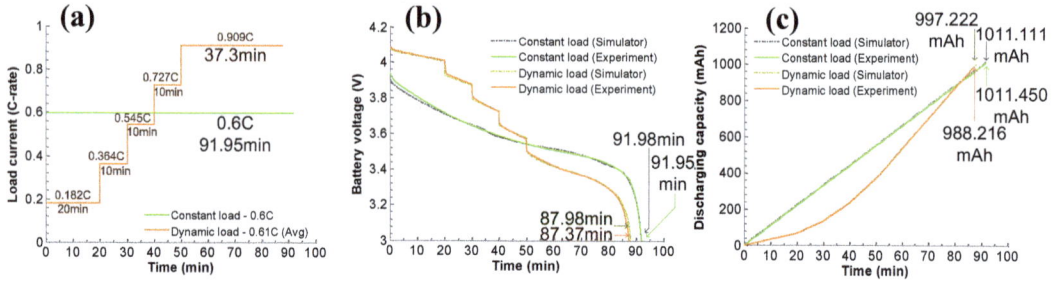

Figure 17. Experimental data and simulation results of the proposed battery simulator: (**a**) load current; (**b**) battery voltage; and (**c**) discharging capacity.

Table 6. Experimental data and simulation results of the proposed battery simulator.

Profile	Runtime		Discharging Capacity	
	Experiment	Simulator	Experiment	Simulator
Dynamic current load	87.37 min	87.98 min	988.216 mAh	997.222 mAh
Constant current load	91.95 min	91.98 min	1011.450 mAh	1011.111 mAh
Growth rate	(+5.24%)	(+4.55%)	(+2.35%)	(+1.39%)

The nonlinear capacity effects can be clearly observed from the experimental data and simulation results. Due to the increasing load current, the rated capacity effect causes the available discharging charge of the dynamic current load to reduce, so the runtime is less than the time of the constant current load. The experimental data show that the runtime and the discharging capacity of the constant current load increase 5.24% and 2.35%, respectively. The simulation results show that the runtime and the discharging capacity of the constant current load increase 4.55% and 1.39%. If the average current of the current profile rises, the runtime differences between a dynamic current load and a constant current load will also increase. This example definitely illustrates the superiority of the proposed battery simulator over the standard ECM. The proposed battery simulator has the capability of reflecting the nonlinear capacity effects in a dynamic discharging current load. However, the standard ECM cannot make the difference between a constant current load and a dynamic current load with the same average current rate. Thus, the standard ECM cannot reflect the runtime and discharging capacity reductions under a dynamic current load.

7. Conclusions

This paper has presented a new battery simulator, which can predict battery dynamic behavior and runtime in existing electronic simulation environments. A hybrid battery model that combines the diffusion model for enhancing the ability to capture the nonlinear capacity effects and a switching overpotential model for simulating the battery terminal voltage have been presented. The linear extrapolation technique has also been introduced to extract the parameters of the two models. A Nokia BL-5C polymer Li-ion battery cell with the 1020 mAh nominal capacity has been used to validate the proposed simulator. In total, 11 constant discharging current, 0.02 C and 0.1 C to 1.0 C, have been tested for extracting parameters, including α and β for the diffusion model and

EMF, overpotential resistance and voltage for the switching overpotential model. The six predefined profiles have been tested in the experiments to check the estimation performances of the simulator under a heavy load, a light load, an interrupted load, increasing and decreasing loads and a varying load. The same profiles have also been tested in MATLAB and PSIM simulations. The results confirm the usefulness and accuracy of the simulator to predict dynamic discharging behavior and battery runtime in electronic design environments for co-simulating and co-designing with electrical circuits and systems.

Acknowledgments: The authors would like to give thanks for the research grant (MOST 102-2221-E-009-145) from the Ministry of Science and Technology (MOST), Taipei, Taiwan.

Author Contributions: Lan-Rong Dung developed the essential ideal of the proposed battery simulator, designed some parts of the experiments and gave some suggestions. Chien-Hua She and Ming-Han Lee carried out the experiments and built the simulator in MATLAB and PSIM. Lan-Rong Dung and Hsiang-Fu Yuan checked the experiment and simulation results. Hsiang-Fu Yuan and Jieh-Hwang Yen wrote the manuscript. The final review and revision were done by Lan-Rong Dung and Hsiang-Fu Yuan.

Conflicts of Interest: The authors declare no conflict of interest.

References

1. Gao, J.; Zhang, Y.; He, H. A real-time joint estimator for model parameters and state of charge of lithium-ion batteries in electric vehicles. *Energies* **2015**, *8*, 8594–8612.
2. Xia, B.; Wang, H.; Tian, Y.; Wang, M.; Sun, W.; Xu, Z. State of charge estimation of lithium-ion batteries using an adaptive cubature kalman filter. *Energies* **2015**, *8*, 5916–5936.
3. Uddin, K.; Picarelli, A.; Lyness, C.; Taylor, N.; Marco, J. An acausal Li-ion battery pack model for automotive applications. *Energies* **2014**, *7*, 5675–5700.
4. Sepasi, S.; Roose, L.R.; Matsuura, M.M. Extended kalman filter with a fuzzy method for accurate battery pack state of charge estimation. *Energies* **2015**, *8*, 5217–5233.
5. Lee, J.; Yi, J.; Shin, C.B.; Yu, S.H.; Cho, W.I. Modeling the effects of the cathode composition of a lithium iron phosphate battery on the discharge behavior. *Energies* **2013**, *6*, 5597–5608.
6. Dees, D.W.; Battaglia, V.S.; Bélanger, A. Electrochemical modeling of lithium polymer batteries. *J. Power Sources* **2002**, *110*, 310–320.
7. Newman, J.; Thomas, K.E.; Hafezi, H.; Wheeler, D.R. Modeling of lithium-ion batteries. *J. Power Sources* **2003**, *119–121*, 838–843.
8. Gomadam, P.M.; Weidner, J.W.; Dougal, R.A.; White, R.E. Mathematical modeling of lithium-ion and nickel battery systems. *J. Power Sources* **2002**, *110*, 267–284.
9. Klein, R.; Chaturvedi, N.; Christensen, J.; Ahmed, J.; Findeisen, R.; Kojic, A. Electrochemical model based observer design for a lithium-ion battery. *IEEE Trans. Control Syst. Technol.* **2013**, *21*, 289–301.
10. Ahmed, R.; El Sayed, M.; Arasaratnam, I.; Tjong, J.; Habibi, S. Reduced-order electrochemical model parameters identification and state of charge estimation for healthy and aged Li-ion batteries-Part II: Aged battery model and state of charge estimation. *IEEE J. Emerg. Sel. Top. Power Electron.* **2014**, *2*, 678–690.
11. Pedram, M.; Wu, Q. Design considerations for battery-powered electronics. In Proceedings of the 36th Annual ACM/IEEE Design Automation Conference, New Orleans, LA, USA, 21–25 June 1999; pp. 861–866.
12. Chiasserini, C.; Rao, R. Energy efficient battery management. *IEEE J. Sel. Areas Commun.* **2001**, *19*, 1235–1245.
13. Linden, D.; Reddy, T.B. *Handbook of Batteries*, 3rd ed.; McGraw-Hill: New York, NY, USA, 2002.
14. Manwell, J.F.; McGowan, J.G. Lead acid battery storage model for hybrid energy systems. *Sol. Energy* **1993**, *50*, 399–405.
15. Rakhmatov, D.; Vrudhula, S.; Wallach, D. A model for battery lifetime analysis for organizing applications on a pocket computer. *IEEE Trans. VLSI Syst.* **2003**, *11*, 1019–1030.
16. Rong, P.; Pedram, M. An analytical model for predicting the remaining battery capacity of lithium-ion batteries. *IEEE Trans. VLSI Syst.* **2006**, *14*, 441–451.
17. Agarwal, V.; Uthaichana, K.; DeCarlo, R.; Tsoukalas, L. Development and validation of a battery modelu useful for discharging and charging power control and lifetime estimation. *IEEE Trans. Energy Convers.* **2010**, *25*, 821–835.

18. Schweighofer, B.; Raab, K.; Brasseur, G. Modeling of high power automotive batteries by the use of an automated test system. *IEEE Trans. Instrum. Meas.* **2003**, *52*, 1087–1091.

19. Chen, M.; Rincon-Mora, G. Accurate electrical battery model capable of predicting runtime and I-V performance. *IEEE Trans. Energy Convers.* **2006**, *21*, 504–511.

20. Szumanowski, A.; Chang, Y. Battery management system based on battery nonlinear dynamics modeling. *IEEE Trans. Veh. Technol.* **2008**, *57*, 1425–1432.

21. Hu, X.; Li, S.; Peng, H. A comparative study of equivalent circuit models for Li-ion batteries. *J. Power Sources* **2012**, *198*, 359–367.

22. Thirugnanam, K.; Ezhil Reena, J.; Singh, M.; Kumar, P. Mathematical modeling of Li-ion battery using genetic algorithm approach for V2G applications. *IEEE Trans. Energy Convers.* **2014**, *29*, 332–343.

23. Sánchez, L.; Blanco, C.; Antón, J.; García, V.; González, M.; Viera, J. A variable effective capacity model for LiFePO$_4$ traction batteries using computational intelligence techniques. *IEEE Trans. Ind. Electron.* **2015**, *62*, 555–563.

24. Yang, H.C.; Dung, L.R. An accurate Lithium-ion battery gas gauge using two-phase STC modeling. In Proceedings of the IEEE 16th International Symposium on Industrial Electronics (ISIE), Vigo, Spain, 4–7 June 2007; pp. 866–871.

25. Kim, T.; Qiao, W. A hybrid battery model capable of capturing dynamic circuit characteristics and nonlinear capacity effects. *IEEE Trans. Energy Convers.* **2011**, *26*, 1172–1180.

26. Zhang, J.; Ci, S.; Sharif, H.; Alahmad, M. An enhanced circuit-based model for single-cell battery. In Proceedings of the 25th Annual IEEE Applied Power Electronics Conference and Exposition (APEC), Palm Springs, CA, USA, 21–25 February 2010; pp. 672–675.

27. Zhang, J.; Ci, S.; Sharif, H.; Alahmad, M. Modeling discharge behavior of multicell battery. *IEEE Trans. Energy Convers.* **2010**, *25*, 1133–1141.

28. Hentunen, A.; Lehmuspelto, T.; Suomela, J. Time-domain parameter extraction method for Thévenin-equivalent circuit battery models. *IEEE Trans. Energy Convers.* **2014**, *29*, 558–566.

29. Yao, L.W.; Aziz, J.; Kong, P.Y.; Idris, N. Modeling of lithium-ion battery using MATLAB/simulink. In Proceedings of the 39th Annual Conference of the IEEE Industrial Electronics Society, Vienna, Austria, 10–13 November 2013; pp. 1729–1734.

30. Abu-Sharkh, S.; Doerffel, D. Rapid test and non-linear model characterisation of solid-state lithium-ion batteries. *J. Power Sources* **2004**, *130*, 266–274.

31. Kim, J.; Seo, G.S.; Chun, C.; Cho, B.H.; Lee, S. OCV hysteresis effect-based SOC estimation in extended Kalman filter algorithm for a LiFePO$_4$/C cell. In Proceedings of the 2012 IEEE International Electric Vehicle Conference (IEVC), Greenville, SC, USA, 4–8 March 2012; pp. 1–5.

32. Baronti, F.; Femia, N.; Saletti, R.; Visone, C.; Zamboni, W. Hysteresis modeling in Li-ion batteries. *IEEE Trans. Magn.* **2014**, *50*, 1–4.

33. Guena, T.; Leblanc, P. How depth of discharge affects the cycle life of Lithium-Metal-Polymer batteries. In Proceedings of the 28th Annual International Telecommunications Energy Conference, Providence, RI, USA, 10–14 September 2006; pp. 1–8.

4

Enhanced Predictive Current Control of Three-Phase Grid-Tied Reversible Converters with Improved Switching Patterns

Zhanfeng Song [1,*,†], Yanjun Tian [2,†], Zhe Chen [2] and Yanting Hu [3]

Academic Editor: Frede Blaabjerg

[1] Department of Electrical Engineering and Automation, Tianjin University, Tianjin 30072, China
[2] Department of Energy Technology, Aalborg University, Aalborg 9220, Denmark; yti@et.aau.dk (Y.T.); zch@et.aau.dk (Z.C.)
[3] Faculty of Science and Engineering, University of Chester, Chester CH1 4BJ, UK; y.hu@chester.ac.uk
* Correspondence: zfsong@tju.edu.cn
† These authors contributed equally to this work.

Abstract: A predictive current control strategy can realize flexible regulation of three-phase grid-tied converters based on system behaviour prediction and cost function minimization. However, when the predictive current control strategy with conventional switching patterns is adopted, the predicted duration time for voltage vectors turns out to be negative in some cases, especially under the conditions of bidirectional power flows and transient situations, leading to system performance deteriorations. This paper aims to clarify the real reason for this phenomenon under bidirectional power flows, *i.e.*, rectifier mode and inverter mode, and, furthermore, seeks to propose effective solutions. A detailed analysis of instantaneous current variations under different conditions was conducted. An enhanced predictive current control strategy with improved switching patterns was then proposed. An experimental platform was built based on a commercial converter produced by Danfoss, and moreover, relative experiments were carried out, confirming the superiority of the proposed scheme.

Keywords: predictive control; bidirectional power flow; three-phase grid-tied converters; improved switching pattern; cost function minimization

1. Introduction

During the past few decades, a large variety of active systems have been connected to the electric power system, including geothermal generators, photovoltaic systems, wind turbines, and so on. Most of these installations are connected to the grid with the assistance of grid-tied converters. Consequently, flexible interconnections of different installations depend largely on the design and control of grid-tied converters. In the past few years, model predictive control (MPC) of converters has received significant attention and experienced sustained development [1]. Various types of model predictive controllers have emerged as effective strategies and promising alternatives for the control of power converters and electrical machines [2–12]. One feasible solution for the application of MPC is known as continuous set model predictive control [13,14]. Continuous voltage reference signals are obtained by means of online optimization and cost function minimization. A modulator is normally required to generate the switching signals. With continuous set model predictive strategies adopted, control actions are considered to be continuous. By contrast, another approach called finite set model predictive control regards these discrete switching actions as the constraints of the system inputs [15,16]. During the optimizing process, the control actions are constrained to the limited available switching states, instead of continuous sets [17–19]. Until recently, finite set model predictive control with one-step prediction horizons has been applied to various converter topologies and also with a wide range of control objectives, including speed control, current control, power control, torque and flux control [20–24].

As demonstrated by recent research, finite set model predictive control outperforms classical linear solutions with regard to transient behaviours and control flexibility [25,26]. Meanwhile, a relative amount of research has been conducted in order to overcome some of the shortcomings possessed by the finite set model predictive control technique, including unsatisfactory steady-state performances, variable switching frequency and the wide distributed spectra of output waveforms. A large number of novel solutions to these practical problems have been reported. Several constant-switching frequency MPC strategies were developed, by means of introducing frequency information to the cost function or the adoption of a modulation stage. However, even though the desired switching patterns and spectrum distribution can both be achieved, transient performances were deteriorated due to the low-pass filter characteristics possessed by these methods. In [27,28], two active vectors were firstly selected during each sampling interval. An optimal vector sequence was then symmetrically built up on the basis of one zero vector and two selected vectors. With the assistance of cost function minimization, the application times of each vector element in the optimal sequence were finally determined. This is quite different from the working principle of finite set model predictive control techniques, in which the optimal switching action is selected and, in particular, remains fixed within each sampling interval once it is determined. Note that when the method proposed in [27,28] is adopted, the predicted application times of each vector element in the optimal sequence might be smaller than zero. In reality, the application times of voltage vectors or switching states should never be negative. This item should be taken into consideration when the algorithm is implemented. Recently, this has been paid special attention to, and several improved versions of this control methodology have been proposed [4,28–32]. In [4], a generalized predictive direct power control strategy was proposed. The cost function was calculated for each sector. Based on cost function comparisons, the optimal voltage vectors and their time sequences are finally determined. Even though a corresponding simplified scheme was further proposed, the cost function still needs to be calculated six times within each sampling period [19]. In [30], the calculated values of the predicted application time are firstly examined. Whenever the application time is less than zero, the optimal voltage vector sequences were to be modified according to a pre-defined switching table. Nevertheless, even though steady-state performance deteriorations can be successfully suppressed, those transient performance deteriorations caused by the incorrect selection of voltage vectors were not paid attention to. Besides, transient performance deterioration caused by the negative duration times also was not mentioned or analysed in previous studies [30–32]. Besides, the research conducted in [31,32] only refers to the control actions when the converter is operating in the rectifier mode. The corresponding control problem under the situation of reversible power flows was not properly solved.

This paper aims to reveal the reason for the operating characteristics under bidirectional power flows, i.e., rectifiers and inverters, and, furthermore, seeks to propose effective solutions. The following sections are organized as follows. In Section 2, the operating principle of conventional predictive current control for three-phase grid-tied reversible converters is briefly depicted. A detailed analysis is then conducted in Section 3 with its shortcomings emphasized, and an enhanced strategy with improved switching patterns is developed. In order to demonstrate the main advantage of the proposed enhanced predictive current control strategy, Section 4 presents comparative experimental results based on a commercial grid-tied converter produced by Danfoss Company. Conclusions are finally presented in Section 5.

2. Operating Principle of Predictive Current Control for Three-Phase Grid-Tied Reversible Converters

In this section, it is assumed that the three-phase grid-tied reversible converter operates in the rectifier mode. In the synchronous reference frame, the instantaneous variations of the converter current within each sampling period can be written as:

$$\Delta I_g = \frac{dI_g}{dt} = \frac{1}{L_g}(U_g - R_g I_g - j\omega L_g I_g - U_c) \tag{1}$$

where U_g represents the grid voltage vector and U_c is the converter output voltage vector; I_g is the converter current vector; ω, L_g and R_g denote the angular frequency of grid voltage, the line inductance and its equivalent resistance, respectively.

Normally, two adjacent voltage vectors should be firstly selected as active vectors according to the angular location of U_g [33]. These two selected active vectors will be applied together with the zero vector $V_{0,7}$. For the purposes of convenience, V_m and V_n are adopted to denote the first and second vectors, respectively. Their application times t_m and t_n will be further determined by minimizing a pre-defined cost function.

Taking Sector 6, for example, V_6 and V_1 should be selected when U_g is located in this sector. These two vectors will then be applied along with the zero vector $V_{0,7}$. When different voltage vectors are applied, the instantaneous current variation differs accordingly. The instantaneous current variations resulting from the application of V_1, V_6 and $V_{0,7}$ can be predicted as:

$$\Delta I_{g\text{-}V1} = \frac{1}{L_g}(U_g - R_g I_g - j\omega L_g I_g - V_1) \tag{2}$$

$$\Delta I_{g\text{-}V6} = \frac{1}{L_g}(U_g - R_g I_g - j\omega L_g I_g - V_6) \tag{3}$$

$$\Delta I_{g\text{-}V0,7} = \frac{1}{L_g}(U_g - R_g I_g - j\omega L_g I_g - V_{0,7}) \tag{4}$$

Within each sampling period T_s, the current variation $\Delta I_{g\text{-}Ts}$ produced by the joint actions of two active vectors and the zero vector $V_{0,7}$ can be predicted as:

$$\Delta I_{g\text{-}Ts} = \Delta I_{g\text{-}V1} t_1 + \Delta I_{g\text{-}V6} t_6 + \Delta I_{g\text{-}V0,7} t_0 \tag{5}$$

where t_1, t_6 and t_0 denote the duration times of different vectors within each sampling interval T_s.

Under the joint actions of V_1, V_6 and $V_{0,7}$, the current tracking error vector E_g at the end of the k-th sampling period can thus be computed as follows:

$$E_g = I_g^* - I_g - \Delta I_{g\text{-}Ts} \tag{6}$$

where $I_g{}^*$ is the current reference vector and I_g represents the current vector at the initial instant of the k-th sampling period.

It is hoped that the converter current evolves from the initial vector I_g toward the reference vector $I_g{}^*$ within one sampling period. In other words, the duration times of two active voltage vectors should be determined with the aim of eliminating the current tracking error vector E_g at the end of each sampling period [34]. Conventionally, the optimal duration values of t_1 and t_6 satisfy the following minimum value condition:

$$\frac{\partial W}{\partial t_1} = \frac{\partial W}{\partial t_6} = 0 \tag{7}$$

where **W** denotes the cost function given by:

$$\mathbf{W} = \left| E_g \right|^2$$

Based on Equation (7), the optimal duration times for two active vectors can be calculated and applied accordingly. It should be noted that Sector 6 is taken as an example here. When U_g is located in other sectors, the selected active voltage vectors will change accordingly.

3. Performance Analysis of the Predictive Current Control Strategy

3.1. Rectifier Mode Analysis

Assuming that the three-phase grid-tied reversible converter is operating in rectifier mode, the instantaneous current variations caused by the application of V_6, V_1 and $V_{0,7}$ when U_g lies in Sector 6 is depicted in Figure 1a. After vector movement and rearrangement, instantaneous current variations in the dq-reference frame can be depicted as in Figure 1b.

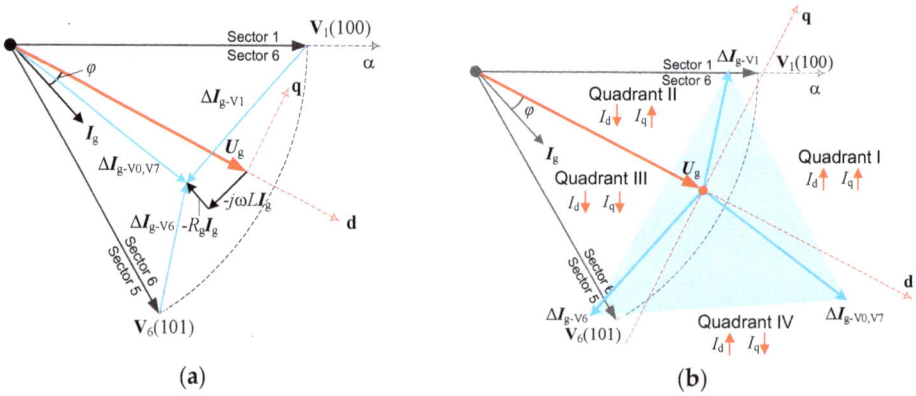

(a)   (b)

Figure 1. Instantaneous current variations caused by the application of V_6, V_1 and $V_{0,7}$ when U_g is located in the middle part of Sector 6 (Rectifier mode). (a) Original diagram; (b) Diagram after vector movement and rearrangement.

As depicted in Figure 1b, the instantaneous current variation vector $\Delta I_{g\text{-}V0,V7}$ resulting from the application of zero vector $V_{0,7}$ lies in Quadrant IV of the dq-reference frame. This indicates that the actions of $V_{0,7}$ will surely result in the increase of i_d, as well as the decrease of i_q. Besides, as $\Delta I_{g\text{-}V1}$ is located in Quadrant II, the application of V_1 will lead to the decrease of i_d and the increase of i_q. Similarly, i_d and i_q will both be decreased under the action of V_6, due to the location of $\Delta I_{g\text{-}V6}$ in Quadrant III. Generally speaking, joint actions of V_6, V_1 and $V_{0,7}$ can result in both an increase and a decrease of the d- and q-axes currents. In other words, the d- and q-axes currents can be properly regulated by combining the actions of V_6, V_1 and $V_{0,7}$. Actually, this is due to the fact that the light blue area composed of three vectors $\Delta I_{g\text{-}V6}$, $\Delta I_{g\text{-}V1}$, $\Delta I_{g\text{-}V0,V7}$ covers four quadrants in the dq-reference frame, as can be clearly seen in Figure 1b. However, this is not the case when U_g is located in the initial part of Sector 6, as shown in Figure 2.

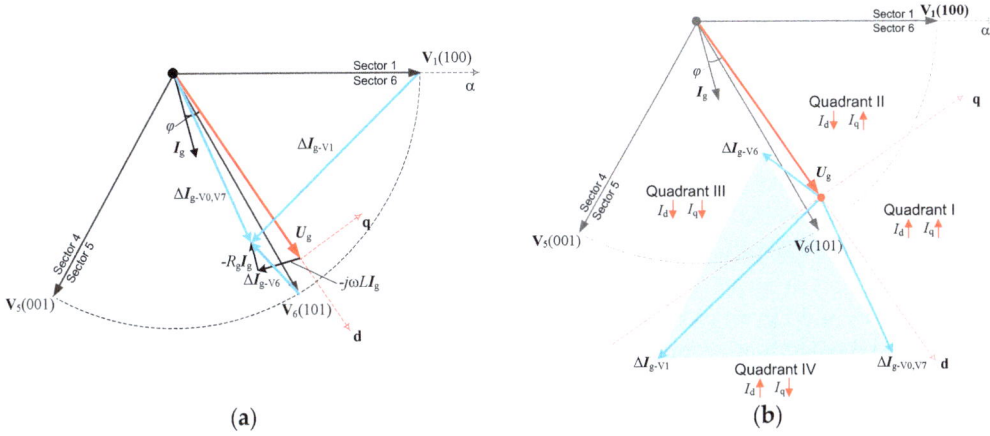

Figure 2. Instantaneous current variations caused by the application of V_6, V_1 and $V_{0,7}$ when U_g is located in the initial part of Sector 6 (rectifier mode). **(a)** Original diagram; **(b)** Diagram after vector movement and rearrangement.

As shown in Figure 2, the light blue area composed of three vectors $\Delta I_{g\text{-}V6}$, $\Delta I_{g\text{-}V1}$, $\Delta I_{g\text{-}V0,V7}$ only covers Quadrants III and IV in the dq-reference frame. The actions of $V_{0,7}$ and V_1 both result in the increase of i_d and the decrease of i_q. The application of V_6 will lead to the decrease of the d- and q-axes currents. Obviously, i_q can only be decreased by the joint actions of V_6, V_1 and $V_{0,7}$. It is found that, under this situation, the duration time t_1 of the selected voltage vector V_1, which is calculated based on cost function minimization, will turn out to be negative. With common sense, the duration time of voltage vectors should not be a negative value and will be forced to zero. Under this situation, the flexible regulation of currents cannot be successfully achieved. Actually, V_1 and V_5 should be chosen and applied under this situation, instead of V_1 and V_6. In order to clarify this remarkable conclusion, Figure 3 demonstrates the instantaneous current variations caused by the application of V_5, V_6 and $V_{0,7}$ when U_g is located in the initial part of Sector 6.

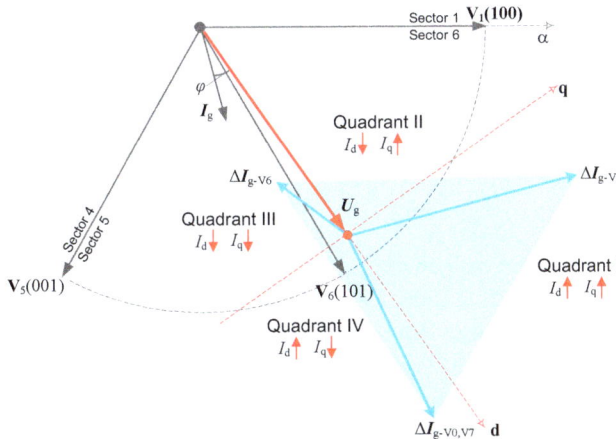

Figure 3. Instantaneous current variations caused by the application of V_5, V_6 and $V_{0,7}$ when U_g is located in the initial part of Sector 6 (rectifier mode).

As shown in Figure 3, the light blue area composed of $\Delta I_{g\text{-}V5}$, $\Delta I_{g\text{-}V6}$, $\Delta I_{g\text{-}V0,V7}$ covers four quadrants in the dq-reference frame. Consequently, both an increase and a decrease of the dq-axes currents can be expected under the joint actions of V_5, V_6 and $V_{0,7}$. Flexible regulation of currents can thus be achieved. Therefore, when U_g is located in the position as shown in Figure 3,

V_5 and V_6 should be selected as active vectors, instead of V_1 and V_6. A similar conclusion can be easily obtained when U_g lies in other sectors.

Furthermore, incorrect selection of voltage vectors and negative duration time also come out when there is a step change in the reference current. Taking Sector 6 for example, the duration time t_1 is negative, while t_6 remains positive when the q-axis reference current is step-up changed. The reason for this phenomenon is that the selected voltage vectors V_6 and V_1 could not produce large current variations as desired. Under this situation, V_5 and V_6 should be selected and applied, instead of V_6 and V_1. Figure 4 depicts the instantaneous current variations caused by the application of V_5 and V_6 when U_g lies in Sector 6. By comparing Figure 4 with Figure 1a, it is clearly visible that the joint actions of V_5 and V_6 can produce much larger current variations than those of V_6 and V_1.

It should be noted that, as pointed out previously, even though the analysis in this section is carried out based on the example of Sector 6, the obtained conclusion can be also extended to other sectors. Besides, the calculated duration times for selected voltage vectors turn out to be negative under the situation of step change in d-axis currents. Similar analysis and the corresponding modifications of vector selection rules can be easily obtained based on previous contents.

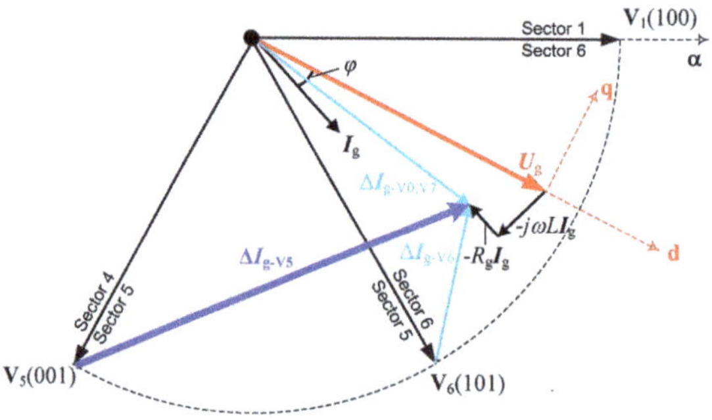

Figure 4. Instantaneous current variations caused by the application of V_5, V_6 and $V_{0,7}$ when U_g lies in the middle part of Sector 6 (rectifier mode).

3.2. Inverter Mode Analysis

In some applications, the three-phase grid-tied converter may operate in the inverter mode. In this case, the instantaneous current variation resulting from the application of voltage vector V_i ($i = 0, \ldots, 7$) should be rewritten as:

$$\Delta I_{g\text{-}Vi} = \frac{1}{L_g}(V_i - R_g I_g - j\omega L_g I_g - U_g) \tag{8}$$

As argued previously, when the converter is operating in rectifier mode and U_g is located in the initial part of Sector 6, the predictive controller will lose its effectiveness if V_1 and V_6 are selected as active voltage vectors. However, this is not the case when the converter operates in the inverter mode. Figure 5 depicts instantaneous current variations when U_g lies in the initial part of Sector 6.

It can be clearly observed in Figure 5 that, when the converter is operating in the inverter mode, the highlighted blue area composed of three vectors $\Delta I_{g\text{-}V1}$, $\Delta I_{g\text{-}V6}$, $\Delta I_{g\text{-}V0,V7}$ covers four quadrants in the dq-reference frame. The increase and decrease of the dq-axes currents can both be achieved. Therefore, when the converter is operating in the inverter mode and U_g is located in the initial part of Sector 6, selection of V_6 and V_1 as active vectors can realize flexible regulation of both d- and q-axis currents. This phenomenon is quite different from that when the converter is operating in the rectifier mode, as mentioned previously.

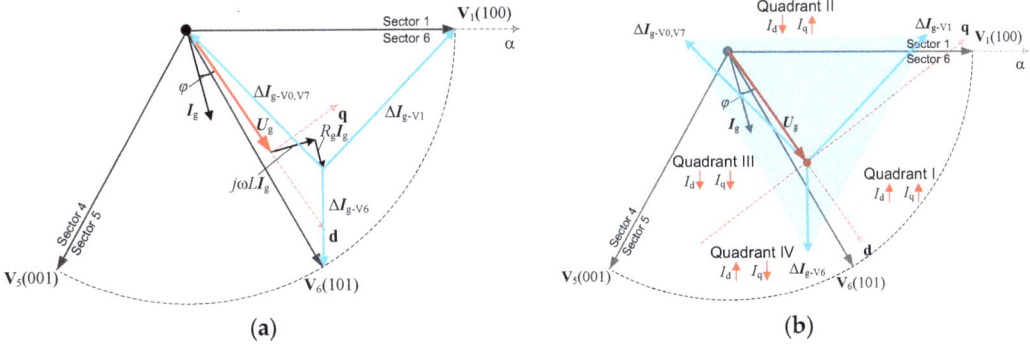

Figure 5. Instantaneous current variations $\Delta I_{g\text{-}V6}$, $\Delta I_{g\text{-}V1}$, $\Delta I_{g\text{-}V0,V7}$ in the dq-reference frame when U_g is located in the initial part of Sector 6 (inverter mode). (**a**) Original diagram; (**b**) Diagram after vector movement and rearrangement.

However, when U_g is located in the latter part of each sector, incorrect selection of active vectors will come out. Taking Sector 5 for example, when U_g is located in the latter part of this sector, V_5 and V_6 should normally be selected and applied. Figure 6 shows the instantaneous current variations by the application of V_5, V_6 and $V_{0,7}$.

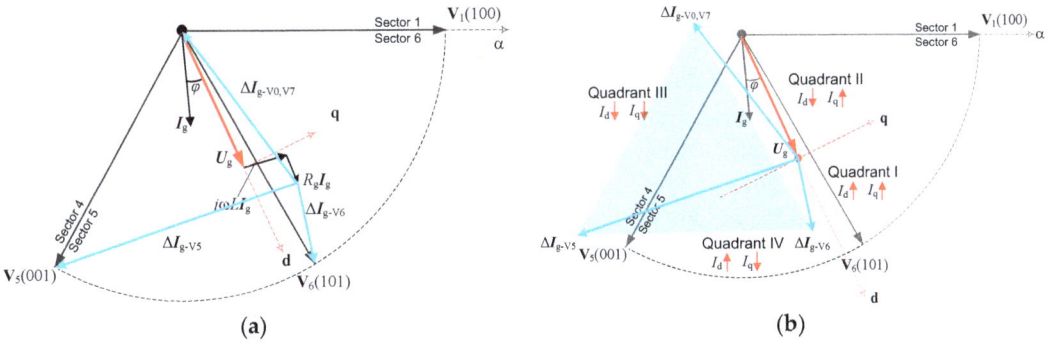

Figure 6. Instantaneous current variations caused by the application of V_5, V_6 and $V_{0,7}$ when U_g is located in the latter part of Sector 5 (inverter mode). (**a**) Original diagram; (**b**) diagram after vector movement and rearrangement.

It is clearly visible in Figure 6 that the light blue area composed of three vectors $\Delta I_{g\text{-}V5}$, $\Delta I_{g\text{-}V6}$, $\Delta I_{g\text{-}V0,V7}$ only covers Quadrants III and IV in the dq-reference frame. Even though the increase and decrease of the d-axis current can both be achieved, the q-axis current will only be decreased under the joint actions of these voltage vectors. When V_5 and V_6 are still selected and applied in this case, the minimization of the cost function will lead to negative duration time t_5. A similar phenomenon can also be observed when U_g is located in other sectors. In general, under inverter mode, the duration time t_m of the voltage vector V_m will become negative when the grid voltage vector is located in the latter part of each sector. The negative duration time will be forced to zero during the application process, resulting in performance deterioration. This will be further validated by the experimental waveforms to be presented in the following sections. It should be noted that, different from the situation of rectifier mode, as shown in Figure 2, incorrect selections of active vectors come out when U_g lies in the latter part of different sectors, instead of their initial part.

Actually, V_5 should be replaced by V_1 when U_g is located in the latter part of Sector 5. Subsequently, V_1 is applied together with V_6. In order to clarify this conclusion, Figure 7 depicts instantaneous current variations under the joint actions of V_1, V_6 and $V_{0,7}$ when U_g is located in the latter part of Sector 5.

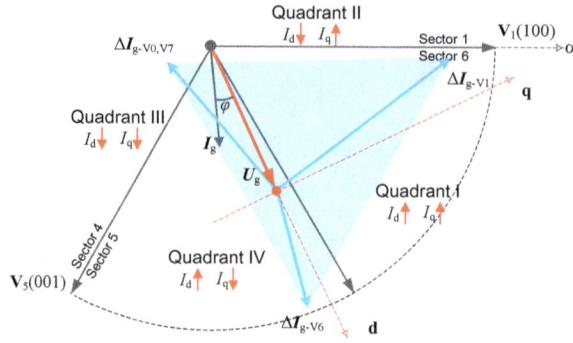

Figure 7. Instantaneous current variations caused by the application of V_1, V_6 and $V_{0,7}$ when U_g is located in the latter part of Sector 5 (inverter mode).

Obviously, when V_1 and V_6 are selected as active vectors in this case, instantaneous current variation vectors can reach all four quadrants in the dq-reference frame. It can be inferred that the dq-axes currents can both be increased and decreased with the joint actions of V_1, V_6 and $V_{0,7}$. Flexible regulation of converter currents can thus be achieved under this situation.

Moreover, with a step change in the reference value of currents, the selection of two adjacent vectors according to the angular location of U_g will also result in negative duration time. Again, Sector 6 is taken as an example here. When the q-axis reference current is step-up changed, the duration time t_6 for voltage vector V_6 is negative, while t_1 for voltage vector V_1 remains positive. This is because large current variations are normally desired when the reference signals are step changed. However, the joint actions of V_6 and V_1 could not produce such large current variations as desired. Under this situation, other combinations of voltage vectors that can generate larger current variations should be chosen and applied, instead of V_6 and V_1. For the purpose of illustration, Figure 8 shows instantaneous current variations caused by the joint actions of V_1, V_2 and $V_{0,7}$ when U_g is located in Sector 6.

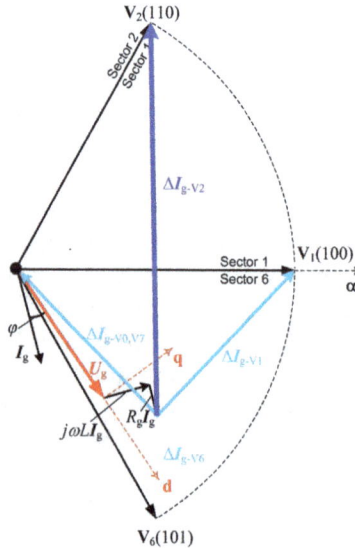

Figure 8. Instantaneous current variations caused by the application of V_1, V_2 and $V_{0,7}$ when U_g is located in Sector 6 (inverter mode).

In fact, V_1 and V_2 should be selected when the q-axis reference current is step-up changed. This is due to fact that the combination of V_1 and V_2 can generate much larger positive variations of q-axis currents than the joint actions of V_6 and V_1. This conclusion can be easily drawn by comparing Figure 8 to Figure 5. As argued previously, the analysis in this section is based on the example of Sector 6. Similar conclusions can be easily obtained when the grid voltage vector is located in other sectors. The corresponding modifications of vector selection rules can be easily obtained based on previous contents.

3.3. Modified Switching Patterns for Reversible Converters

Based on the analysis mentioned above, modified switching patterns for reversible converters can be summarized as Table 1. For clarity and better understanding, the improved switching patterns mentioned above can be explained as follows. Taking Sector I for example, when the grid voltage U_g is located in Sector I, two adjacent voltage vectors V_1 and V_2 are firstly selected. Their duration times t_m (t_1) and t_n (t_2) are then determined by minimizing the cost function. If t_1 and t_2 are both positive ($t_m > 0$ and $t_n > 0$), V_1 and V_2 are directly applied, as well as the zero vector $V_{0,7}$. If t_1 is negative while t_2 keeps positive ($t_m < 0$ and $t_n > 0$), V_2 and V_3 should be chosen to replace the combination of V_1 and V_2. If t_2 is negative while t_1 remains positive ($t_m > 0$ and $t_n < 0$), V_6 and V_1 are to be applied. When t_1 and t_2 are both negative ($t_m < 0$ and $t_n < 0$), the corresponding opposite vectors to V_4 and V_5 should be selected in this case. Finally, the application times of re-selected vectors are then calculated based on cost function minimization.

The illustrative block diagram of the proposed control strategy is depicted in Figure 9, where the current references are obtained based on the outer voltage control loop. The pulse width modulation (PWM) block is used to generate switching signals for the converter.

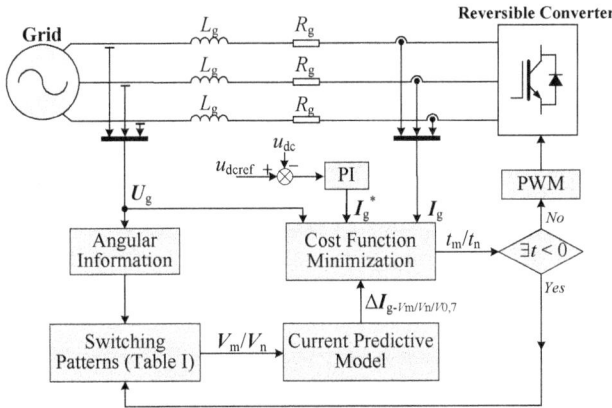

Figure 9. Block diagram of the proposed control strategy.

Table 1. Improved switching patterns for the proposed strategy.

Sector	$t_m > 0$ and $t_n > 0$	$t_m > 0$ and $t_n < 0$	$t_m < 0$ and $t_n > 0$	$t_m < 0$ and $t_n < 0$
1	V_1, V_2	V_6, V_1	V_2, V_3	V_4, V_5
2	V_2, V_3	V_1, V_2	V_3, V_4	V_5, V_6
3	V_3, V_4	V_2, V_3	V_4, V_5	V_6, V_1
4	V_4, V_5	V_3, V_4	V_5, V_6	V_1, V_2
5	V_5, V_6	V_4, V_5	V_6, V_1	V_2, V_3
6	V_6, V_1	V_5, V_6	V_1, V_2	V_3, V_4

4. Experimental Validations

With the aim of comparing the performances of the proposed improved predictive control strategy with those of the conventional controller, extensive experimental research on steady-state behaviours and transient responses was carried out on a three-phase grid-tied reversible converter prototype. The parameters are listed in Table 2. Throughout the experiments, a Danfoss FC302 converter for industrial applications was used to realize reversible power conversion. Besides, a variac was adopted to connect the AC supply and the experimental prototype. The algorithm processing, AD input, DA output and pulse width modulation (PWM) generation are realized by a dSPACE DS1006 board, a DS 2004 ADC board, a DS2102 DAC board and a DS5101 PWM board, respectively. During the tests, the sampling period is chosen as 100 μs. It should be noted that, for the purposes of convenience, the conventional predictive current controller is referred to as Method I, while the proposed strategy is referred to as Method II.

Table 2. Parameters of the experimental platform.

Parameter	Value
Rated power	11 kW
Line-to-line voltage (Root mean squar value)	245 V
Grid frequency	50 Hz
Filter inductance	7.8 mH
Filter resistance	0.1 Ω
DC-link capacitor	950 μF
Sampling period	100 μs

4.1. Steady-State Behaviours

When Method I is adopted, two adjacent voltage vectors are normally selected as active vectors according to the angular position of the grid voltage vector. However, this might lead to negative duration time when the cost function is minimized. Conventionally, the duration time of voltage vectors should be larger than zero. The negative duration time will thus be forced to zero. In order to evaluate its influence on system steady-state behaviours when the converter is operating in rectifier and inverter modes, relative experiments were conducted. Here, i_d^* and i_q^* are used to denote the reference signals for d- and q-axis currents, respectively. As is clearly visible in Figures 10 and 11 severe periodic fluctuations can be observed in both d- and q-axes currents. In other words, when the negative duration time comes out and is forced to zero, the phenomenon of control failure comes out, as shown in Figures 10b and 11b. Under this situation, the dq-axis currents could not be regulated in a smooth way, therefore generating significant harmonic distortions in thephase current.

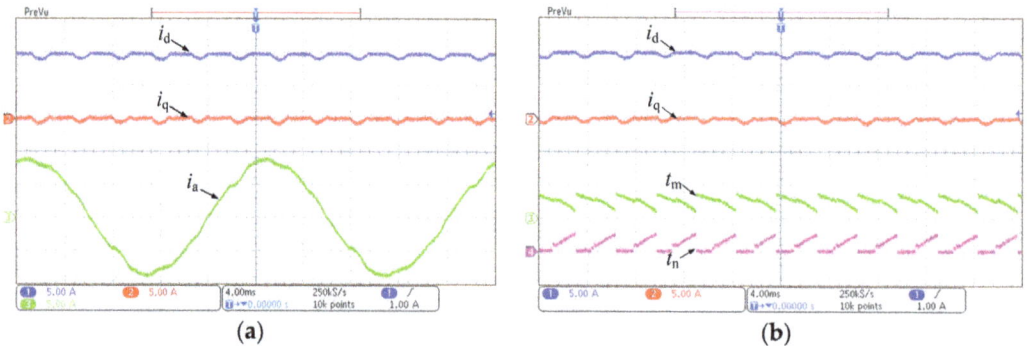

(a)   (b)

Figure 10. Experimental results: dq-axis currents, Phase A current, t_m and t_n (rectifier mode, Method I, $i_d^* = 10$ A, $i_q^* = 0$ A). (a) dq-axis and Phase A currents; (b) t_m and t_n (50 μs/division).

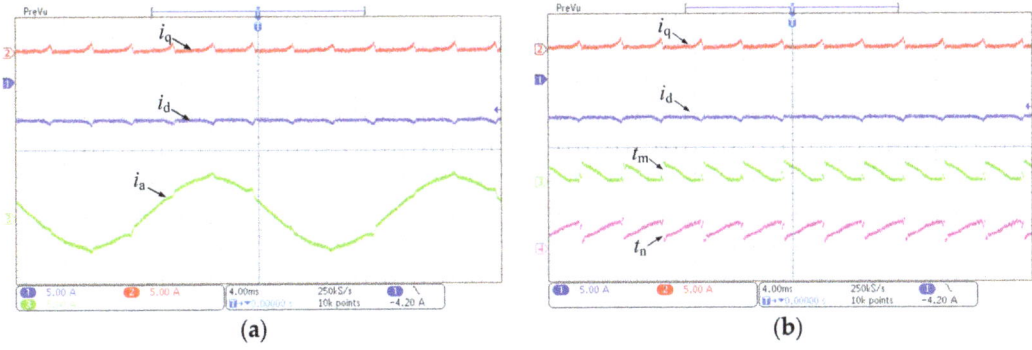

Figure 11. Experimental results: dq-axis currents, Phase A current, t_m and t_n (inverter mode, Method I, $i_d^* = 6$ A, $i_q^* = 0$ A). (a) dq-axis and Phase A currents; (b) t_m and t_n (50 μs/division).

As argued in previous sections, the control failure and performance deterioration demonstrated in Figures 10 and 11 result from the incorrect selection of voltage vectors. Those two voltage vectors, which are adjacent to the grid voltage vectors, are not always the optimum vectors. Figure 12 presents experimental waveforms when the proposed Method II is adopted and the converter is operating in rectifier mode. These experimental results were obtained under the same conditions as those of Figure 10. It can be seen that the proposed Method II presents superior performances to the conventional predictive control strategy. Due to the modified switching pattern shown in Figure 12c, both t_m and t_n stay positive at all times when the proposed strategy is used, as shown in Figure 12b. Periodic fluctuations in the dq-axes currents are successfully eliminated as expected, and as a result, dq-axes currents can both be regulated to be smooth and constant. The sinusoidal AC current is further obtained. Similar results can be obtained when the converter is operating in inverter mode, as shown in Figure 13.

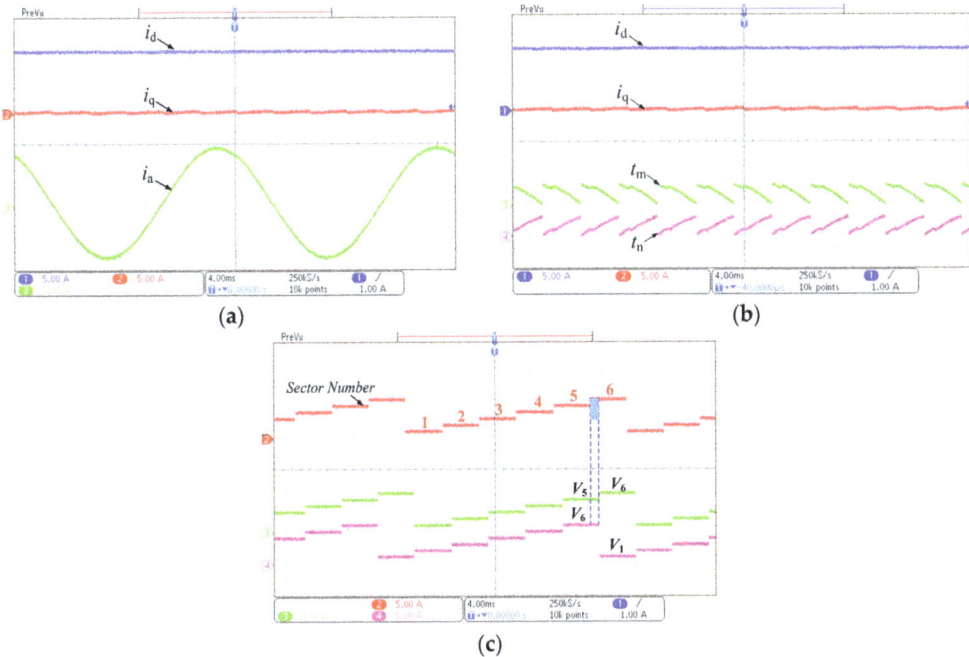

Figure 12. Experimental results: dq-axis currents, Phase A current, t_m and t_n (rectifier mode, Method II, $i_d^* = 10$ A, $i_q^* = 0$ A). (a) dq-axis and Phase A currents; (b) t_m and t_n (50 μs/division); (c) selected voltage vectors.

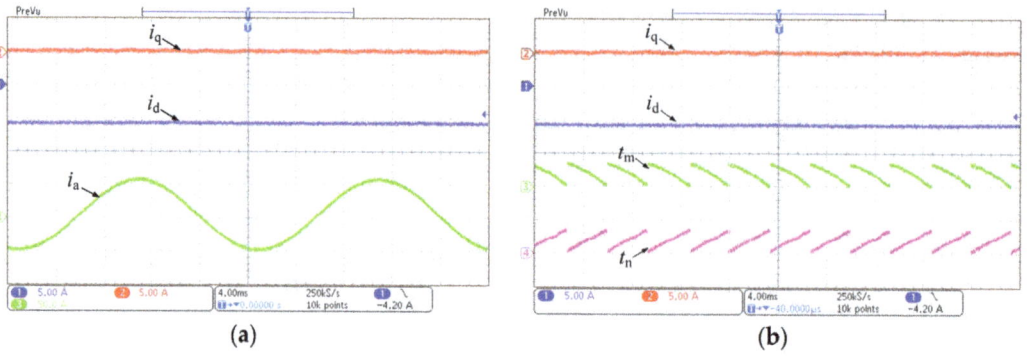

Figure 13. Experimental results: dq-axis currents and Phase A current (inverter mode, Method II, $i_d{}^* = 6$ A, $i_q{}^* = 0$ A). (a) dq-axis and Phase A currents; (b) t_m and t_n (50 μs/division).

4.2. Transient Responses

For the purpose of further illustrating the superiority of Method II with regard to the transient responses, experimental studies were carried out. Figures 14 and 15 present the experimental waveforms with Method I adopted. As pointed out previously, when the reference current is step changed, the conventional Method I will lead to incorrect selection of the voltage vector and negative duration time, which as a result, deteriorates the transient behaviours of the converter. As shown in Figures 14 and 15 a slow transient response can be observed for both rectifier and inverter modes, and furthermore, the cross-coupling effect arises. The d-axis is significantly influenced by the step change of the q-axis current.

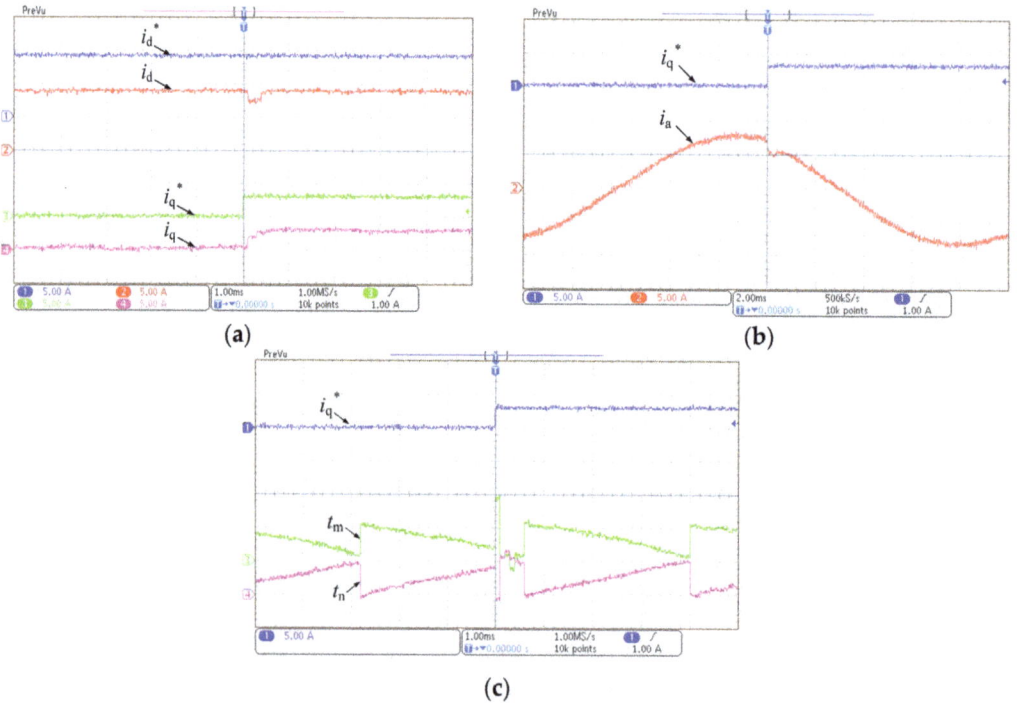

Figure 14. Experimental results: dq-axis currents and Phase A current with $i_q{}^*$ step changed from 0 to 3 A (rectifier mode, Method I, $i_d{}^* = 9$ A). (a) dq-axis currents; (b) Phase A current; (c) t_m and t_n (25 μs/division).

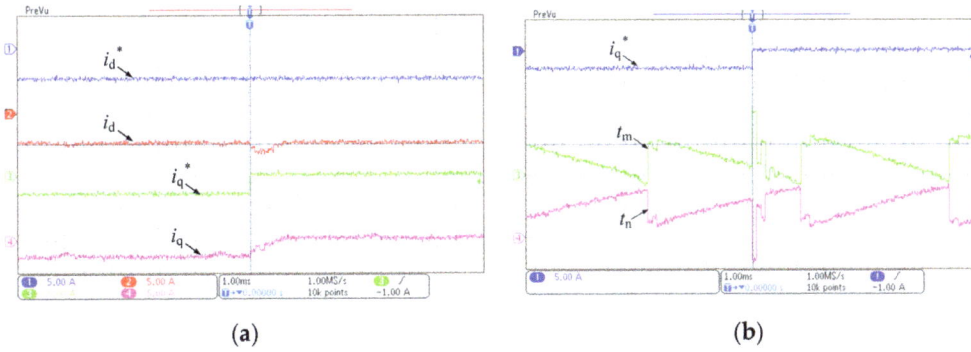

Figure 15. Experimental results: dq-axis currents and Phase A current with i_q^* step changed from 0 A to 3 A (inverter mode, Method I, $i_d^* = 0$ A). (**a**) dq-axis currents; (**b**) t_m and t_n (25 μs/division).

With the aim of performance comparison, Figure 16 demonstrates the results when Method II is used and the converter is operating in rectifier mode. These figures were obtained under the same situation as those of Figure 14. By comparing Figures 14 and 16 the improvements are evident with respect to dynamic responses. Highly dynamic behaviours can be expected with Method II. Actually, these excellent behaviours presented by Method II result from the improved switching patterns, which avoid the negative application times, as shown in Figure 16c,d. A similar conclusion can be made when system behaviours in the case of inverter mode were compared, as show in Figures 15 and 17.

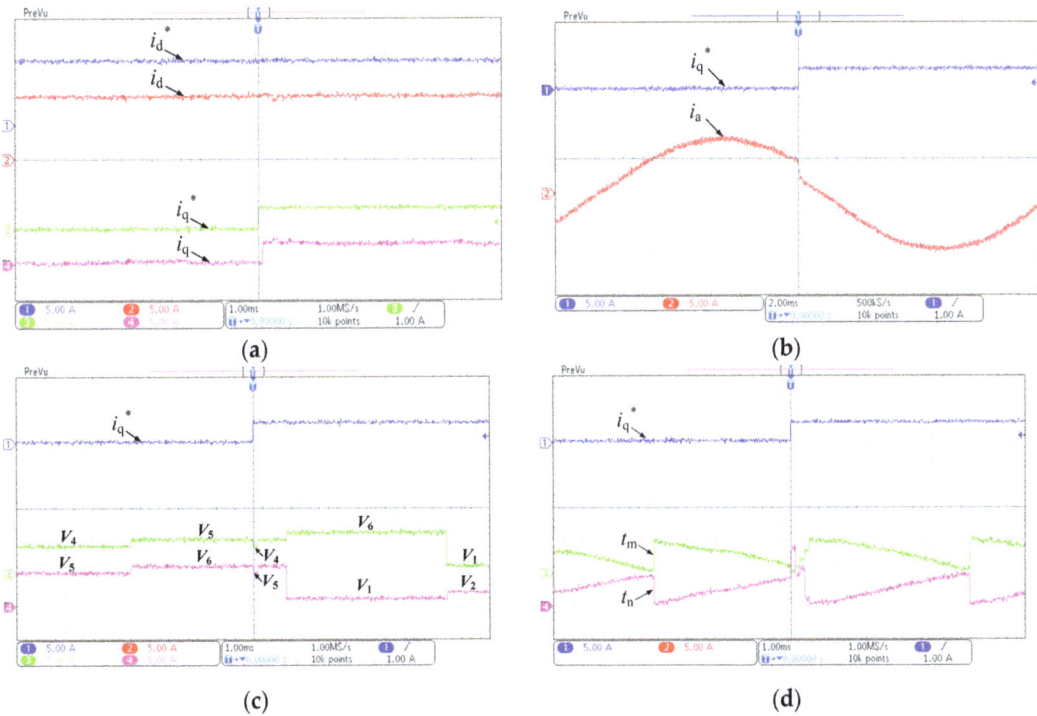

Figure 16. Experimental results: dq-axis currents and Phase A current with i_q^* step changed from 0 A to 3 A (rectifier mode, Method II, $i_d^* = 0$ A). (**a**) dq-axis currents; (**b**) Phase A current; (**c**) selected voltage vectors; (**d**) t_m and t_n (25 μs/division).

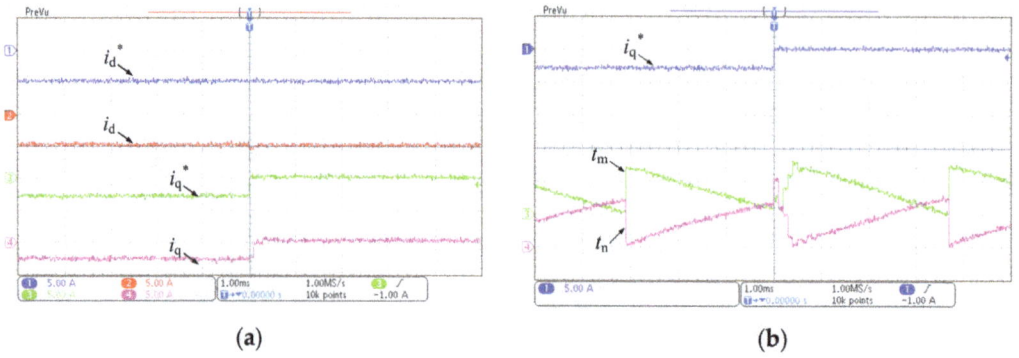

(a)   (b)

Figure 17. Experimental results: dq-axis currents and Phase A current with $i_q{}^*$ step changed from 0 A to 3 A (inverter mode, Method II, $i_d{}^* = 0$ A). (a) dq-axis currents; (b) t_m and t_n (25 µs/division).

Besides, system transient behaviours with an outer DC-voltage control loop were also tested. Figure 18 demonstrates system transient responses when the converter was operating in rectifier mode and a disturbance was introduced to the resistance load of the DC-link. The control strategy for the DC-link voltage was the typical PI controller, which generated the reference signals for the d-axis current. Additionally, the q-axis current remained zero. As can be seen in the waveforms, the d-axis current can track the reference current in a good way, and the q-axis was not influenced by this transient process, demonstrating the decoupling performances of the proposed controller.

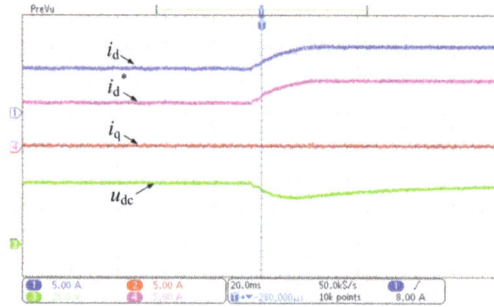

Figure 18. Experimental results: dq-axis currents and dc voltage with load disturbances (Method II; reference DC voltage: 445 V; offset voltage of DC voltage signal: 400 V).

5. Conclusions

When the predictive current control strategy is adopted, three-phase grid-tied converters can be flexibly regulated based on system behaviour prediction and cost function minimization. However, due to the adoption of conventional switching patterns, system performance deteriorations often take place under the conditions of both bidirectional power flows and transient situations. This paper analysed the switching patterns of three-phase grid-tied reversible converters under different conditions and investigated the predicted duration time of voltage vectors under bidirectional power flows, i.e., rectifier mode and inverter mode. A detailed analysis of switching patterns was further conducted. An enhanced predictive current control strategy with improved switching patterns was then proposed. According to the obtained experimental results, the system performances of three-phase grid-tied reversible converters were improved in terms of both steady-state and transient behaviours.

Acknowledgments: This work was supported by the National Science Foundation of China under Grants 51477113 and 51107084.

Author Contributions: Zhanfeng Song and Yanjun Tian contributed to experimental validations. Zhe Chen and Yanting Hu contributed to the result analysis.

Conflicts of Interest: The authors declare no conflict of interest.

References

1. Vazquez, S.; Leon, J.I.; Franquelo, L.G.; Rodriguez, J.; Young, H.A.; Marquez, A.; Zanchetta, P. Model predictive control: A review of its applications in power electronics. *IEEE Ind. Electron. Mag.* **2014**, *8*, 16–31. [CrossRef]

2. Min, R.; Chen, C.; Zhang, X.; Zou, X.; Tong, Q.; Zhang, Q. An optimal current observer for predictive current controlled buck DC-DC converters. *Sensors* **2014**, *14*, 8851–8868. [CrossRef] [PubMed]

3. Tong, Q.; Chen, C.; Zhang, Q.; Zou, X. A sensorless predictive current controlled boost converter by using an EKF with load variation effect elimination function. *Sensors* **2015**, *15*, 9986–10003. [CrossRef] [PubMed]

4. Zhang, Y.; Xie, W.; Li, Z.; Zhang, Y. Low-complexity model predictive power control: Double-vector-based approach. *IEEE Trans. Ind. Electron.* **2014**, *61*, 5871–5880. [CrossRef]

5. Aguilera, R.P.; Quevedo, D.E.; Vazquez, S.; Franquelo, L.G. Generalized predictive direct power control for AC/DC converters. In Proceeding of the IEEE Annual International Energy Conversion Congress and Exhibition (ECCE Asia), Melbourne, VIC, Australia, 3–6 June 2013.

6. Yaramasu, V.; Rivera, M.; Narimani, M.; Wu, B.; Rodriguez, J. Model predictive approach for a Simple and effective load voltage control of four-leg inverter with an output LC filter. *IEEE Trans. Ind. Electron.* **2014**, *61*, 5259–5270. [CrossRef]

7. Riveros, J.A.; Barrero, F.; Levi, E.; Durán, M.J.; Toral, S.; Jones, M. Variable-speed five-phase induction motor drive based on predictive torque control. *IEEE Trans. Ind. Electron.* **2013**, *60*, 2957–2968. [CrossRef]

8. Tarisciotti, L.; Zanchetta, P.; Watson, A.; Bifaretti, S.; Clare, J.C. Modulated model predictive control for a seven-level cascaded h-bridge back-to-back converter. *IEEE Trans. Ind. Electron.* **2014**, *61*, 5375–5383. [CrossRef]

9. Zhang, Q.; Min, R.; Tong, Q.; Zou, X.; Liu, Z.; Shen, A. Sensorless predictive current controlled DC-DC converter with a self-correction differential current observer. *IEEE Trans. Ind. Electron.* **2014**, *61*, 6747–6757. [CrossRef]

10. Tong, Q.; Zhang, Q.; Min, R.; Zou, X.; Liu, Z.; Chen, Z. Sensorless Predictive peak current control for boost converter using comprehensive compensation strategy. *IEEE Trans. Ind. Electron.* **2014**, *61*, 2754–2766. [CrossRef]

11. Vu, T.; Chen, C.K.; Hung, C.W. A model predictive control approach for fuel economy improvement of a series hydraulic hybrid vehicle. *Energies* **2014**, *7*, 7017–7040. [CrossRef]

12. Yang, J.; Zeng, Z.; Tang, Y.; Yan, J.; He, H.; Wu, Y. Load Frequency control in isolated micro-grids with electrical vehicles based on multivariable generalized predictive theory. *Energies* **2015**, *8*, 2145–2164. [CrossRef]

13. Geyer, T.; Papafotiou, G.; Morari, M. Model predictive direct torque control—Part I: Concept, algorithm and analysis. *IEEE Trans. Ind. Electron.* **2009**, *56*, 1894–1905. [CrossRef]

14. Papafotiou, G.; Kley, J.; Papdopoulos, K.; Bohren, P.; Morari, M. Model predictive direct torque control—Part II: Implementation and experimental evaluation. *IEEE Trans. Ind. Electron.* **2009**, *56*, 1906–1915. [CrossRef]

15. Karamanakos, P.; Pavlou, K.; Manias, S. An enumeration-based model predictive control strategy for the cascaded H-bridge multilevel rectifier. *IEEE Trans. Ind. Electron.* **2014**, *61*, 3480–3489. [CrossRef]

16. Choi, D.; Lee, K. Dynamic performance improvement of AC/DC converter using model predictive direct power control with finite control set. *IEEE Trans. Ind. Electron.* **2015**, *62*, 757–767. [CrossRef]

17. Lim, C.; Levi, E.; Jones, M.; Rahim, N.A.; Hew, W. A fault-tolerant two-motor drive with FCS-MP-Based flux and torque control. *IEEE Trans. Ind. Electron.* **2014**, *61*, 6603–6614. [CrossRef]

18. Ramírez, R.O.; Espinoza, J.R.; Villarroel, F.; Maurelia, E.; Reyes, M.E. A novel hybrid finite control set model predictive control scheme with reduced switching. *IEEE Trans. Ind. Electron.* **2014**, *61*, 5912–5920. [CrossRef]

19. Davari, S.S.; Khaburi, D.; Kennel, R. An improved FCS–MPC algorithm for an induction motor with an imposed optimized weighting factor. *IEEE Trans. Power Electron.* **2012**, *27*, 1540–1551. [CrossRef]

20. Calle-Prado, A.; Alepuz, S.; Bordonau, J.; Nicolas-Apruzzese, J.; Cortés, P.; Rodriguez, J. Model predictive current control of grid-connected neutral-point-clamped converters to meet low-voltage ride-through requirements. *IEEE Trans. Ind. Electron.* **2015**, *62*, 1503–1514. [CrossRef]

21. Yaramasu, V.; Wu, B. Model predictive decoupled active and reactive power control for high-power grid-connected four-level diode-clamped inverters. *IEEE Trans. Ind. Electron.* **2014**, *61*, 3407–3416. [CrossRef]

22. Yaramasu, V.; Wu, B.; Alepuz, S.; Kouro, S. Predictive control for low-voltage ride-through enhancement of three-level-boost and NPC-Converter-Based PMSG wind turbine. *IEEE Trans. Ind. Electron.* **2014**, *61*, 6832–6843. [CrossRef]

23. Zhang, Z.; Xu, H.; Xue, M.; Chen, Z.; Sun, T.; Kennel, R.; Hackl, C.M. Predictive control with novel virtual-flux estimation for back-to-back power converters. *IEEE Trans. Ind. Electron.* **2015**, *62*, 2823–2834. [CrossRef]

24. López, M.; Rodriguez, J.; Silva, C.; Rivera, M. Predictive torque control of a multidrive system fed by a dual indirect matrix converter. *IEEE Trans. Ind. Electron.* **2015**, *62*, 2731–2741. [CrossRef]

25. Lim, C.; Levi, E.; Jones, M.; Rahim, N.A.; Hew, W. A comparative study of synchronous current control schemes based on FCS-MPC and PI-PWM for a two-motor three-phase drive. *IEEE Trans. Ind. Electron.* **2014**, *61*, 3867–3878. [CrossRef]

26. Cortes, P.; Rodriguez, J.; Quevedo, D.E.; Silva, C. Predictive current control strategy with imposed load current spectrum. *IEEE Trans. Power Electron.* **2008**, *23*, 612–618. [CrossRef]

27. Song, Z.; Xia, C.; Liu, T. Predictive current control of three-phase grid-connected converters with constant switching frequency for wind energy systems. *IEEE Trans. Ind. Electron.* **2013**, *60*, 2451–2464. [CrossRef]

28. Xia, C.; Wang, M.; Song, Z.; Liu, T. Robust model predictive current control of three-phase voltage source PWM rectifier with online disturbance observation. *IEEE Trans. Ind. Inf.* **2012**, *8*, 459–471. [CrossRef]

29. Vazquez, S.; Marquez, A.; Aguilera, R.; Quevedo, D.; Leon, J.I.; Franquelo, L.G. Predictive Optimal switching sequence direct power control for grid-connected power converters. *IEEE Trans. Ind. Electron.* **2015**, *62*, 2010–2020. [CrossRef]

30. Song, Z.; Xia, C.; Liu, T.; Dong, N. A modified predictive control strategy of three-phase grid-connected converters with optimized action time sequence. *Sci. China Technol. Sci.* **2013**, *56*, 1017–1028. [CrossRef]

31. Hu, J.; Zhu, Z. Investigation on switching patterns of direct power control strategies for grid-connected DC–AC converters based on power variation rates. *IEEE Trans. Power Electron.* **2011**, *26*, 3582–3598. [CrossRef]

32. Song, Z.; Chen, W.; Xia, C. Predictive direct power control for three-phase grid-connected converters without sector information and voltage vector selection. *IEEE Trans. Power Electron.* **2014**, *29*, 5518–5531. [CrossRef]

33. Sergio, A.; Rodriguez, M.A.; Oyarbide, E.; Torrealday, J.R. Predictive direct power control—A new control strategy for DC/AC converters. In Proceeding of the IECON 2006–32nd Annual Conference on IEEE Industrial Electronics, Paris, France, 6–10 November 2006; pp. 1661–1666.

34. Larrinaga, S.A.; Vidal, M.A.R.; Oyarbide, E.; Apraiz, J.R.T. Predictive control strategy for DC/AC converters based on direct power control. *Ind. Electron. IEEE Trans.* **2007**, *54*, 1261–1271. [CrossRef]

5

Assessing the Potential of Plug-in Electric Vehicles in Active Distribution Networks

Reza Ahmadi Kordkheili [1,*], Seyyed Ali Pourmousavi [2], Mehdi Savaghebi [1], Josep M. Guerrero [1] and Mohammad Hashem Nehrir [3]

Academic Editor: K. T. Chau

[1] Department of Energy Technology, Aalborg University, Pontoppidanstraede 101, Aalborg 9220, Denmark; mes@et.aau.dk (M.S.); joz@et.aau.dk (J.M.G.)

[2] NEC Laboratories America Incorporations, Cupertino, CA 95014, USA; s.pourmousavikani@msu.montana.edu

[3] Electrical and computer engineering department, Montana State University, Bozeman, MT 59717, USA; hnehrir@ece.montana.edu

[*] Correspondence: rak@et.aau.dk

Abstract: A multi-objective optimization algorithm is proposed in this paper to increase the penetration level of renewable energy sources (RESs) in distribution networks by intelligent management of plug-in electric vehicle (PEV) storage. The proposed algorithm is defined to manage the reverse power flow (PF) from the distribution network to the upstream electrical system. Furthermore, a charging algorithm is proposed within the proposed optimization in order to assure PEV owner's quality of service (QoS). The method uses genetic algorithm (GA) to increase photovoltaic (PV) penetration without jeopardizing PEV owners' (QoS) and grid operating limits, such as voltage level of the grid buses. The method is applied to a part of the Danish low voltage (LV) grid to evaluate its effectiveness and capabilities. Different scenarios have been defined and tested using the proposed method. Simulation results demonstrate the capability of the algorithm in increasing solar power penetration in the grid up to 50%, depending on the PEV penetration level and the freedom of the system operator in managing the available PEV storage.

Keywords: optimization; plug-in electric vehicle (PEV); photovoltaic (PV) panels; state of charge (SoC); vehicle to grid (V2G)

1. Introduction

Inevitable presence of renewable energy sources (RESs) in modern power system has led to significant changes in different operation, protection, and management aspects of power systems. The distribution network plays a major role in the system overall performance, as it is the interface between the customers and the electric network. The main purpose of the network in the traditional power system was to respond to the customers' demand anytime, no matter the circumstances. However, the improving and wide acceptance of RESs technologies in the last decade required new concepts for the design and operation of distribution networks, their functionality, and the role of customers in the new electric network paradigm [1]. The availability of RES for household application has changed the traditional customers to "prosumers", where a customer can be a producer as well [2]. This paradigm shift introduces new opportunities as well as new challenges for the grid operation. In the future power system, each prosumer will be able to sell the excess available power to the upper grid, if requested, as a part of ancillary services [2]. However, new functionalities lead to certain challenges for grid operators, such as the unidirectional structure of the existing distribution networks [3,4], and voltage issues in the grid [5,6]. Therefore, optimal sizing and siting of RESs in distribution networks has received significant interest in recent years. An optimal placement of distributed generation (DG) units is discussed in [7,8] to reduce network losses. Lin *et al.* [9] proposes an optimization algorithm for photovoltaic (PV) penetration in distribution systems in order to maximize the net present value (NPV) of PV panels. An optimization algorithm based on a genetic algorithm (GA) was proposed in [10], with the main objective of maximizing the savings in the system upgrade investment deferral, and minimizing cost of annual energy loss. On the other hand, increasing interest in plug-in electric vehicles (PEVs), especially PEVs with lithium-ion (Li-ion) storage, has led to a huge interest in application of PEVs for grid support and ancillary services. The main role of ancillary service is to enable the system operator to deal with balancing issues. PV generation is variable and non-dispatchable. Therefore, its application in ancillary service depends on the availability and application of storage devices. The application of PEVs can be beneficial to overcome a part of inherent variability of PV generation, and also to improve the applicability of distributed RES for ancillary services [11]. Different research works have addressed the PEV potential for grid support [11–20]. On the other hand, due to technology improvement and cost reduction, PEVs are becoming popular, imposing new loads on power system. The grid features are normally designed considering current (and the traditional future forecast) customers' consumption. Therefore, the low voltage (LV) grid would face new challenges as PEVs add on to the grid. As a result, PEVs charging scheduling becomes important for system operators, and has attracted attention from many research groups [13].

In this paper, an algorithm is proposed based on optimal management of PEV storage, in order to maximize PV participation in ancillary services. This way, part of the required ancillary services will be provided by the PV panels installed in distribution networks to alleviate the system operational costs. In addition, the total power loss of the system is included in the optimization procedure, so that the system energy losses will also be minimized. Although an optimal placement of DGs has been proposed in [7,8], but the main objective in these literature was to minimize the network losses, and providing ancillary service has not been considered in their optimization. On the other hand, the proposed algorithm in [9] aims at maximizing NPV by optimizing the placement of PV panels in the network. Therefore, it does not consider ancillary service as its major objective, and network losses are not a big concern in this study. In addition, the proposed GA optimization in [10] aims at minimizing the system investment cost, as well as minimizing annual energy loss. On the other hand, most of the studies on PEVs, such as [13,15], are focused on optimal charging strategies, regardless of the potential of PEVs for grid support. In [16], a vehicle to grid (V2G) strategy has been discussed for PEVs, but the role of DGs has not been a major focus in [16].

A multi-objective optimization procedure is proposed to obtain the best viable solution for this purpose [21–23]. Appropriate voltage profile at different buses and the quality of service (QoS) for the PEV owners are among the major constraints in the proposed algorithm. Considering the physical constraints of electric equipment in the system, *i.e.*, transformer nominal power and nominal currents of the cables, the problem formulation should take these limits into account as well. On the other hand, since the PEVs are utilized by the algorithm for grid support, the proposed algorithm should include a proper charging algorithm for PEVs to avoid any inconvenience for the PEV owners. The rest of the paper is organized as follows. Section 2 provides details of the proposed optimization method, together with algorithms and driving patterns proposed for electric vehicles (EVs). Section 3 explains the system modeling approach, including load, PV panels, and PEVs. Section 4 is dedicated to simulation results. Conclusions of the work results are presented in Section 5.

2. Proposed Intelligent Algorithm

The proposed optimization algorithm is presented in this section. A general overview of the LV grid is presented in Figure 1. In general, LV grids have a radial layout, and each distribution transformer is connected to four to seven LV distribution feeders. In this work, the 10 kV side of the grid is referred to as the "upper grid", while the 0.4 kV side is considered the "LV grid".

An overview of the proposed algorithm is shown in Figure 2. At the beginning, the "Objective Function" block generates optimization variables, "*x*", *i.e.*, number and locations of PV panels based on the optimization data. The generated PV sizes and locations along with solar irradiation and ambient temperature data will be utilized to prepare the power flow (PF) matrices. Then, the first PF study will be carried out to calculate the voltage magnitudes of different buses (PF-1 Block). The PF studies are based on Newton-Raphson load flow. The calculated voltages will be used in the PEV Block to compute PEV charging pattern, and PEV for grid support. Then, the load flow matrices will be modified based on the impact of the PEVs, and the load flow runs again (PF-3 Block). Finally, the required data for the Optimization Block will be calculated based on the recent load flow. In the following sub-sections, each block will be discussed in detail.

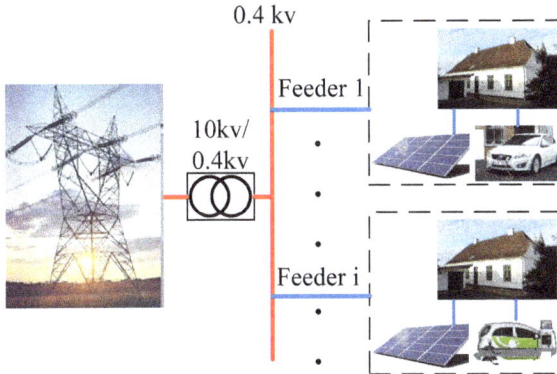

Figure 1. General layout of low voltage (LV) grid.

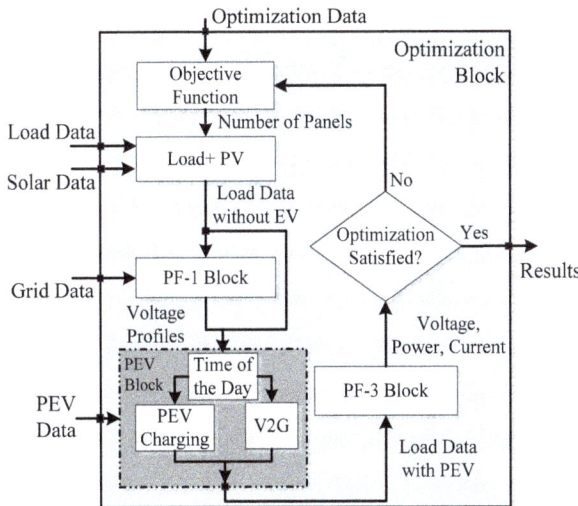

Figure 2. Proposed optimization method.

2.1. Objective Function

This block is responsible for finding "*x*", *i.e.*, the optimal number and location of the PV panels. Optimization data include the maximum and minimum number of panels that can be installed at each bus, and optimization technique parameters. The maximum number of panels is obtained by assuming that each household cannot install more than 2 units of "3 kW PV panel". This is a fair assumption, considering the available roof area of a house. As a standard procedure for optimization problems, the first step is to define an appropriate objective function. Increasing the penetration of PV panels in the grid will lead to higher PF from the LV grid to the upper grid. Therefore, the main objective is to maximize the PF to the upper grid. This objective is defined as "f_1" in Equation (3). Two other objectives are defined as minimizing the voltage deviation, represented by "f_2", and minimizing power losses, shown as "f_3". The voltage profile is known as an important operational index in the LV grids. In general, a multi-objective problem can be expressed as follows [21,22]:

$$\text{minimize } F(x) : \quad F(x) = [f_1(x), f_2(x), ..., f_k(x)] \qquad K = 3$$
$$\text{subject to} : \quad h_j(x) \leqslant 0 ; \quad ; \quad j = 1, 2, ..., n \tag{1}$$

where "n" is the number of inequality constraints, and K is the number of objective functions. Here, "x" represents the optimization variables. In this work, the optimization variables are the number of PV panels on different grid buses, as presented in Equation (2):

$$x = [x_1, x_2, ..., x_q] \tag{2}$$

In Equation (2), "q" represents the total number of buses in the grid. As mentioned above, the number of PV panels on each household is limited between 0 and 2, due to the available rooftop area of a common house. Therefore, the number of PV panels on each bus is limited to the maximum number of PV panels that can be installed on all the households on the bus. If four households are connected to a bus, then the maximum number of PV panels on the bus would be limited to eight.

As mentioned above, three different objective functions are considered in this paper, *i.e.*, $K = 3$. The first objective function, *i.e.*, maximizing the PF from the LV grid to the upper grid, can be equivalently replaced by minimizing the power from the upper grid to the LV grids, as follows:

$$f_1(x) = \min P_T \tag{3}$$

In Equation (3), P_T represents the PF through the transformer to the LV grid for the whole simulation period. In this work, the study was performed in 15 min time intervals, and the simulation was performed for a 24 h period. In other words, to obtain P_T, PF through transformer is obtained for each simulation interval. Then, the average PF through the transformer for the whole simulation period (24 h) was obtained, *i.e.*,

$$P_T = \text{mean } P_{\text{trafo}}(x) \tag{4}$$

The second objective is the voltage deviation among the grid buses, which can be obtained as follows:

$$f_2(x) = \min \Delta V_{\text{grid}} \tag{5}$$

where "ΔV_{grid}" is the maximum voltage deviation in the grid for the whole simulation time, as presented in Equation (6):

$$\Delta V_{\text{grid}} = \max [\Delta V_1, \Delta V_2, \Delta V_3, ..., \Delta V_b] \quad b = 1 : q \tag{6}$$

In Equation (6), "ΔV_b" represents the maximum voltage deviation of bus "b" during the whole simulation period. In this study, the study is performed for a 24 h time period, with 15 min time intervals, meaning that 96 time intervals have been defined for the study. Therefore, to calculate the maximum voltage deviation of bus "b" in Equation (6), "ΔV_b", voltage deviation of bus "b" is calculated in each time interval. Then, the maximum (*i.e.*, worst) voltage deviation of bus "b" among the whole simulation time, "ΔV_b", is obtained, as presented in Equation (7):

$$\begin{cases} \Delta V_b = \max [A_{b1}, A_{b2}...., A_{bz}] ; \\ A_{bz} = |V_{bz} - 1| \end{cases} \begin{cases} z = 1 : 96 \\ b = 1 : q \end{cases} \tag{7}$$

In Equation (7), "*b*" represents the number of bus in the grid, and "*z*" represents the time interval of the study. As mentioned, this study is performed in 15 min time intervals, giving 96 time intervals for a 24 h case study. "A_{bz}" is the voltage deviation of bus "*b*" (in p.u.) in simulation interval "*z*". Minimizing the maximum voltage deviation amongst all buses guarantees that voltage deviations of other buses in the grid will also be minimized.

The last objective function is defined to minimize power losses (copper losses) in the grid, as expressed in Equation (8):

$$f_3(x) = \min\ P_{loss,total} \tag{8}$$

To obtain $P_{loss,total}$, the power loss of the whole LV grid is calculated at the end of each simulation interval. This value is obtained from the Newton-Raphson load flow of the grid. Then, total power losses of the grid for a day of simulation (*i.e.*, $P_{loss,total}$) is computed by adding power loss values of all intervals together.

Three inequality constraints are considered in this study, as explained below:

(1)   Voltage of all buses must stay within a specific limit:

$$A_{bz} \leqslant \Delta V_{standard} \quad \begin{cases} z = 1:96 \\ b\ = 1:q \end{cases} \tag{9}$$

where "A_{bz}", "*b*", and "*z*" are defined in Equation (7). On the other hand, $\Delta V_{standard}$ is the standard (maximum) voltage deviation of grid buses, which is defined by the grid codes and regulations [24].

(2)   PF through the transformer must be less than transformer nominal power (transformer capacity), as presented in Equation (10):

$$P_T \leqslant P_{nominal} \tag{10}$$

where P_T is already presented in Equation (4). Here, $P_{nominal}$ is the nominal power (capacity) of transformer.

(3)   Line currents should never exceed the nominal currents of the cables:

$$i_{line} \leqslant i_{cable} \tag{11}$$

In Equation (11), i_{cable} is the nominal current of the cables in the network, which is obtained from the network data. In addition, the line current, i_{line}, is calculated from the Newton-Raphson load flow.

The multi-objective optimization can be solved in different ways, such as Pareto optimality methods [21], and weighting methods. In weighting methods, different objective functions can be combined into a single function with different weights. One of the weighting methods, known as the "Rank Order Centroid" (ROC) approach, is used in this work [22]. In this method, different objective functions are weighted and added to each other to form the single objective function as follows:

$$\min \sum_{s=1}^{K} w_s f_s(x) \quad \text{subject to:} \sum_{s=1}^{K} w_s = 1 \tag{12}$$

where "$f_s(x)$" is the sth objective function and "*K*" is the total number of objective functions, "*K* = 3". In this work, the weight factors for functions "f_1", "f_2", and "f_3" are 11/18, 5/18, and 2/18, respectively. In this paper, a heuristic method (specifically a GA) is used since the objective functions and constraints are nonlinear [23]. Details of the GA parameters are presented in Table A1.

2.2. "Load + Photovoltaic" Block

This block prepares bus and line data matrices based on grid topology and grid demand data. Bus data, which contain generation and load data at each bus, will be modified based on the generated

PV sizes and locations by the Optimization Block. Therefore, it is primarily required to calculate PV generation based on irradiation and ambient temperature.

2.3. Power Flow-1 Block

"PF-1 Block" represents a Newton-Raphson load flow study for the grid. In this study, the voltage level of different grid buses, in presence of PV panels, is used to design a charging/discharging algorithm for the PEVs in order to increase PV penetration and support the upper grid. Once the PV generation is too much, voltage at the bus of connection (and probably at the neighbor buses) shows undesired increase. On the other hand, LV at different buses shows deficiency of power, which can be compensated by PEV discharging, if available. Thus, the voltage magnitudes at different buses should be obtained before the utilization of PEVs. The "PF-1 Block" takes the data and calculates voltage magnitudes for the next time interval, using Newton-Raphson load flow [25].

2.4. Plug-in Electric Vehicle Block

The "PEV Block" is responsible for computing the charging/discharging pattern for the PEVs based on their distance profile (DP) and the grid condition at each interval. The block receives PEV data and voltage magnitudes at different buses as the input. The PEV data includes: (1) the number and type of PEVs; (2) PEVs' battery size; (3) initial state of charge (SoC) of PEVs at the beginning of simulation; (4) average consumption of different PEVs (Wh/km); (5) DP of PEVs; (6) nominal power of PEV charger; and (7) PEV buses with the number and types of PEVs on each bus. The SoC of PEVs changes during the day with respect to their DP and their average energy consumption. Therefore, in order to define a proper algorithm for PEVs, it is necessary to calculate the SoC of each PEV at the end of each interval. If the PEV is moving, the new SoC at the end of the interval will be calculated based on the PEV's driving distance for that interval, type of PEV, its previous SoC, and its battery capacity. Otherwise, when the PEV is idle, its SoC does not change. The SoC will be calculated within "PEV Block" in Figure 2. From Figure 2, the "PEV Block" includes two individual blocks, "PEV Charging" and "V2G". The reason is that a day of simulation is divided into two different blocks: (1) high irradiation and (2) low irradiation. The reason for the division is to find appropriate set points for the algorithm. The high solar irradiation and solar production during the daytime increases grid voltage levels, especially during the daytime. However, the solar production reaches zero during the evening and night hours. Therefore, different set points are defined for the algorithm to efficiently charge the PEV battery for customer requirements in the morning, while enabling the grid to utilize PEV storage for grid support.

2.4.1. Plug-in Electric Vehicle Charging

Considering the number of PEVs in the LV grid and their significant power demand, especially at night when most PEV owners plug their car for charging, a huge amount of load will be imposed on the grid which can result in voltage issues, as well as overcurrent in the lines and transformer. Therefore, a proper algorithm is required to handle PEV charging appropriately. Figure 3 presents the flowchart of the proposed algorithm for PEV charging. From Figure 3, all the PEVs connected to the grid will be sorted in ascending order based on their available SoC from the last interval. Sorting PEVs will enable the algorithm to charge certain PEVs with the least SoCs for simultaneous charging. After selecting the PEVs, each selected PEV will be examined for charging. If the ith PEV is moving (i.e., $DP(i,t) \neq 0$), it cannot be charged and the algorithm only calculates its SoC at the end of the current interval. Otherwise, the PEV's SoC at the end of the last interval will be evaluated, i.e., $SoC(i, t - 1)$. If the PEV's SoC is above SoC_{max}, the algorithm does not charge the PEV. This way, the SoC at the end of the current interval will be equal to its SoC from the last interval, i.e., $SoC(i, t) = SoC(i, t - 1)$, because internal battery discharging is neglected in this study. Once the SoC of the ith PEV is less than SoC_{max}, the PEV will be charged. Here, the PEV's SoC at the end of the charging period will be calculated and the "load data" matrix will be modified, due to PEV charging impact.

Defining a maximum value for battery charging (SoC_{max}) is due to battery lifetime issues [17]. Furthermore, the random displacement of different PEVs in different feeders and different buses enhance the algorithm performance and voltage profile at different buses, as it minimizes the probability of simultaneous charging of too many PEVs at the same feeder. The random placement of PEVs is performed only once at the initialization of the whole optimization program. Additionally, when the number of PEVs increases, the random selection nature of the proposed algorithm helps to keep the diversity for PEV selection, which results in higher probability of available PEV for charging at different buses at any time.

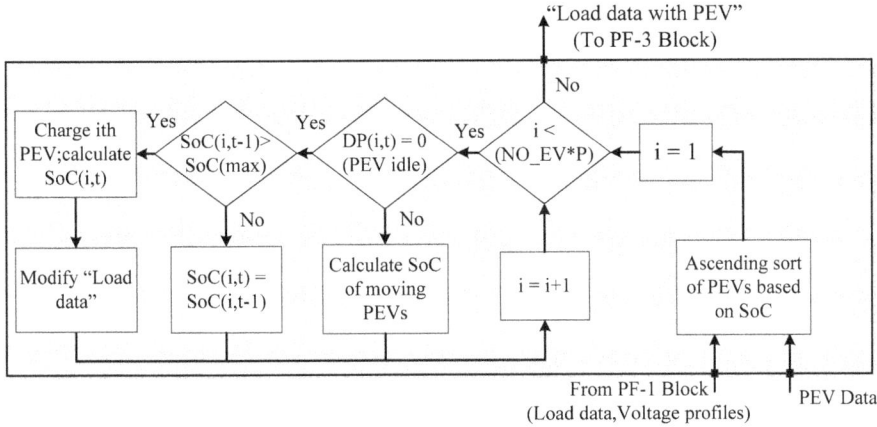

Figure 3. Charging pattern for plug-in electric vehicles (PEVs) for each interval of simulation.

2.4.2. Vehicle-to-Grid

PEVs utilization is a random behavior because of human interaction as the driver. This way, the PEV's SoC at the end of each trip, charging and discharging preferences of the owner, and connectivity to the grid during the idle periods brings a new level of uncertainty in the power system. This results in higher operational cost due to the required ancillary services. In this study, the PEV behavior is represented by generating random driving patterns, selecting the type and their random displacement in the grid. The random driving patterns are produced with respect to statistical driving data provided by authorities. Two types of cars are considered in this study: commuters and family cars. Details of the PEVs are presented in Section 3. Commuters are mainly used in the morning and evening time, for travelling between home and work. Therefore, their driving pattern is mainly in the morning and evening time, and they are idle for most of the daytime. On the other hand, family cars have a more diverse driving pattern, as they will be used during the day more frequently. However, family cars are mainly used for short trips. Therefore, their battery charge does not change drastically. Further details on the driving patterns of PEVs are presented in Section 3.

Considering the driving patterns of PEVs and their idle time, when the PEV is idle and connected to the grid, the grid can benefit in two ways: (1) positive balance, where the storage capacity of PEV can be used to store extra energy [12]; and (2) negative balance, where the PEV can support the upper grid [17]. It should be noted that a PEV can be used for grid support only when: (1) the PEV is idle; (2) the PEV is connected to the grid; and (3) the PEV battery is available for such function. Due to the battery lifetime issues, batteries should not be charged more than a certain level (90%) [17]. On the other hand, batteries should always have a minimum level of charge to respond to the owners' requirements. Thus, battery availability is limited to certain "minimum" and "maximum" values. Determining "minimum" value can be done considering PEVs' DP and their daily distance. Figure 4 presents the proposed algorithm for such functions. For each interval, PEVs' data (including locations, DP, and initial SoC) are given to the PEV Block (Figures 2 and 4). Then, each PEV will be examined for its potential for grid support based on different rules and constraints. Since charging and discharging of PEV could affect the voltage of the whole feeder, the proposed algorithm is designed to consider voltage of the feeder to which PEV is connected.

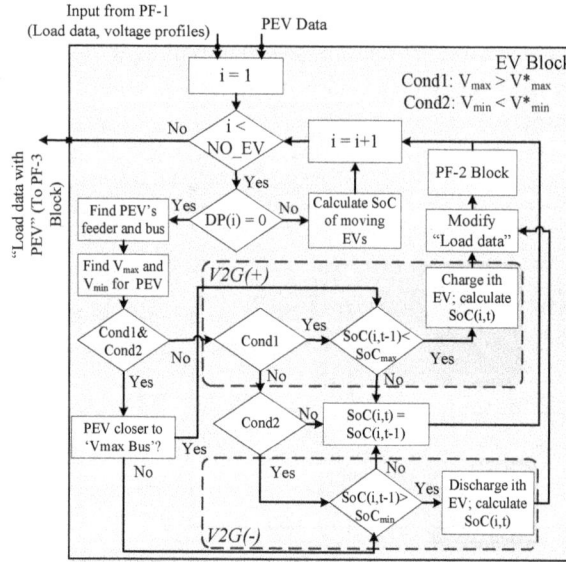

Figure 4. Proposed algorithm for vehicle to grid (V2G) application of electric vehicles (EVs).

In the first iteration, the maximum and minimum voltage of the feeder (V_{max} and V_{min}, respectively) will be determined based on the data from PF-1 Block. However, for the next iterations these values will be determined by "PF-2 Block", shown in Figure 4. "PF-2 Block" is also based on Newton-Raphson load flow calculations. Then, the obtained V_{max} and V_{min} will be compared with predefined set points, "V^*_{max}" and "V^*_{min}". If "$V_{max} > V^*_{max}$", named as "Cond1" in Figure 4, then the "V2G(+)" scenario applies. On the other hand, if "$V_{min} < V^*_{min}$", which is shown by "Cond2" in Figure 4, then the "V2G(−)" scenario is valid. For cases where both "Cond1" and "Cond2" criteria could happen, the decision is made based on the distance of the PEV from the buses. These scenarios are further explained below:

Scenario 1 (charging and discharging): If both maximum and minimum voltages of the feeder at hand are beyond the thresholds, which might happen if the feeder is too long and there are many PV panels installed at certain buses, priority between discharging and charging is given to the one that the PEV bus is closer to. As a result, discharging is preferred if the bus with minimum voltage is closer to the current PEV ($D\_V_{min} < D\_V_{max}$). Conversely, if the bus with maximum voltage is closer to the current PEV bus, then charging will be preferred ($D\_V_{max} < D\_V_{min}$). The desired operation on the PEV, however, depends on its availability and its SoC level.

Scenario 2 (only charging): If the maximum voltage is higher than a threshold (V_{max}) in the whole feeder, there is an excess power generated by the PVs in that feeder which can be possibly stored in the PEVs. Thus, the current PEV will be examined to store the excess power. However, charging happens only if the PEV is idle and its battery SoC is less than the technical upper limit, *i.e.*, 90% of its nominal capacity.

Scenario 3 (discharging): If the minimum voltage of the feeder is below V_{min}, a shortage in the generation will be recognized. In this condition, the PEV will be examined for discharging based on its availability and SoC level.

Scenario 4: If the maximum and minimum voltages are within the pre-defined values, and the PEV is idle, the SoC of the PEV does not change.

In all scenarios (except Scenario 4), the bus matrix should be modified for the new Newton-Raphson PF in the "PF-3 Block". This procedure will be done for all intervals of all PEVs.

2.5. Power Flow-3 Block

In this block, the modified "load data" matrix, considering charging and discharging operation of PEVs for the length of simulation, along with line matrix will be utilized to calculate new steady-state operating points. A similar PF structure, similar to PF-1 Block (*i.e.*, Newton-Raphson-based PF), is used in this block; and only the loads have changed. Final load flow results will be utilized to calculate the required data for the Optimization in order to generate new solutions for the PV sizes and locations.

3. Grid Layout and Photovoltaic Modeling

The configuration of a Danish LV grid under study is shown in Figure 5. The LV grid has six feeders which are connected to a 20 kV/0.4 kV, 630 kVA distribution transformer. The number of households on each feeder is given in Table 1. The data of the cables are obtained using grid data presented in Table A2 in Appendix A [17,26]. For modeling PV panels, the model presented in [27] is applied. The only available datum in this study was the transformer load curve. Therefore, load modeling is done using the "Velander method". This method is explained in [28,29]. This method is mainly used for studies where no measurement of single households in the grid is available. In this method, the power demand of the households will be obtained from the house's "annual energy demand".

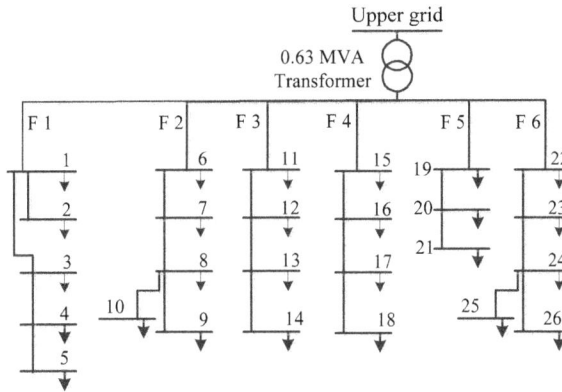

Figure 5. Overall view of the LV grid.

Based on battery type and details required for a certain study, different models are proposed for PEVs [17,30]. In this study, PEVs with Li-Ion battery are considered since they are widely considered recently [31]. Besides, the PEVs' DP, types, and their allocations in the grid should be defined. Although these parameters are not deterministic, a typical scenario is developed in this study to simplify it. Two types of PEVs are considered here: (a) commuters; and (b) family cars.

As mentioned in Section 2.4.2, two types of PEVs are considered in this study: commuters and family cars. In this study, the commuters and family PEVs have a 30 kW battery pack [32]. The required data for the PEVs are reported in Table 1 [32]. The main feature which differentiates PEVs is their average daily distance and owner behavior. To generate DP for PEVs, a random normal distribution is used. Based on the statistical data, commuters are mainly on the road in the morning, from 7:00 a.m. to 9:00 a.m., and during the evening time, from 5:00 p.m. until 7:00 p.m. Therefore, to provide a proper DP for commuters, their average daily distance (mentioned in Table 1) is split between morning and evening time intervals. The DPs of commuters are similar, except for a time shift for DPs of different commuters. On the other hand, family cars are used for short trips, while they have a more diverse moving time, depending on the needs of the car owner [32].

Table 1. Details of feeders and PEVs.

Number of households on grid feeders	Feeder No. of household	1	2	3	4	5	6
		20	33	27	28	17	42

Details of PEVs	PEV type	% of PEVs	Battery (kWh)	Average consumption (Wh/km)	Charger power (kW)	Daily distance (km)
	Commuter	80	30	150	7.2	40
	Family car	20	30	150	7.2	25

The distribution of PEVs among feeders and buses depends on the number of the households on different buses, *i.e.,* buses with higher number of households are likely to have higher number of PEVs. As a result, feeders with higher number of households will have higher number of PEVs connected to them. Such distribution of PEVs provides a more realistic scenario, since all the households are equally likely to own a PEV, regardless of their location on the LV grid.

4. Simulation Results

In this work, the nature of the optimization problem and variables (*i.e.,* number of PVs) is an integer optimization. Usually, heuristic approaches are more effective than gradient-based techniques when it comes to nonlinear integer optimization. As a result, GA method is chosen for this study.

4.1. Case Studies

4.1.1. CASE I: Original Grid without Photovoltaic and Plug-in Electric Vehicle

First, the original grid is simulated with actual load demand without PV panels and PEVs. The data used in this simulation belong to the first day of July 2010. The reason for picking a summer day is the fact that solar irradiation becomes maximum in the summer. Therefore, it is better to determine the penetration level of solar panels in the summer. This way, it will be guaranteed that the solar power production never exceeds the grid limits. The solar irradiance also belongs to the same day of July 2010. The irradiation is obtained from a measurement device in the laboratory. As mentioned above, load modeling is performed using "Velander method". The only available measurement for the grid was the power profile of transformer. Therefore, to provide a demand profile for the grid households, it is assumed that the demand profile of the grid households is similar to the demand profile of transformer.

The simulation reveals the possible issues with the grid which further can be utilized for comparison purposes. Voltages at some critical buses (*i.e.,* at the end of the feeders) are shown in Figure 6.

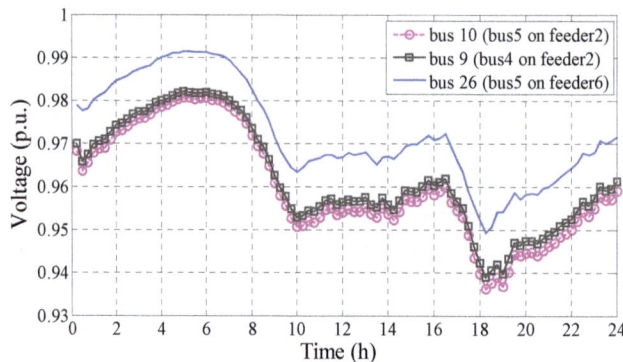

Figure 6. Voltage profile of grid critical buses (no photovoltaic (PV) panels and no PEV).

Considering the standard voltage deviation limit [24], voltage violations at some buses can be recognized. Voltage violation is more severe for the end buses of feeders 2 and 6. The reason is the number of household on these feeders. It should be mentioned that, despite such voltage drop, such voltage profile is acceptable for Danish grid [24]. Since the "Velander" method is used for load modeling in this study, the voltage profiles obtained from simulations include more extremes comparing to the real situation. In reality, the voltage profiles in the grid have lower deviation and the voltage drop is also less than what is shown here.

4.1.2. CASE II: Optimization without Grid Support

In this case, it is assumed that the PEVs are considered as loads. PEVs are charged during the night time, using the smart charging algorithm, as presented in Figure 3. During the daytime, PEVs can only be used for positive balancing (from grid to vehicle). The goal of the optimization-based simulation is to find the optimal number and location of PV panels considering the new PEV loads. Typically, PEVs are under charge during the night hours, where the household demand is normally minimal. However, connection of a large number of PEVs to the feeder imposes a significant amount of load and leads to a significant voltage drop. To overcome this issue, the charging algorithm presented in Figure 3 is utilized. The lower and upper SoC of PEVs' batteries are 20% and 90%, respectively.

4.1.3. CASE III: Optimization Considering Grid Support

CASE III is similar to CASE II, except that the PEVs are utilized for V2G purposes (both positive and negative balance). Thus, CASE III contains the features of the proposed method in Section 2.4.2. Considering customer's driving requirements, the lower limit of SoC is set to 35%, while the upper limit is 90% for night time charging, similar to CASE II.

4.1.4. CASE IV: Optimization Considering Grid Support

Case IV is similar to case III, but with a lower SoC limit during the night time. In other words, in order to increase the available PEV battery capacity for absorbing PV power during the daytime, it is assumed that the grid operator is allowed to restrict the upper limit of SoC for night-time charging to 50%. In such case, the available PEV storage for the daytime is higher.

4.2. Optimization Results

In this section, simulation results of CASE II, CASE III, and CASE IV are presented. The number of PEVs in the LV grid was increased gradually to have the maximum PEV penetration. Here, the results with 16 PEVs (10% PEV penetration) and 40 PEVs (25% penetration) in the grid are presented. The optimization results are shown in Table 2.

To ease the comparison, the total number of PV panels for each case study is presented in Figure 7 as well. Considering the total number of panels for each case, the effectiveness of the proposed algorithm can be realized. Activating the negative balance option (*i.e.*, discharging PEVs for grid support), for both CASE III and CASE IV, has led to a higher number of panels in the grid. Besides, in CASE IV, where the grid operator is allowed to limit the maximum charge of batteries (maximum 50% charging for the night time in this study), the number of panels has increased compared to CASE III. Such increase can be realized both with 10% PEV penetration, and with 25% PEV penetration, although the rate of increase is not the same. The reason for different rates is the fact that the placement of PEVs on different buses and the number of PEVs on different buses are not similar for the two cases. The maximum increase in the number of panels is in CASE IV with 40 PEVs in the grid (205 panels), which shows more than 32.7% increase compared to CASE II (138 panels). Since each panel's nominal capacity is 3 kW, such increase is equal to 201 kW more PV panels.

Table 2. Number of PV panels for cases II–IV considering two PEV penetration levels: 10% and 25% PEV penetration.

Bus	Case II		Case III		Case IV	
	10% (16 PEV)	25% (40 PEV)	10%	25%	10%	25%
1	2	2	2	3	2	3
2	6	6	6	7	8	9
3	1	1	3	4	3	6
4	4	4	4	5	4	7
5	3	3	4	4	4	4
6	6	6	6	6	6	7
7	9	9	10	12	11	12
8	2	2	3	5	6	10
9	4	4	4	6	6	7
10	1	1	2	4	4	6
11	3	3	5	8	6	8
12	10	10	10	10	10	10
13	4	4	5	6	5	8
14	2	2	3	5	4	6
15	6	6	6	6	6	6
16	8	8	9	10	10	10
17	8	8	9	9	9	10
18	8	8	8	8	8	8
19	10	10	10	10	10	10
20	8	8	8	8	8	8
21	8	8	8	8	8	8
22	5	5	6	8	7	11
23	6	6	7	8	8	9
24	8	8	8	8	8	10
25	5	5	5	6	6	8
26	1	1	1	3	2	4
Total	138	138	152	177	169	205

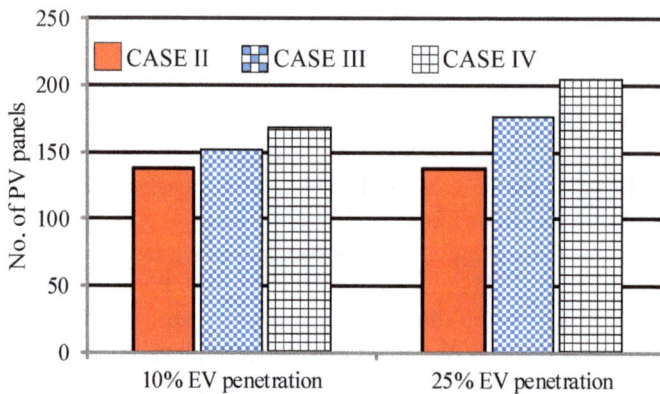

Figure 7. Number of PV panels in different cases.

4.3. Analysis of Plug-in Electric Vehicle Impact

To show the impact of PEVs on the voltage profile at different buses, voltage profile of bus 10 in Figure 5 (bus 5 on feeder 2), is depicted in Figure 8 for CASE II and CASE IV. As mentioned earlier, bus 10 is the worst bus of the system. The role of PEVs for both negative and positive balance is demonstrated in Figure 8. In CASE II, the PEV is fully charged during the night, whereas in CASE IV the PEV charging is limited to 50%. The difference in settings causes the voltage difference between the two cases as well. PEV is on the road around 7:30 a.m. until 8:00 a.m., and it loses part of its charge. During the day time, as the solar irradiation increases, the PEV battery starts charging until around 2:00 p.m. However, after this time, although the solar power is still high, the battery reaches its maximum capacity (90%) and cannot participate in positive balance. This causes an increase in voltage. The PEV is on the road around 5:30 p.m. until 6:00 p.m. as well. As PEV becomes idle in the evening, it starts supporting the grid in CASE IV. The provided support by PEVs in this scenario is quite significant, and it clearly elevates the voltage profile of the bus. It can be seen that, despite the higher PV penetration in CASE IV compared to CASE II (32.7% higher as mentioned above), the algorithm keeps voltage deviation in the grid less than 10%.

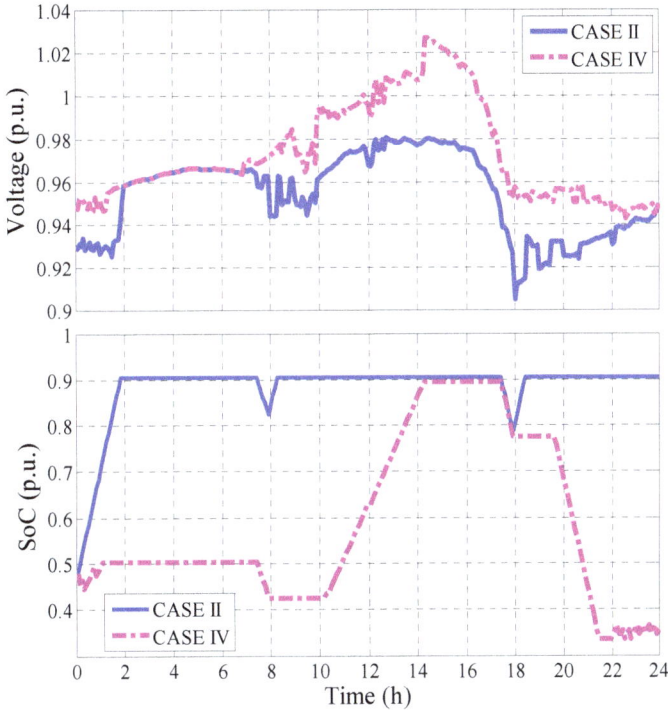

Figure 8. Voltage profile and SoC for bus 10: CASE II and CASE IV.

Figure 9 compares the voltage profile of bus 10, and the SoC of PEV for this bus, for CASE III and CASE IV. For CASE III, the night charging is not limited, while it is limited to 50% for CASE IV. Therefore, voltage profile of bus 10 has lower values for night hours. On the other hand, when the PEV is idle during the day, the grid uses PEV storage for storing the solar production. However, since the available PEV storage is limited in CASE III compared to CASE IV, the PEV provides less grid support (positive balance) for the grid. Such scenario results in lower solar penetration, as presented in Table 2. Therefore, solar penetration is higher in CASE IV (205 panels) compared to CASE III (177 panels). Transformer power profile is presented in Figure 10 for 40 PEVs (25% penetration) in the grid. Considering the results of CASE II as the base case, it can be seen that PEV charging during the night time for CASE II occupies almost all the transformer capacity.

PV penetration provides extra power which can support the upper grid during the day. However, PV production doesn't exist during evening time, when the grid has a significant peak in demand. Therefore, the grid is forced to absorb power from the upper grid to support its consumption. Using PEVs for grid support, the excess available energy can be stored in PEVs batteries during the daytime. The stored energy can then be used during the evening hours to respond to the peak demand of the grid. Such opportunity provides a great flexibility for the grid to not only provide some extra power to the upper grid, but the grid can also deal with its peak demand problem.

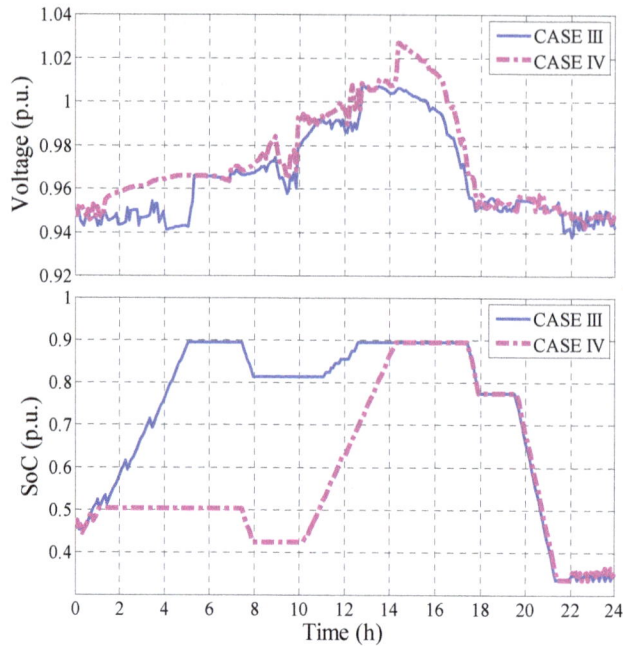

Figure 9. Voltage profile and SoC for bus 10: CASE III and CASE IV.

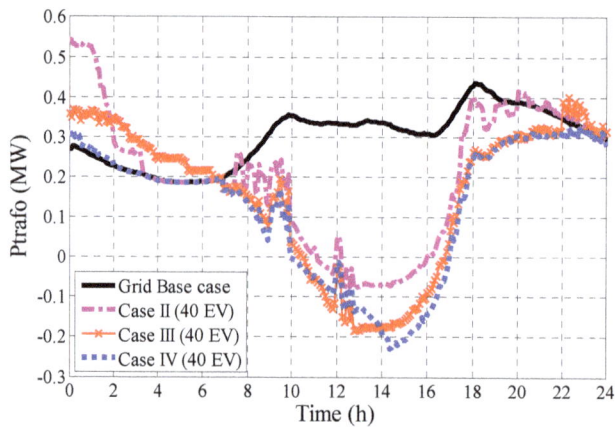

Figure 10. Transformer power profile for different cases.

5. Conclusions

This work focuses on the challenges and opportunities of a grid with the presence of PV panels and PEVs. The idea is to find a proper solution for the new challenges in the grid without changing the grid existing conditions, as well as benefiting from the new potentials. The focus is to utilize the PEV storage to provide maximum ancillary service. Different scenarios are evaluated to support the proposed method. Simulation results quantify the potential and effect of PEVs for the grid and the grid operator. From the results in Table 2, compared to CASE II where only positive balance is available, enabling both positive balance and negative balance increases the PV penetration by 50% with 40 PEVs in the grid (25% PEV penetration), and by 22.4% for 16 PEVs in the grid (10% PEV penetration). In addition, PV panels are optimally allocated through different grid feeders and buses to maximize the ancillary service. The results verify the capability of PEV storage in increasing the penetration of renewable energy in the grid, which leads to a more sustainable power system.

Author Contributions: Reza Ahmadi Kordkheili is the main researcher who initiated and organized research reported in the paper. He contributed to the sections on load modeling and grid modeling, EV modeling, designing intelligent algorithm, and optimization. In addition, as the first author, he is responsible for writing main parts of the paper, including paper layout and results. S. Ali Pourmousavi contributed in designing algorithms and optimization method, as well as drafting the article. Mehdi Savaghebi and Josep M. Guerrero assisted with the vehicle-to-grid application and algorithm, and contributed jointly to analysis, and the writing of this article. M. Hashem Nehrir contributed towards the optimization algorithm, and his comments on the paper draft have had a big impact on improving its quality. All authors were involved in preparing the manuscript.

Conflicts of Interest: The authors declare no conflict of interest.

Appendix A

Table A1. Parameters of genetic algorithm (GA).

Parameter	Population	Number of Generation	Stall Generation	Tolerance Function
Value	40	200	100	1×10^{-8}

Table A2. Data of grid cables.

Bus Number	R (Ω/km)	X (Ω/km)	R	X	R (p.u.)	X (p.u.)
Trafo →1 (feeder 1)	0.2080	0.086	0.0859	0.0355	53,703	22,204
1 to 2	0.6420	0.0870	0.0399	0.0054	24,958	3382
1 to 3	0.6420	0.0870	0.0658	0.0089	41,128	5573
3 to 4	0.6420	0.0870	0.0642	0.0087	40.125	5438
4 to 5	0.3210	0.0860	0.0551	0.0148	34,467	9234
Trafo →6 (feeder 2)	0.2080	0.0840	0.0539	0.0217	33,657	13,592
6 to 7	0.2080	0.0840	0.0186	0.0075	11,609	4688
7 to 8	0.2080	0.0840	0.0448	0.0181	28,028	11,319
8 to 9	0.3210	0.0840	0.0544	0.0142	34,006	8899
9 to 10	0.3210	0.0840	0.0258	0.0068	16,130	4221
Trafo →11 (feeder 3)	0.2080	0.0840	0.0347	0.0140	21,671	8752
11 to 12	0.2080	0.0840	0.0233	0.0094	14,560	5880
12 to 13	0.3210	0.0840	0.0261	0.0068	16,331	4274
13 to 14	0.3210	0.0840	0.0546	0.0143	34,106	8925
Trafo →15 (feeder 4)	0.2080	0.0840	0.0613	0.0247	38,285	15,461
15 to 16	0.2080	0.0840	0.0223	0.0090	13,949	5633
16 to 17	0.2080	0.0840	0.0267	0.0108	16,718	6752
17 to 18	0.2080	0.0840	0.0132	0.0053	8268	3339
Trafo →19 (feeder 5)	0.2080	0.0860	0.0908	0.0376	56,771	23,473
19 to 20	0.3210	0.0840	0.0414	0.0108	25,901	6778
20 to 21	0.6420	0.0870	0.0545	0.0074	34,066	4616
Trafo →22 (feeder 6)	0.2080	0.0840	0.0590	0.0238	36,855	14,884
22 to 23	0.2080	0.0840	0.0146	0.0059	9100	3675
23 to 24	0.3890	0.0830	0.0204	0.0044	12,764	2723
24 to 25	0.3890	0.0830	0.0311	0.0066	19,450	4150
24 to 26	0.5260	0.0830	0.0395	0.0062	24,656	3891

References

1. *Guide for Smart Grid Interoperability of Energy Technology Operation with Electric Power System (EPS), and End-Use Applications and Loads*; IEEE P2030; IEEE Standards Association: Piscataway, NJ, USA, 2011.
2. *CEN-CENELEC-ETSI Smart Grid Coordination Group-Sustainable Processes*; European Committee for Electrotechnical Standardization-European Telecommunications Standards Institute (CENELEC-ETSI): Brussels, Belgium, 2012.

3. Shafiullah, G.M.; Oo, A.M.T.; Jarvis, D.; Ali, A.B.M.S.; Wolfs, P. Potential Challenges: Integrating Renewable Energy with the Smart Grid. In Proceedings of the 20th Australasian Universities Power Engineering Conference, Christchurch, New Zealand, 5–8 December 2010; pp. 1–6.

4. Omran, W.A.; Kazerani, M.; Salama, M.M.A. Investigation of Methods for Reduction of Power Fluctuations Generated from Large Grid-Connected Photovoltaic Systems. *IEEE Trans. Energy Convers.* **2011**, *26*, 318–327. [CrossRef]

5. Ueda, Y.; Kurokawa, K.; Tanabe, T.; Kitamura, K.; Sugihara, H. Analysis Results of Output Power Loss Due to the Grid Voltage Rise in Grid-Connected Photovoltaic Power Generation Systems. *IEEE Trans. Ind. Electron.* **2008**, *55*, 2744–2751. [CrossRef]

6. Tonkoski, R.; Turcotte, D.; EL-Fouly, T.H.M. Impact of High PV Penetration on Voltage Profiles in Residential Neighborhoods. *IEEE Trans. Sustain. Energy* **2012**, *3*, 518–527. [CrossRef]

7. Hung, D.Q.; Mithulananthan, N. Multiple Distributed Generator Placement in Primary Distribution Networks for Loss Reduction. *IEEE Trans. Ind. Electron.* **2013**, *60*, 1700–1708. [CrossRef]

8. Al-Sabounchi, A.; Gow, J.; Al-Akaidi, M. Simple Procedure for Optimal Sizing and Location of a Single Photovoltaic Generator on Radial Distribution Feeder. *IET Renew. Power Gener.* **2014**, *8*, 160–170. [CrossRef]

9. Lin, C.H.; Hsieh, W.L.; Chen, C.S.; Hsu, C.T.; Ku, T.T. Optimization of Photovoltaic Penetration in Distribution Systems Considering Annual Duration Curve of Solar Irradiation. *IEEE Trans. Power Syst.* **2012**, *27*, 1090–1097. [CrossRef]

10. Shaaban, M.F.; Atwa, Y.M.; El-Saadany, E.F. DG Allocation for Benefit Maximization in Distribution Networks. *IEEE Trans. Power Syst.* **2013**, *28*, 639–649. [CrossRef]

11. Alsokhiry, F.; Lo, K.L. Distributed Generation Based on Renewable Energy Ancillary Services. In Proceedings of the 2013 Fourth International Conference on Power Engineering, Energy and Electrical Drives, Istanbul, Turkey, 13–17 May 2013; pp. 1200–1205.

12. Kempton, W.; Tomić, J. Vehicle-to-Grid Power Implementation: From Stabilizing the Grid to Supporting Large-Scale Renewable Energy. *J. Power Sources* **2005**, *144*, 280–294. [CrossRef]

13. Sortomme, E.; El-Sharkawi, M.A. Optimal Charging Strategies for Unidirectional Vehicle-to-Grid. *IEEE Trans. Smart Grid* **2011**, *2*, 131–138. [CrossRef]

14. Hong, Y.-Y.; Lai, Y.-M.; Chang, Y.-R.; Lee, Y.-D.; Liu, P.-W. Optimizing Capacities of Distributed Generation and Energy Storage in a Small Autonomous Power System Considering Uncertainty in Renewables. *Energies* **2015**, *8*, 2473–2492. [CrossRef]

15. Gao, S.; Chau, K.T.; Liu, C.; Wu, D.; Chan, C.C. Integrated Energy Management of Plug-in Electric Vehicles in Power Grid with Renewables. *IEEE Trans. Veh. Technol.* **2014**, *63*, 3019–3027. [CrossRef]

16. Escudero-Garzas, J.J.; Garcia-Armada, A.; Seco-Granados, G. Fair Design of Plug-in Electric Vehicles Aggregator for V2G Regulation. *IEEE Trans. Veh. Technol.* **2012**, *61*, 3406–3419. [CrossRef]

17. Rottondi, C.; Fontana, S.; Verticale, G. Enabling Privacy in Vehicle-to-Grid Interactions for Battery Recharging. *Energies* **2014**, *7*, 2780–2798. [CrossRef]

18. Diaz, N.L.; Dragicevic, T.; Vasquez, J.C.; Guerrero, J.M. Intelligent Distributed Generation and Storage Units for DC Microgrids—A New Concept on Cooperative Control Without Communications Beyond Droop Control. *IEEE Trans. Smart Grid* **2014**, *5*, 2476–2485. [CrossRef]

19. Wu, D.; Tang, F.; Dragicevic, T.; Vasquez, J.C.; Guerrero, J.M. A Control Architecture to Coordinate Renewable Energy Sources and Energy Storage Systems in Islanded Microgrids. *IEEE Trans. Smart Grid* **2015**, *6*, 1156–1166. [CrossRef]

20. Latvakoski, J.; Mäki, K.; Ronkainen, J.; Julku, J.; Koivusaari, J. Simulation-Based Approach for Studying the Balancing of Local Smart Grids with Electric Vehicle Batteries. *Systems* **2015**, *3*, 81–108. [CrossRef]

21. Marler, R.T.; Arora, J.S. Survey of Multi-Objective Optimization Methods for Engineering. *Struct. Multidiscip. Optim.* **2004**, *26*, 369–395. [CrossRef]

22. Kordkheili, R.A.; Pourmousavi, S.A.; Pillai, J.R.; Hasanien, H.M.; Bak-Jensen, B.; Nehrir, M.H. Optimal Sizing and Allocation of Residential Photovoltaic Panels in a Distribution Network for Ancillary Services Application. In Proceedings of the 2014 International Conference on Optimization of Electrical and Electronic Equipment, Bran, Romania, 22–24 May 2014; pp. 681–687.

23. Liang, H.; Zhuang, W. Stochastic Modeling and Optimization in a Microgrid: A Survey. *Energies* **2014**, *7*, 2027–2050. [CrossRef]

24. *Technical Regulation 3.2.1 for Electricity Generation Facilities with a Rated Current of 16 A per Phase or Lower*; Energinet.dk: Fredericia, Denmark, 2011.

25. Saadat, H. *Power System Analysis*, 2nd ed.; Mc-Graw Hill: New York, NY, USA, 2002.

26. Pillai, J.R.; Huang, S.; Bak-Jensen, B.; Mahat, P.; Thogersen, P.; Moller, J. Integration of Solar Photovoltaics and Electric Vehicles in Residential Grids. In Proceedings of the 2013 IEEE Power and Energy Society General Meeting, Vancouver, BC, Canada, 21–25 July 2013.

27. Thompson, M.; Infield, D.G. Impact of Widespread Photovoltaics Generation on Distribution Systems. *IET Renew. Power Gener.* **2007**, *1*, 33–40. [CrossRef]

28. Neimane, V. On Development Planning of Electricity Distribution Networks. Ph.D. Thesis, Kungliga Tekniska högskolan (KTH), Stockholm, Sweden, 2001.

29. Kordkheili, R.A.; Bak-Jensen, B.; Pillai, J.R.; Mahat, P. Determining Maximum Photovoltaic Penetration in a Distribution grid Considering Grid Operation Limits. In Proceedings of the 2014 IEEE PES General Meeting/Conference & Exposition, National Harbor, MD, USA, 27–31 July 2014; pp. 1–5.

30. Alimisis, V.; Hatziargyriou, N.D. Evaluation of a Hybrid Power Plant Comprising Used EV-Batteries to Complement Wind Power. *IEEE Trans. Sustain. Energy* **2013**, *4*, 286–293. [CrossRef]

31. Affanni, A.; Bellini, A.; Franceschini, G.; Guglielmi, P.; Tassoni, C. Battery Choice and Management for New-Generation Electric Vehicles. *IEEE Trans. Ind. Electron.* **2005**, *52*, 1343–1349. [CrossRef]

32. Sundstrom, O.; Binding, C. Flexible Charging Optimization for Electric Vehicles Considering Distribution Grid Constraints. *IEEE Trans. Smart Grid* **2012**, *3*, 26–37. [CrossRef]

6

Methods for Global Survey of Natural Gas Flaring from Visible Infrared Imaging Radiometer Suite Data

Christopher D. Elvidge [1,*], Mikhail Zhizhin [2,3], Kimberly Baugh [2], Feng-Chi Hsu [2] and Tilottama Ghosh [2]

Academic Editor: Richard B. Coffin

[1]   Earth Observation Group, National Centers for Environmental Information,
       National Oceanic and Atmospheric Administration, 325 Broadway, Boulder, CO 80205, USA
[2]   Cooperative Institute for Research in the Environmental Sciences, University of Colorado,
       Boulder, CO 80303, USA; mikhail.zhizhin@noaa.gov (M.Z.); kim.baugh@noaa.gov (K.B.);
       feng.c.hsu@noaa.gov (F.-C.H.); tilottama.ghosh@noaa.gov (T.G.)
[3]   Russian Space Research Institute, Moscow 117997, Russia
*   Correspondence: chris.elvidge@noaa.gov

Abstract: A set of methods are presented for the global survey of natural gas flaring using data collected by the National Aeronautics and Space Administration/National Oceanic and Atmospheric Administration NASA/NOAA Visible Infrared Imaging Radiometer Suite (VIIRS). The accuracy of the flared gas volume estimates is rated at $\pm 9.5\%$. VIIRS is particularly well suited for detecting and measuring the radiant emissions from gas flares through the collection of shortwave and near-infrared data at night, recording the peak radiant emissions from flares. In 2012, a total of 7467 individual flare sites were identified. The total flared gas volume is estimated at 143 (± 13.6) billion cubic meters (BCM), corresponding to 3.5% of global production. While the USA has the largest number of flares, Russia leads in terms of flared gas volume. Ninety percent of the flared gas volume was found in upstream production areas, 8% at refineries and 2% at liquified natural gas (LNG) terminals. The results confirm that the bulk of natural gas flaring occurs in upstream production areas. VIIRS data can provide site-specific tracking of natural gas flaring for use in evaluating efforts to reduce and eliminate routine flaring.

Keywords: Visible Infrared Imaging Radiometer Suite (VIIRS); Nightfire; gas flaring; carbon intensity; carbon dioxide emissions

1. Introduction

Flaring is widely used to dispose of natural gas produced at oil and gas facilities that lack sufficient infrastructure to capture all of the gas that is produced (Figure 1). The term "associated gas" refers to natural gas that emerges when crude oil is brought to the Earth's surface. This is the largest source of gas flaring. Smaller quantities of gas flaring occur at oil refineries and natural gas processing facilities. Because flaring is a waste disposal process, there is no systematic reporting of the flaring locations and flared gas volumes. Additionally, where flare volume data are reported, the data are typically self-reported by the flare operators, estimated from the difference between the natural gas volume produced and the quantity used or sold. It is therefore difficult to assess the reliability and accuracy of the reported data.

There are four distinct applications for site-specific estimates of flared gas volumes. First, there are carbon cycle analyses that rely on site-specific knowledge of the locations and magnitudes of greenhouse gas emissions to the atmosphere [1]. Second is the tracking of activities to reduce gas flaring [2]. Third is the identification of potentially attractive locations for gas utilization. Fourth is the calculation of the carbon intensity of fuels, such as the California Low Carbon Fuel Standard [3].

In this paper, we present a series of methods that produce site-specific estimation of flared gas volumes worldwide using data collected by the National Aeronautics and Space Administration/National Oceanic and Atmospheric Administration NASA/NOAA Visible Infrared Imaging Radiometer Suite (VIIRS). Using data collected in 2012, we conducted a global survey of gas flaring sites and separated these into upstream (production sites) and downstream refineries and liquefied natural gas (LNG) terminals. We developed a calibration for estimating flared gas volumes and applied this to the individual sites. The results have been aggregated to the national level, with tallies of the number of flaring sites and estimates for the total flared gas volume in 2012.

Figure 1. Large quantities of radiant energy are produced by gas flares.

2. Satellite Observation of Gas Flares

Because of the lack of systematic reporting from flare operators and the remote nature of many flare locations, satellite sensors are an attractive option for global monitoring of gas flares. However, none of the existing sensors have been designed specifically for the detection and monitoring of gas flares. Systems that collect at high spatial resolution are not well suited to collect global data on large numbers of flare sites, lacking a repeat cycle suitable for cloud clearing and capturing the variability in flaring activity. In addition, the high spatial resolution sensors need to be tasked to collect data at specific sites, and the data are sold commercially, significantly raising the cost and complexity of any potential effort for global gas flaring monitoring. For global monitoring of gas flares, the option to analyze data from moderate spatial resolution (~1 km^2) polar orbiting sensors has considerable merit. Here, the data are free and the coverage is global every 24 h. The challenge with this style of data is that flares are subpixel sources requiring specialized analysis to identify flares and to extract the flaring radiant emissions.

There have been several published studies describing gas flaring detection using satellite systems. NASA's moderate-resolution imaging spectroradiometer (MODIS) data were used to inventory gas flares with the 4 μm spectral band and to estimate flared gas volumes in Nigeria [4]. Nighttime shortwave infrared (SWIR) data from advanced along track scanning radiometer (AATSR) have been used to map gas flares globally based on the temporal persistence of gas flares [5]. A survey of gas flaring sites was developed for a portion of Canada using daytime data collected by Landsat 8 [6].

To date, the most extensive time series of global gas flaring, with national estimates of flared gas volumes, comes from low light imaging nighttime data acquired by the Operational Linescan System (OLS) operated by the U.S. Air Force Defense Meteorological Satellite Program (DMSP) [7]. Gas flares were identified visually in DMSP data because the sensor detects electric lights from cities and towns, as well as gas flares. Estimation of gas flaring volumes using DMSP data ended in 2012 due to orbit degradation, resulting in solar contamination.

This study was conducted using VIIRS data on the Suomi National Polar Partnership (SNPP) satellite, launched in 2011. The VIIRS is operated in an unusual way that offers a substantial advantage for the observation of gas flaring. At night, the VIIRS continues to record data in three near- to short-wave infrared channels designed for daytime imaging (Figure 2). At night, the only features detected in these channels are combustion sources [8]. The SWIR channel, at 1.6 μm, is at the wavelength region where peak radiant emissions from gas flares occur. The 4 μm channel, widely used in fire detection [9], only detects large flares due to the fact that it falls on the trailing edge of gas flare radiant emissions and observes a mixture of flare plus background radiant emissions. Typically, the flare radiant emissions in the 4 μm channel are about a third of the emissions at 1.6 μm. This has a dramatic effect, limiting the detection of smaller flares in standard satellite fire products based on channels set at the 4 μm wavelength.

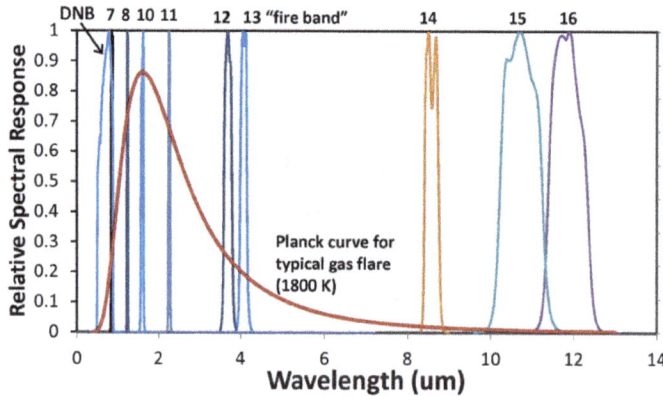

Figure 2. Relative spectral response of visible infrared imaging radiometer suite (VIIRS) bands and the Planck curve of typical gas flares at 1800 K. Day night band: DNB.

By detecting flare radiances in multiple spectral bands, it is possible to model Planck curves for gas flares. The temperature of the hot source is calculated using Wien's displacement law:

$$T = b/\lambda_{max} \tag{1}$$

where T is temperature in kelvin (K), b is Wien's displacement constant = 2897.8 K*μm and λ_{max} is the wavelength of peak radiant emissions.

The gas flares appear as graybodies because they are sub-pixel sources. The emission scaling factor (ε) is defined as the ratio between the observed radiances and the radiances for an object at that temperature filling the entire field of view. The source area (S) is calculated by multiplying ε by the size of the pixel footprint. Radiant heat (RH) is calculated using the Stefan–Boltzmann law:

$$RH = \sigma T^4 S \tag{2}$$

where RH = radiant heat in megawatts (MW), σ = the Stefan–Boltzmann constant, T is temperature in K and S = source area in square meters.

Figure 2 shows spectral bands collected by VIIRS at night. Bands M7-10 are daytime spectral bands that continue to be collected at night, enabling unambiguous detection of combustion sources at 0.87 µm, 1.24 µm and 1.6 µm. Nighttime collection with the M11 band at 2.2 µm, which will improve detection and quantification of gas flares, has been approved for VIIRS and is expected to commence in 2016. The solid red line is the Planck curve for an 1800 K object, typical of a gas flare. The M10 spectral band records the peak radiant emissions from the typical gas flare.

3. Methods

3.1. Visible Infrared Imaging Radiometer Suite Nightfire Processing

The VIIRS Nightfire (VNF) algorithm [8] was applied to all of the usable nighttime VIIRS data from 2012. The usable record started 1 March 2012 and ran to the end of December, 2012. This processing included the detection of subpixel hot sources in five spectral bands (M7, M8, M10, M12 and M13). Redundant "bow-tie" pixels found along the outer portions of the swath were marked for exclusion in the output. The location (latitude, longitude) of the hot pixels, spectral radiances, satellite zenith angle and cloud state were recorded in the output. While all detected pixels were recorded in the output, local maxima were marked. As cloud cover falsely provides a 'no flaring' signal, pixel cloud states were extracted from VIIRS using a cloud mask (VCM) [10], so that these non-detections can be ignored. While the VCM generally successfully identifies cloud cover, it sometime identifies gas flares as clouds due to spectral confusion in the mid-wave infrared. To address this, VNF removes isolated patches of cloud where these coincide with M10 detections.

Planck curve fitting was applied to the detected pixels. Pixels with detection in M12 and M13 were processed with dual Planck curve fitting, with one Planck curve for the background and one for the hot source. From the Planck curve fits, temperature (K) and source area (m²) were calculated. A single hot source Planck curve fit is developed for pixels that lack detection in the M12 and M13 spectral bands.

Planck curve fitting cannot be performed on the pixels with detection in only a single spectral band. This is a common occurrence for small gas flares detected only in the M10 spectral band. The M10-only pixel detections were treated using a different set of processing steps for incorporation in the flared gas estimates (Section 3.5).

Examination of the results on a temperature *versus* source area basis reveals that there were two primary data clusters (Figure 3). There is a low temperature detection set, peaking in the 800–1200 K range, dominated by biomass burning. The high temperature set (above 1450 K) is dominated by gas flares, with peak numbers of flares in the 1700–1800 K range.

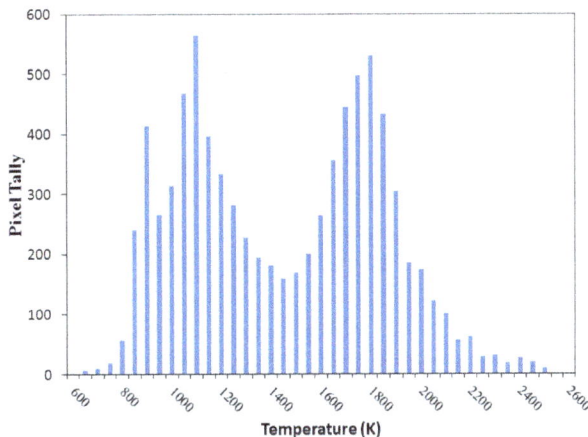

Figure 3. Temperature histogram for a single day of VIIRS Nightfire (VNF) data. The distribution is bimodal, with the majority of gas flares falling in the range from 1500 K to 2000 K. Biomass burning, industrial sites and volcanoes have temperatures in the 600–1300 K range. The range from 1300 K to 1500 K is a crossover zone between gas flares and biomass burning.

3.2. Analyzing Atmospheric Effects

The VIIRS measures top-of-atmosphere (TOA) radiances. In some spectral bands, there can be substantial losses in radiance from the Earth's surface to the TOA from atmospheric absorption and scatter. A study was conducted to analyze the effects of atmospheric variations on the VNF data. The study was based on a flare in Iraq (dry atmosphere) and two flares in Nigeria (moist atmosphere, onshore and offshore). Data included in the analysis spanned from 1 March 2012 to 31 December 2014. The analysis included all available satellite zenith angles. An atmospheric correction was developed for each of the VIIRS observations using the MODerate Resolution Atmospheric TRANsmission (MODTRAN) model [11] parameterized by atmosphere temperature, pressure and water vapor profiles, as well as ground temperature derived from simultaneously-acquired advanced technology microwave sounder (ATMS) data and a surface elevation model. For each observation, RH (MW) was calculated with and without the atmospheric correction. It was found that there is a strongly coherent linear relationship between the TOA and atmospherically-corrected RH data (Figure 4). We attribute this to the fact that the M10 spectral band is in a clear atmospheric window. Based on these results, the study was conducted with the uncorrected TOA radiances.

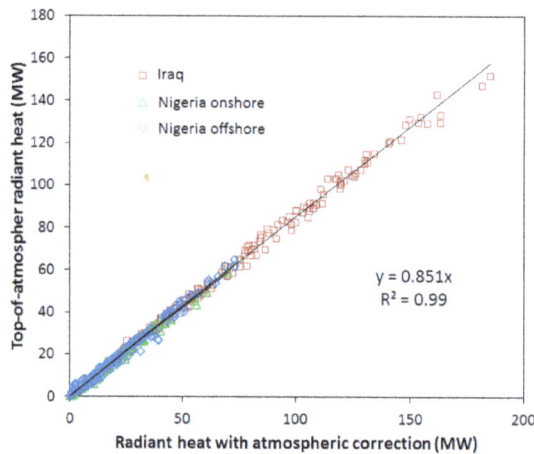

Figure 4. Radiant heat (RH) for three gas flares, with and without atmospheric correction. Sites include onshore and offshore flares in Nigeria (moist atmosphere) and a flare in Iraq (dry atmosphere).

3.3. Identifying Gas Flaring Sites

It is not possible to adequately separate gas flares from other hot sources based solely on temperature due to the overlap between high temperature biomass burning and low temperature gas flaring (Figure 3). To separate gas flares from fires, we use both temperature and persistence. To accomplish this, we built a global 15 arc second grid tallying the number of times an M10 detection occurred during the year. The vast majority of biomass burning events could be filtered out by excluding single and double detections. Manual editing was used to mask out the few remaining biomass burning events.

The remaining sites were divided into three classes based on their temperature records: (1) sites where the maximum temperature exceeded 1400 K; these were taken to be gas flares; (2) sites where temperatures never exceeded 1400 K; these are primarily industrial sites; and (3) sites with single band, M10 detections, where temperature could not be calculated. If the "no temperature" sites were within 10 kilometers of a gas flare, they were classed as potential gas flares and were subsequently resolved by visual inspection of satellite images. Finally, a water-shedding algorithm was used to separate conjoined gas flare features. A raster to vector algorithm was used to draw vectors defining the outlines of the individual gas flaring sites.

3.4. Building a Global Gas Flaring Database

A database was built to hold the time series of gas flaring detections from the individual VIIRS observations. The purpose of the database is to enable rapid extraction of the time series of VNF data for individual sites. The database has three tables, each with the spatial extensions. Table 1 is used to store all VNF detections as individual points. This includes the latitude and longitude of the pixel center, radiances from the individual spectral bands, view geometry and specification of local maxima within clusters of adjacent pixels. Table 2 stores boundaries for individual sites with the polygon geometry. The third table stores the date/time and the cloud state conditions for all satellite observations for the centroid points of the individual sites. A typical database query will select all of the VNF detections within an individual site polygon and fill the gaps when the flare was not detected with the dates and the cloud state observed at the site center.

3.5. Assigning Temperatures to M10-Only Flare Detections

Small flares often have detection only in a single band, M10, the location of peak radiant emissions for typical gas flares. In these cases, it is not possible to model a Planck curve to derive temperature and source area, and there is insufficient information therefore to calculate the *RH* with the Stefan–Boltzmann law. We develop two methods for assigning temperatures to M10-only gas flare detections. The first method is used in cases where the flare site had observations on other nights with Planck curve fits. In this method, the M10-only detections were assigned the average temperature for flare observations from the same site. The second method is used for cases where the flare site never has observations with Planck curve fits. In this case, a temperature is assigned from the nearest flare site having Planck curve fits. Using these methods for assigning temperatures, it was possible to calculate source areas and *RH* values for weak flare detections based on the M10 radiance.

3.6. Adjusting Flare Area Estimates for View Angle Differences

VIIRS observes the Earth at satellite zenith angles ranging from zero (nadir) to 70 degrees (edge of scan). In examining the observed signals from flares, we found that flares tend to have higher radiance when viewed at high satellite zenith angles. Our interpretation of this phenomena is that flares are typically taller than they are wide due to the buoyancy of the hot gas relative to the surrounding air (Figure 1). Thus, flare footprints appear larger when viewed from the side and smaller when viewed from straight above (the nadir view). The expression of this in VIIRS data is that the flares have higher radiance when viewed at an oblique angle when compared to the nadir, yet the temperature remains stable across all viewing angles. This results in larger source areas for flares when viewed at the edge-of-scan.

The three-dimensional shape of flares can be modeled as an ellipsoid, based on the apparent size of flares *versus* the satellite zenith angle (Figure 5). Using the approach developed by Jekrard [12], it is possible to derive the ellipticity or height (H) *versus* width (R) ratios of individual flares. For a vertically-standing flare, the footprint viewed by satellite from a zenith angle α will be:

$$S(\alpha, H, R) = \pi R \sqrt{(H^2 + R^2 - (H^2 - R^2)\cos 2\alpha)/2} \qquad (3)$$

Non-linear regression of Equation (3) to the set of average flare footprints $S(\alpha_i)$ from different satellite zenith angles α_i can estimate the flare shape H/R. Typical flares have ellipticities in the range of 1–4, with an average of 1.6.

Testing indicated that the calibration to flared gas volumes had a higher coefficient of determination (R^2) if the source size was adjusted to the side view. For frequently-observed flares, it is generally possible to calculate the ellipticity, in which cases, the source sizes were adjusted to a horizontal side view or satellite zenith angle of 90 degrees. For flares with infrequent detection, we used the average ellipticity of 1.6.

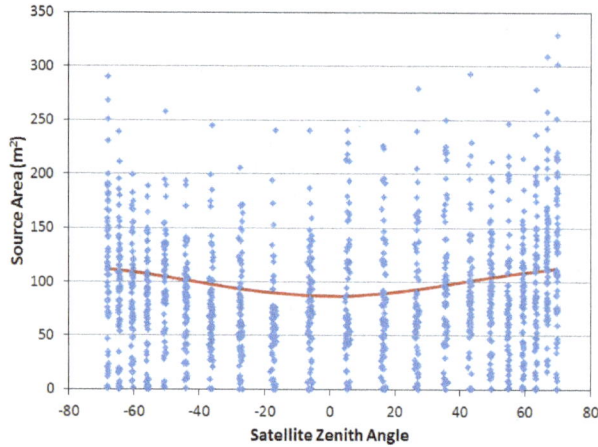

Figure 5. Variation in flare area estimates as a function of satellite zenith angle. The red line is the best fit line based on an elliptical model of the flare shape.

3.7. Discrimination of Flare Types

Flare sites were divided into upstream (or production) sites, downstream processing sites and flares at LNG terminals. This was based on spatial databases [13] containing 658 refineries and 35 LNG terminals combined with visual inspection of high spatial resolution images available in Google Earth. The upstream sites include oil and natural gas production facilities. Downstream sites are primarily refineries, identified either from the database or based on the spatial extent of infrastructure and large numbers of circular storage tanks. In some cases, oil refineries are adjacent to LNG terminals, with each site having flares. In this case, the LNG section was identified based on the presence of LNG carrier loading infrastructure and LNG storage tanks.

3.8. Calibration to Estimate Gas Flaring Volumes

RH with units of MW, is calculated from temperature and source size using the Stefan–Boltzmann law (Equation (2)). Although the efficiency of gas combustion at the flare and variations in gas composition (gas heating value) somewhat affect the relationship between gas volume entering a flare and the RH emitted, there should be a reasonably consistent relationship between the reported flare volumes and the estimated RH. We attempted to develop a calibration for estimating flared gas volumes based on the monthly sum of RH estimates (normalized for cloud cover and the number of valid nighttime observations) *versus* reported flaring from individual sites in Nigeria, Texas and North Dakota. Over the limited range of data available, the calibration appeared linear. When applied globally, however, the linear calibration resulted in unrealistically high flared gas volumes for the largest ~100 flares out of the >7000 detected globally. This result implied that there is a non-linear relationship between RH and flared gas volumes, with RH growing in a logarithmic fashion relative to flared gas volume.

To address this issue, we developed a non-linear calibration using national-level reporting of upstream flaring (plus venting) for 47 countries provided by Cedigaz [14], plus state-level reporting for Texas and North Dakota. Non-linearity was introduced by applying an exponent to the source area in the calculation of a modified RH estimate, RH', for each flare. As part of the calibration process, the value of the exponent was tuned to achieve the highest possible R^2 coefficient of determination between reported flare volumes and RH' (Figure 6). As vented gas does not contribute to RH emissions, we assumed that the quantity of reported venting was negligible. This was reasonable, as gas venting is rare and generally not reported by oil field operators. For the calibration, annual RH' estimates from all of the upstream gas flaring sites within the national boundaries were summed, with normalization for cloud cover and the number of valid nighttime observations.

Cedigaz includes only flare volumes at oil fields in its reported data. In Russia, in particular, there is a substantial volume of flaring at non-associated gas and gas condensate fields, which is not included in the Cedigaz estimates. Indeed, the RH' for flaring in all of Russia is high relative to the Cedigaz reported number (Figure 7). To remedy this, the Russian flares were associated with vector maps of oil fields, natural gas fields and gas condensate fields, and only the RH' for flares related to oil fields were used for the calibration.

To determine the optimal exponent for modulating the source areas for estimating RH', source areas were modulated using exponents ranging from 0.4 to 1.0 and evaluating the coefficient of determination (R^2) between RH' and the reported data. The highest R^2 occurred for an exponent of 0.7 (Figure 6).

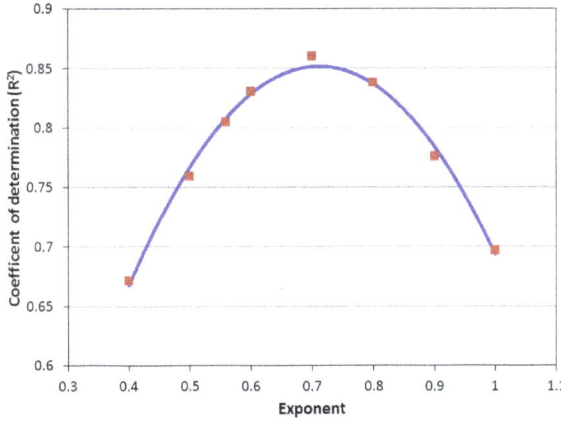

Figure 6. The exponent applied to the flare source area was tuned to 0.7 to yield the highest coefficient of determination (R^2).

The resulting calibration is shown in Figure 7. To estimate the slope in the linear equation $BCM = slope \times RH'$ we use a standard linear regression through the origin. The confidence intervals for the total sum $\pm$ BCM are derived from the 95% confidence intervals of the regression slope multiplied by the total sum of the observed RH'.

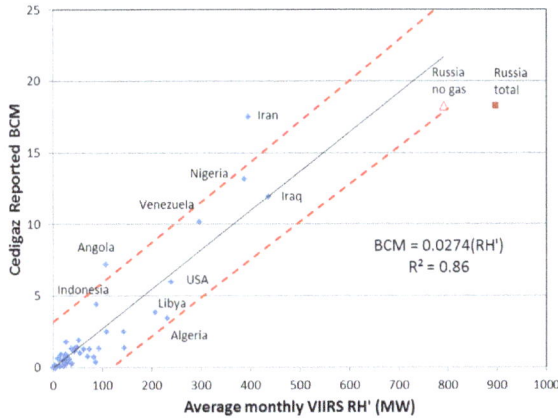

Figure 7. Calibration for estimating flared gas volumes from VIIRS-derived RH based on Cedigaz reported data. The dashed red lines indicate the positions of the 95% confidence intervals for the billion cubic meter (BCM) prediction errors. Note that the Russian data used in the regression was filtered to remove flares in natural gas and gas condensate fields, since the flaring in these areas is not represented in the Cedigaz gas flaring estimates.

3.9. Calculation of Flared Gas Volumes

The calibration from Figure 7 was applied to each individual flare site worldwide. In addition to the gas flare volume, the location, average temperature, average source area and percent frequency of detection were calculated.

4. Results

The analysis produced results identifying the locations of flares sites and flared gas volume estimates for 2012. These can be aggregated to national level to understand the global distribution of flaring and national potential for reducing CO_2 emissions through reductions in flaring.

4.1. Number of Flaring Sites

A total of 7467 flare sites were identified in 2012 (Figure 8). Of these, 6802 were upstream flares, 628 were downstream sites (predominantly refineries) and 37 flares were found at LNG terminals. The USA had the largest number of flare sites, with 2399 (Figure 9). Russia had the second largest number of flare sites (1053), less than half the USA tally. Flare numbers by country trail off rapidly below Russia, with Canada (332), Nigeria (325) and China (309).

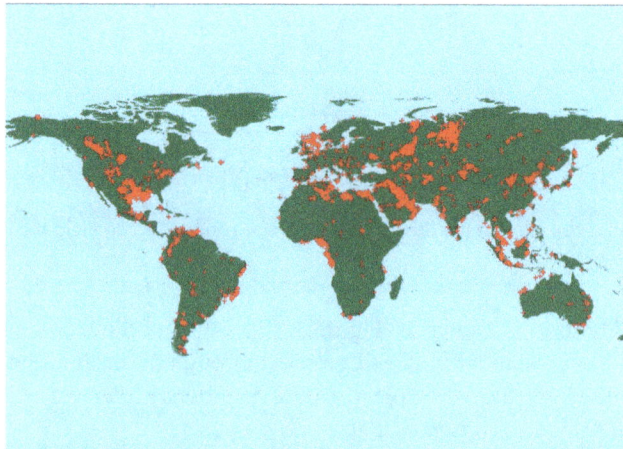

Figure 8. Spatial distribution of natural gas flaring in 2012.

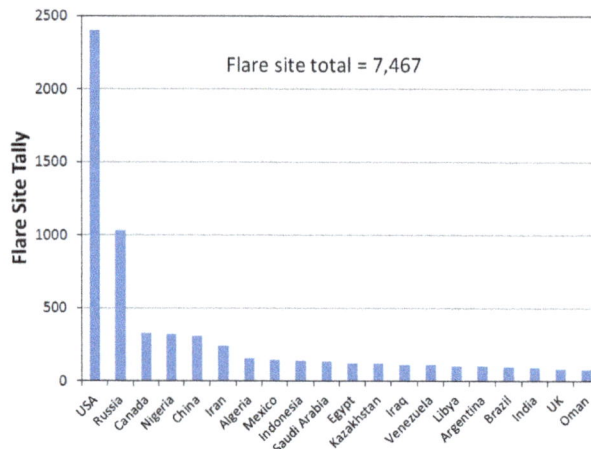

Figure 9. Flare tally by country for 2012.

4.2. Flared Gas Volume Estimates for Individual Flaring Sites

An output file was generated listing annual flared gas volume estimates for all of the detected flares. The output lists each flare site's latitude and longitude, average flare temperature, the percent frequency of detection, the normalized sum of *RH'* and the annual flared gas volume estimate. In addition to the tabular data, a Keyhole Markup Language (KML) file was produced to enable a review of the results in Google Earth. The KML for 2012 is included as a supplemental file with this publication.

Figure 10 shows the estimated flare volumes for each of the flare sites identified with VIIRS data. It is a classic exponential distribution, with high flare volumes concentrated in a relatively small percentage of the flare sites and large numbers of sites with small flared gas volumes. Half of the flared gas volume is concentrated at fewer than 400 flares, and 90% of the flaring occurs at just 30% of the sites. The largest flare found (Figure 10), located 7 km southeast of Punta de Mata in Venezuela, had an estimated flared gas volume of 1.13 billion cubic meters (BCM). The smallest flare had a 2012 gas flaring volume of 28,431 cubic meters, located in the Lekhwair oil field, Oman. Thus, from smallest to largest, the flared gas volumes mapped by VIIRS span five orders of magnitude.

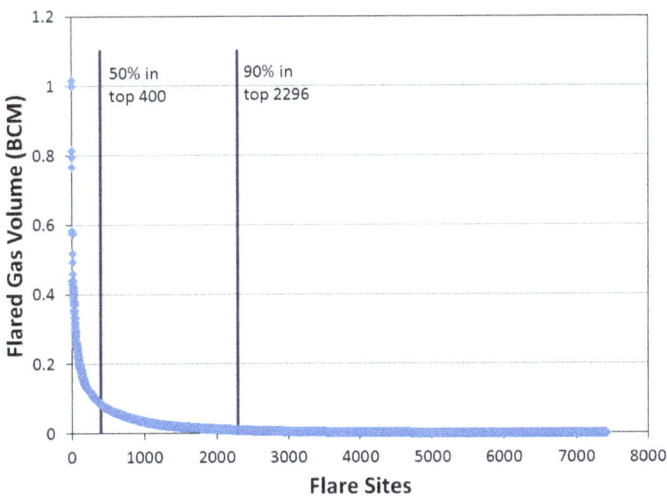

Figure 10. VIIRS estimate flared gas volumes for 7467 gas flares, worldwide. This includes both upstream, downstream and LNG terminal flare sites.

4.3. National- and Global-Level Estimates

National-level flared gas volumes were calculated by summing the estimates from individual flare sites within the national boundaries and associated exclusive economic zones. The global total flared gas volume is estimated at 143 ± 13.6 BCM. The results can be further divided into upstream, downstream and LNG terminals. Upstream flaring was found in 88 countries; with total flared gas volume in 2012 estimated at 129.4 ± 12.2 BCM (Figure 11). Russia leads in estimated upstream flaring with 24.6 BCM, followed by Iraq (11.9), Iran (10.7), Nigeria (10.5), Venezuela (8.1) and the USA (6.5). Downstream flaring (Figure 12) was found in 85 countries with a total of 10.7 ± 1.0 BCM. Algeria leads in estimated downstream flaring with 1.4 BCM, followed by Iran, Qatar, Mexico and Saudi Arabia. Estimated flaring at LNG liquefaction plants totaled 3.1 ± 0.29 BCM, with Algeria leading with 0.82 BCM (Figure 13).

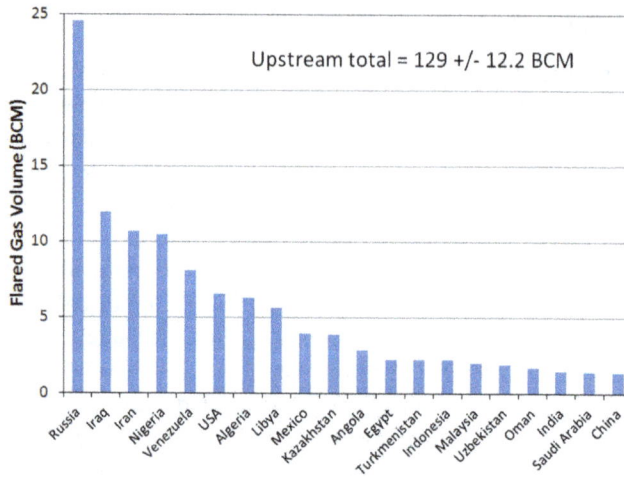

Figure 11. Top 20 countries for upstream gas flaring in 2012.

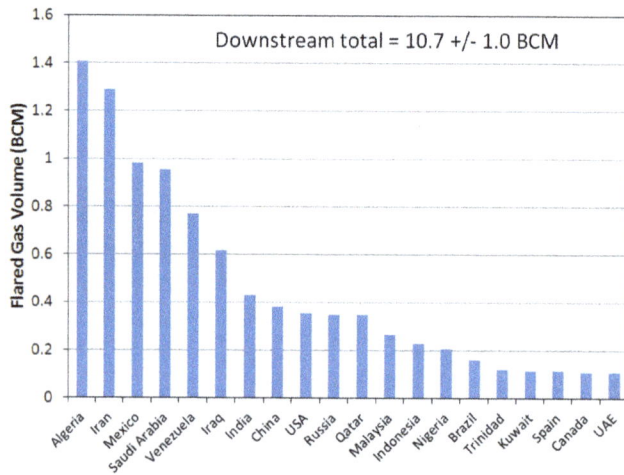

Figure 12. Top 20 countries for downstream gas flaring in 2012.

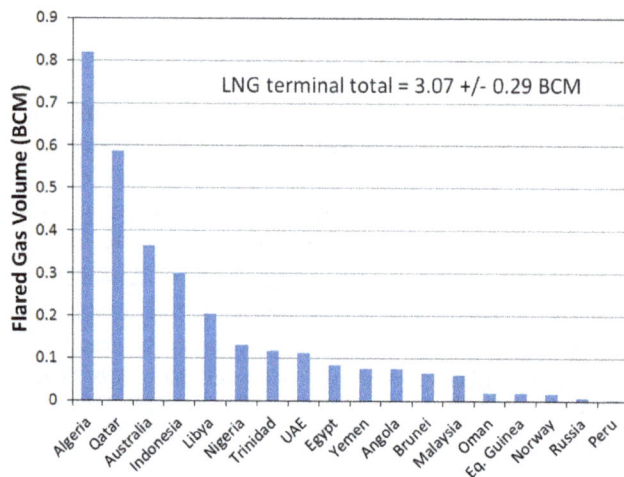

Figure 13. 18 countries with LNG terminal gas flaring in 2012.

4.4. Comparison with Defense Meteorological Satellite Program Satellite Estimates

The last year where DMSP estimates of flared gas volumes were produced was 2012. DMSP orbit degradation from 2013 onward resulted in solar contamination that made it impossible to produce global data for estimating flared gas volumes.

The 2012 global flaring estimate from VIIRS (143 BCM) is 4% higher than the DMSP estimate for the same year (137.4 BCM). The DMSP and VIIRS estimates for individual countries are highly correlated in most cases (Figure 14). There are several factors that may be contributing to the differences between the flared gas volume estimates between DMSP and VIIRS. Firstly, the DMSP sensor sensitivity to electric lighting made it impossible to identify gas flares imbedded in urban areas. VIIRS does not suffer from this drawback and has identified a substantial number of flaring sites that could not be identified in DMSP data. Secondly, the DMSP's inability to distinguish between flare radiant emissions and electric lighting may have resulted in overestimates of flared gas volumes at heavily lit locations. Third, the center core of many gas flaring detections made with DMSP were saturated, meaning that the signal was truncated by the limited dynamic range of the DMSP instrument. In contrast, no saturation was encountered in gas flare observations in the M10 spectral band. Fourth, the DMSP low light imaging data have no in-flight calibration. An empirical intercalibration was used to reduce sensor differences in the flared gas estimation [7]. This issue is resolved with VIIRS, which is widely regarded as a well calibrated instrument. Fifth, the DMSP calibration for estimating flared gas volumes excluded Russian data, since it was impossible to separate flaring in oil *versus* gas areas at that time due to the lack of field-specific vectors. Finally, the multispectral VIIRS data provides samples across 99% of the Planck curve modelling of a flare's radiant energy, as compared to 2% with the single DMSP band.

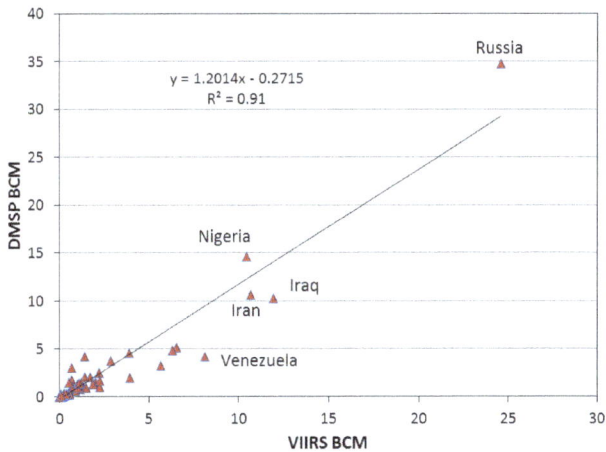

Figure 14. VIIRS *versus* DMSP flared gas estimates for 2012.

5. Conclusions

Using a series of processing steps, it is now possible to conduct global surveys to identify gas flaring sites and estimate flared gas volumes using nighttime VIIRS data. A survey of global gas flaring in 2012 found 7467 flaring sites worldwide, with an estimated 143 ($\pm$13.6) BCM of flared gas volume. The quantity of upstream (production related) flaring was estimated at 129.4 ($\pm$12.2) BCM, downstream flaring at 10.7 ($\pm$1.01) BCM and flaring at LNG liquefaction plants estimated at 3.07 ($\pm$0.29) BCM. While the USA had the largest number of individual flare sites, Russia led in terms of the largest flaring volume.

The VIIRS instrument has substantial advantages over other satellite sensors in terms of global monitoring of gas flaring. VIIRS collects global data every 24 h, providing repeat observations that enable numerous cloud-free observations of each site to be made over the course of a year. VIIRS is unique in the collection of near-infrared and SWIR data at night that has proven to be extremely useful for detecting flares and measuring their radiant output. The VIIRS M10 spectral band, centered at 1.6 μm, covers the peak radiant emissions for gas flares. The other two bands that collect at night are at 0.75 and 0.9 μm, recording radiances on the leading edge of gas flaring radiant emissions. At night, these three spectral bands record unambiguous radiant emissions from gas flares and other hot sources. These are augmented with radiance measurements from two traditional fire detection spectral bands in the 4-μm region. Planck curve fitting of the hot source radiances yields estimates of the temperature (K), source size (m^2) and, hence, RH (MW) of the hot sources. Flares can be distinguished from other hot sources based on their high temperature and persistence.

While VIIRS has a substantial number of favorable characteristics, it has two primary shortcomings of which the data users should be aware. The first shortcoming is the inability to derive temperature, source size and RH for weak detections made on small gas flares. The Planck curve fitting requires detection radiances in at least two spectral bands. There is a class of small flares where detection occurs only in the M10 spectral band, the location of the peak radiant emissions from most flares. These M10-only events can also arise from biomass burning and industrial sites. With no Planck curve fit, there is no direct calculation for RH. We developed methods to assign temperatures to the M10-only detections; however, these observations are not as good as the multispectral Planck curve fit observations. This single band detection problem on small gas flares will be resolved when nighttime M11 collections are added to the VIIRS data stream.

The second shortcoming arises from the temporal sampling limitations of the VIIRS instrument. VIIRS collects global data every night, but the dwell time of VIIRS on a flaring site is a fraction of a second. For steady and continuous flares, this temporal sampling appears to be adequate. However, VIIRS under-samples intermittent or rarely active flares. This under-sampling lowers the probability of detection and decreases the accuracy of flared gas estimates for flares with highly variable flared gas volumes.

We investigated the impact of applying an atmospheric correction using two flares in a humid tropical environment (onshore and offshore) and one flare in a dry desert environment. The atmospheric correction boosted the RH, but the effects across the three sites were highly linear. Our conclusion is that no atmospheric correction is required for gas flare detection made with the Nightfire algorithm. We attribute this to the high transmissivity of the atmosphere in the 1.6 μm window, where the gas flares typically have their highest radiant output.

Gas flares tend to be taller than they are wide due to the buoyancy of the heated air mass. Because VIIRS views gas flaring sites over a wide range of view angles, a substantial portion of the variation in RH estimates can be attributed to the viewed source size varying as a function of view angle. We developed a method for characterizing the average ellipticity of individual flares by analyzing the source size as a function of satellite zenith angle. We use the ellipticity to estimate the source size for all flares as viewed from the horizontal, which presents the largest apparent source area. This increases the coefficient of determination (R^2) in the calibration to estimate flared gas volumes.

A calibration for estimating flared gas volumes has been developed based on national-level data for upstream flaring reported by Cedigaz. For this calibration, the VIIRS data for Russia were filtered to only include flares at oil production facilities, since this is the only type of flaring reported in the Cedigaz data.

Despite the good correlation for many countries, there remains a considerable spread between the Cedigaz reported data and the VIIRS observed RH (Figure 7). For instance, the Cedigaz estimate for Iran (17.55 BCM) is high relative to the observed RH'. As a result, the VIIRS estimated flared gas volume for Iran (12.54 BCM) is 29% lower than the Cedigaz number. Other countries where the VIIRS estimates are lower than the Cedigaz numbers include Nigeria, Venezuela, the USA, Angola and Indonesia. Countries where the VIIRS estimates are higher than the Cedigaz numbers include Russia, Iraq, Libya and Algeria. In the future, it may be possible to improve the VIIRS gas flaring calibration using observations of flaring sites with metered flare volumes spanning a substantial portion of the range found in the VIIRS data.

This paper reports on results for the year 2012. Results from 2013, 2014 and 2015 will be available in the near future. The VIIRS data on flared gas volumes should be useful in carbon cycle studies, identification of sites for natural gas utilization projects, calculating carbon intensities of fuels and tracking the progress of efforts to reduce gas flaring. Such projects can rely on the availability of VIIRS data from NOAA for the next several decades.

Because flaring burns off a potentially valuable fuel commodity, it is one of the obvious places to focus efforts to reduce the carbon loading on the atmosphere as it represents. Making use of natural gas that would otherwise be flared could reduce the consumption of other fuels, thus lowering the total volume of carbon emissions to the atmosphere. Flared gas volume represents about 3.5% of total worldwide natural gas consumption and 19.8% of U.S. natural gas consumption in 2012 [15]. If used to fuel vehicles, it could power 74 million automobiles in the USA based on an average of 25 miles per gallon of gasoline [16] and 13,476 miles per year [17].

In summary, we have developed a systematic set of algorithms and methods that can be used on a repeated basis to identify gas flaring sites worldwide, with estimation of flared gas volumes. With this capability, VIIRS can provide detailed, site-specific data for tracking efforts to reduce natural gas flaring.

Acknowledgments: This study was jointly funded by the NOAA Joint Polar Satellite System (JPSS) proving ground program and the World Bank Global Gas Flaring Reduction partnership (GGFR). Calibration data were provided by Cedigaz.

Author Contributions: Chris Elvidge designed and managed the study and served as the lead author. Mikhail Zhizhin developed the Nightfire software, conducted the water-shedding to identify the individual flaring sites, developed the flare database, the calibration and produced the output files. Kim Baugh managed the data processing and made the cloud-free composite used to identify the gas flaring sites. Feng-Chi Hsu conducted the atmospheric effects analysis. Tilo Ghosh assisted in the download and processing of the data. She also produced Figure 8.

Conflicts of Interest: The authors declare no conflict of interest.

References

1.   Peylin, P.; Law, R.M.; Gurney, K.R.; Chevallier, F.; Jacobson, A.R.; Maki, T.; Niwa, Y.; Patra, P.K.; Peters, W.; Rayner, P.J.; *et al.* Global Atmospheric Carbon Budget: Results from an ensemble of atmospheric CO_2 inversions. *Biogeosciences* **2013**, *10*, 6699–6720. [CrossRef]

2.   Sonibare, J.A.; Akeredolu, F.A. Natural gas domestic market development for total elimination of routine flares in Nigeria's upstream petroleum operations. *Energy Policy* **2006**, *34*, 743–753. [CrossRef]

3.   Sonia, Y.; Witcover, J.; Kessler, J. *Status Review of California's Low Carbon Fuel Standard—Spring 2013 Issue*; Research Report UCD-ITS-RR-13-06; Institute of Transportation Studies, University of California: Davis, CA, USA, 2013.

4.   Anejionu, O.C.D.; Blackburn, G.A.; Whyatt, J.D. Detecting gas flares and estimating flaring volumes at individual flow stations using MODIS data. *Remote Sens. Environ.* **2014**, *158*, 81–94. [CrossRef]

5.   Casadio, S.; Arino, O.; Serpe, D. Gas flaring monitoring from space using the ATSR instrument series. *Remote Sens. Environ.* **2012**, *116*, 239–249. [CrossRef]

6.   Chowdhury, S.; Shipman, T.; Chao, D.; Elvidge, C.D.; Zhizhin, M.; Hsu, F.-C. Daytime gas flare detection using Landsat-8 multispectral data. In Proceedings of the IEEE International Geoscience Remote Sensing Symposium, Quebec City, QC, Canada, 13–18 July 2014; pp. 258–261.

7.   Elvidge, C.D.; Ziskin, D.; Baugh, K.E.; Tuttle, B.T.; Ghosh, T.; Pack, D.W.; Erwin, E.H.; Zhizhin, M. A Fifteen Year Record of Global Natural Gas Flaring Derived from Satellite Data. *Energies* **2009**, *2*, 595–622. [CrossRef]

8.   Elvidge, C.D.; Zhizhin, M.; Hsu, F.-C.; Baugh, K.E. VIIRS Nightfire: Satellite Pyrometry at Night. *Remote Sens.* **2013**, *5*, 4423–4449. [CrossRef]

9.   Giglio, L.; Descloitres, J.; Justice, C.O.; Kaufman, Y.J. An enhanced contextual fire detection algorithm for MODIS. *Remote Sens. Environ.* **2003**, *87*, 273–282. [CrossRef]

10.   Kopp, T.J.; Thomas, W.; Heidinger, A.K.; Botambekov, D.; Frey, R.A.; Hutchison, K.D.; Iisager, B.D.; Brueske, K.; Reed, B. The VIIRS Cloud Mask: Progress in the first year of S-NPP toward a common cloud detection scheme. *J. Geophys. Res. Atmos.* **2014**, *119*, 2441–2456. [CrossRef]

11.   Berk, A.; Anderson, G.P.; Acharya, P.K.; Shettle, E.P. *MODTRAN 5.2. 1 User's Manual*; Spectral Sciences Inc.: Burlington, MA, USA; Air Force Research Laboratory: Hanscom Air Force Base, MA, USA, 2011.

12.   Jekrard, H.G. Transmission of Light through Birefringent and Optically Active Media: The Poincaré Sphere. *J. Opt. Soc. Am.* **1954**, *44*, 634–640. [CrossRef]

13.   *ArcGIS Shapefiles for Global Crude Oil Refinereis and LNG Liquifaction Terminals*; Environmental Systems Research Institute (ESRI): Redlands, CA, USA, 2014.

14.   Cedigaz National Flared Gas Volumes. Available online: http://www.cedigaz.org/products/natural-gas-database.aspx (accessed on 8 September 2015).

15.   Natural Gas Statistics, U.S. Energy Information Administration. Available online: http://www.eia.gov (accessed on 25 October 2015).

16.   Monthly Monitoring of Vehicle Fuel Economy and Emissions. University of Michigan, Transportation Research Institute. Available online: http://www.umich.edu/~umtriswt/EDI_sales-weighted-mpg.html (accessed on 7 December 2015).

17.   Average Annual Miles per Driver by Age Group. Department of Tranportation, Federal Highway Administration. Available online: https://www.fhwa.dot.gov/ohim/onh00/bar8.htm (accessed on 7 December 2015).

7

Optimal Siting and Sizing of Distributed Generators in Distribution Systems Considering Cost of Operation Risk

Qingwu Gong, Jiazhi Lei * and Jun Ye

Academic Editor: João P. S. Catalão

School of Electrical Engineering, Wuhan University, Wuhan 430072, China; qwgong@whu.edu.cn (Q.G.);
yejunz@126.com (J.Y.)
* Correspondence: leijiazhi@126.com

Abstract: With the penetration of distributed generators (DGs), operation planning studies are essential in maintaining and operating a reliable and secure power system. Appropriate siting and sizing of DGs could lead to many positive effects forthe distribution system concerned, such as the reduced total costs associated with DGs, reduced network losses, and improved voltage profiles and enhanced power-supply reliability. In this paper, expected load interruption cost is used as the assessment of operation risk in distribution systems, which is assessed by the point estimate method (PEM). In light with the costs of system operation planning, a novel mathematical model of chance constrained programming (CCP) framework for optimal siting and sizing of DGs in distribution systems is proposed considering the uncertainties of DGs. And then, a hybrid genetic algorithm (HGA), which combines the GA with traditional optimization methods, is employed to solve the proposed CCP model. Finally,the feasibility and effectiveness of the proposed CCP model are verified by the modified IEEE 30-bus system, and the test results have demonstrated that this proposed CCP model is more reasonable to determine the siting and sizing of DGs compared with traditional CCP model.

Keywords: distributed generators; siting and sizing; distribution systems; point estimate method; hybrid genetic algorithm; chance constrained programming

1. Introduction

With the continuous exhaustion of fossil energy, the gradual increase in the global temperature and the limitation of available transmission corridors, rapid development of distributed generators (DGs) has been vigorously developing around the world [1]. The subsequent growth of DGs has enabled distribution systems to actively respond to the dynamics of the main grid [2].

Although the use of DGs is helpful at many aspects, the extensive penetration of DGs could lead to some security and economic risks to active distribution networks. Specifically, the intermittency and variability of renewable-type DGs impose challenges when planning active distribution networks [3–5]. Inappropriate sitingand sizing of DGs could lead to many negative effects on the distribution systems concerned, such as the relay system configurations, voltage profiles, and network losses [6]. Therefore, it is becoming increasingly important of siting and sizing the DGs in active distribution networks planning.

There are various uncertainty handling methods developed for dealing with the uncertainties caused by the randomness of DGs'output [7]. These methods can be categorized into three main categories [8]: information gap decision theory (IGDT), robust restoration method and fuzzy modeling. However, the mathematical models that are obtained are mixed-integer nonlinear programming ones with multiple included variables and constraints. Chance constrained programming in [9] is powerful and robust in cases of severe uncertainty.

Many research papers on the subject of the optimal siting and sizing of DGs have become available in the past decade [10–20]. However, most of them are based on how to establish the objective function to obtain the maximum economic benefits. For example, in [10], an optimal distributed generation allocation on an existing MV distribution network based on genetic algorithm considering the uncertainties of DGs' numbers, locations and capacities was proposed; in [11,12], locations and discrete capacities of DGs are determined in the view of minimizing the distribution network losses; References [13,14] determined the locations and capacities of DGs in case of minimum network losses based on an improved Hereford ranch algorithm and a combination of genetic algorithm and simulated annealing respectively. These papers only considered the network losses with the penetration of DGs, which is not adequate in determining the locations and capacities of DGs. In [15], an optimal investment planning for distributed generation in a competitive electricity market was proposed. In [16,17], a multi-objective evolutionary algorithm for the sizing and siting of distributed generation was proposed. In these papers, many costs including investment cost and network loss cost are considered but chance constraints of the established system operation planning are not considered. Also, in these papersthe only DGs considered arediesel generators and micro-turbines.

In recent years, [18] presented an chance constrained programming (CCP) framework of optimal siting and sizing of DGs with the minimization of the DGs' investment cost, operating cost, maintenance cost, network loss cost, as well as the capacity adequacy cost. The authors of [19,20] determined the optimal location for placing the distributed generators based on loss sensitivity factor and also proposed a successive sizing based algorithm for determining their optimal sizes. In these papers, chance constraints are considered, but none of the papers mentioned above considered the costs of operation risk caused by DGs.

Many other research papers have showed that the presentation of DGs could increase the operation risk in distribution systems such as load curtailment. In [21], a hierarchical method was presented to evaluate the risk of active distribution networks considering the influence of different kinds of DGs. In [22], a novel operation risk assessment method based on credibility theory was used to assess the power system operations. Reference [23] presented a risk assessment approach to analyze power system security for operation planning under high penetration of wind power generation. These papers aimed at calculating the risk assessment of active distribution networks, which can be evaluated by expected energy not supplied (EENS) [24–26]. In [27,28], the unit interruption cost (UIC) is estimated, which can be used to assess the operation risk cost of distribution systems when combined with EENS.

In this paper, expected load interruption cost (ELIC) is used as the assessment of operation risk in distribution systems, which is assessed by point estimate method (PEM). In light of the costs of system operation planning, a novel chance constrained programming (CCP) framework mathematical model for optimal siting and sizing of DGs in distribution systems is proposed considering the uncertainties of DGs. Then, a hybrid genetic algorithm (HGA), which combines the genetic algorithm (GA) with traditional optimization methods, is employed to solve the proposed CCP model. This HGA overcome the drawbacks of both GA and traditional optimization methods.

Finally, the feasibility and effectiveness of the proposed CCP model are verified by the modified IEEE 30-bus system, and the test results have demonstrated that this proposed CCP model is more reasonable to determine the siting and sizing of DGs compared with traditional CCP model. Simulation results also demonstrate that the appropriate siting and sizing of DGs could lead to many positive effects on the distribution system concerned, such as the reduced total costs associated with DGs, reduced network losses, and improved voltage profiles and enhanced power-supply reliability.

2. Output Power of Distributed Generators

The dynamic output power of DGs is intricate with great volatilities and randomness, which mainly depends on the local weather condition. In distribution system, whether the load can be supplied is much related with the output power of DGs when malfunction appears, so the output power of DGs should be calculated for distributed system assessment. In this paper, wind generating units and photovoltaic generation units are considered as DGs, but it should be mentioned that the proposed optimization method in this paper could accommodate other DGs as well.

2.1. Output-Power of Wind Generating Units

The output-power of wind generating units is mainly related with the stochastic wind speed of the surroundings where wind generating units locate. In the past decades, large amounts of researches have demonstrated that the stochastic wind speed in most regions approximately follows the Weibull distribution [18,29,30], and this conclusion is employed in this paper. In general, probability density function of the stochastic wind speed can be depictedin Equation (1):

$$f_v(v) = \frac{k}{c}(\frac{v}{c})\exp[-(\frac{v}{c})^k]$$
(1)

where k and c are, respectively, the shape index and the scale index of the Weibull distribution, which can be calculated by Maximum Likelihood Estimate and the historical data of wind speed.

In light with the known probability distribution function of the wind speed, the output power of wind generating units can be fitted as follows [31]:

$$P_\omega = \begin{cases} 0, & v < v_{in} \\ P_N(a_1v^3 + a_2v^2 + a_3v + a_4), & v_{in} \leqslant v < v_N \\ P_N, & v_N \leqslant v < v_{out} \\ 0, & v \geqslant v_{out} \end{cases}$$
(2)

where, a_1, a_2, a_3 and a_4 are the fitting coefficients. P_ω and P_N are the active output power and the rated outputpower of wind generating units respectively, and v, v_N, v_{in} and v_{out} are the wind speed at the hub height of the wind unit, the rated wind speed, the cut-in wind speed, and the cut-out wind speed respectively.

2.2. Output-Power of Photovoltaic Generating Units

Many factors, such as illumination intensity I_{PV}, array area of photovoltaic cell A, and photoelectric conversion efficiency η, can affect the output-power of photovoltaic generating units P_{PV}. But, the illumination intensity is usually considered the dominant factor of affecting the output power of a solar generating source. In a certain period of time, one hour or several hours, the illumination intensity I_{PV} follows the Beta distribution, and its probability density function can be expressed as follows:

$$f(I) = \frac{\Gamma(\alpha + \beta)}{\Gamma(\alpha)\Gamma(\beta)} \cdot \left(\frac{I_{PV}}{I_{PV}^{max}}\right)^{\alpha-1} \cdot \left(1 - \frac{I_{PV}}{I_{PV}^{max}}\right)^{\beta-1}$$
(3)

Where α and β are the shape indexes of the Betadistribution. I_{PV}^{max} is the maximum value of I_{PV}, Γ is the gamma function.

Based on the known probability distribution function of the illumination intensity, the probability density function of the output-power of photovoltaic generating units can be calculated as follows [32]:

$$f(P_{PV}) = \frac{\Gamma(\alpha + \beta)}{A \cdot \eta \cdot \Gamma(\alpha)\Gamma(\beta)} \cdot \left(\frac{P_{PV}}{P_{PV}^{max}}\right)^{\alpha-1} \cdot \left(1 - \frac{P_{PV}}{P_{PV}^{max}}\right)^{\beta-1}$$
(4)

where, P_{PV}^{max} is the maximum value of P_{PV}. P_{PV} is the output of photovoltaic generating units, which can bedescribed as $P_{PV} = A \cdot \eta \cdot I_{PV}$. As can be shown in Equation (4), the output-power of photovoltaic generating units P_{PV} follows the beta distribution $B(\alpha, \beta)$.

3. Cost of Risk Assessment

3.1. Cost of System Operation Risk Level

The classic definition of power system reliability is related to the existence of sufficient facilities within the system so that it is capable of supplying electric power to its customers with an acceptable assurance of continuity and quality. The reliability performance of a distribution system can be measured at each customer connection point and for the whole system or any group of customers. In [26], SAIFI (system average interruption frequency index), SAIDI (system average interruption duration index) and EENS are used as reliability assessment indexes in active distribution network. In [21], four risk indices, including EENS, PLC (probability of load curtailment), EFLC (expected frequency of load curtailment) and SI (severity index), are adopted to evaluate the system operation risk level. It should be noted that EENS (expected energy not supplied, MWh/year) is one of the most important parameters to reflect the risk level of transmission system. In this paper, the assessment of system operation risk level is measured by the amount of expected load interruption cost (ELIC), which is the product of the unit interruption cost (UIC) in \$/kWh and the expected energy not supplied (EENS).

$$\text{ELIC} = \text{UIC} \times \text{EENS} \tag{5}$$

The UIC can be estimated by a method based on the revenue lost to a utility due to power outages [27]. The index of EENS (expected energy not supplied, MWh/year) is calculated by Equation (6):

$$\text{EENS} = \sum_{i=1}^{N_L} \left(\sum_{s \in Q_i} p_T(s) \cdot C_0(s) \right) \cdot T_i \tag{6}$$

where, N_L is total number of load levels; T_i is the durationof load level i; Q_i is system state set for load level$i$; $p_T(s)$ is occurrence probability of system state s; $C_0(s)$ is total load curtailment in system state s.

3.2. Costs of System OperationPlanning

The costs of system operation planning can consists of many aspects, such as the investment cost, operating cost, maintenance cost, network loss cost, and capacity adequacy cost. Capacity adequacy cost can be reflected by ELIC to some degree, and operating cost for a renewable DG is zero. Furthermore, the installed DGs in a distribution systems will bring a benefit to distribution systems, which is to generate electricity replacing conventional fossil energy. As a result, the generation cost replaced by DGs should be considered in the costs of system operation planning.

Therefore, investment cost C^I, maintenance cost C^M, network loss cost C^L, and the generation cost C^{DG} replaced by DGs are used for the costs of system operation planning, which can be depicted as Equation (7):

$$C_{DG} = C^I + C^M + C^L - C^{DG} \tag{7}$$

In [33–35], many methods have been used to calculate the indexes of investment cost, maintenance cost, and network loss cost. In this paper, network loss cost C^L can be depicted as Equation (8), the investment cost C^I can be depicted as Equation (9), the maintenance cost C^M can be depicted as Equation (10), the generation cost C^{DG} replaced by DGs can be depicted as Equation (11):

$$C^L = \sum_{j=1}^{L} C_e \tau_{jmax} R_j \frac{P_j}{(U_N \lambda_j)^2} \tag{8}$$

In Equation (8), C_e is the unit electricity saleprice ($\cdot kW^{-1}$), τ_{jmax} is the maximum hours of load loss in branch j, and R_j is the resistance of branch j. P_j is the active power that flows through branch j, U_N is the rated voltage of line j, and λ_j is the load power factor of branch j:

$$C^I = \sum_{i=1}^{N} \sum_{m=1}^{2} a_i E \cdot P_{i-max} \cdot C_{OMi}^m \cdot \frac{r(1+r)^{n_{DG}}}{(1+r)^{n_{DG}} - 1} \tag{9}$$

In Equation (9), C_{OMi}^m represents the per-unit capacity investment cost of the DG m at node i, and N is the number of candidate node. $a_i = f(x_1, x_2, x_3)$ is the weighting coefficients of investment cost, which relates to environmental factor x_1, displacement factor x_2, transportation and labor cost factor x_3. r is the discount rate, and n_{DG} is the lifetime of DGs. P_{i-max} is the upper limits of DGs capacities in candidate node i, and E is candidate DG capacity in candidate node i, which will be introduced in Section 4.2:

$$C^M = \sum_{i=1}^{N} \sum_{m=1}^{2} T_{max} \cdot C_{OPi}^m \cdot E \cdot P_{i-max} \tag{10}$$

In Equation (10), T_{max} is the maximum hours of DGs generation, N is the number of DGs, and C_{OPi}^m represents the per-unit capacity maintenance cost of the DG m at node i:

$$C^{DG} = \sum_{i=1}^{N} \sum_{m=1}^{2} E \cdot P_{i-max} \cdot \eta_i \cdot T_{eq} \cdot C_{en}^m \tag{11}$$

In Equation (11), T_{eq} is the equivalent generation hours of DGs generation, and C_{en}^m is the unit on-grid electricity price of DG m. η_i is the efficiency of DG m in node i.

4. Mathematical Model

4.1. Methodological Framework

The developed mathematical model of the CCP-based optimal siting and sizing of DGs in this paper can be formulated as:

$$
\begin{aligned}
\min \quad & f(E, X) \\
s.t. \quad & Pr\{g_j(E, X) \leqslant 0\} \geqslant \lambda \quad (j = 1, 2, \cdots, N_L) \\
& G = 0 \\
& H_{min} \leqslant H \leqslant H_{max}
\end{aligned}
\tag{12}
$$

where, $f(E, X)$ is the objective function, $g_i(E, X)$ are chance constraints, $G = 0$ represents the equality constraints, and $H_{min} \leqslant H \leqslant H_{max}$ is the inequality constraints. E is the decision-making vector; X is a set of stochastic variables with known probability. λ is the given confidence levels; N_L is the number of feeders in the distribution system; H_{min}/H_{max} is the set of the minimal/maximal limits of inequality constraints.

4.2. Objective Function

In order to obtain the optimal locations and capacities of DGs, the optimization variable included in the planning scheme should be constructed.

Suppose that m represents the species of DGs in the distribution system. $m = 1/2$ represent wind generating units and photovoltaic generating units, respectively. If y represents the installed capacities of DGs in the distribution system, then $y_{m1}, y_{m2}, \ldots, y_{mn}$ can represent the candidate locations and installed capacities for DGs in the distribution system. y_{mi} ($i = 1, 2, \ldots, N$) = 0 denotes that there will not be a DG m built at candidate nodes i ($i = 1, 2, \ldots, N$). y_{mi} ($i = 1, 2, \ldots, n$) > 0 denotes that there will be a DG m built at candidate nodes i ($i = 1, 2, \ldots, n$), and the installed capacity is y_{mi}.

Furthermore, the installed capacities of DGs could be normalized, namely $y_{mi} = y_{mi}/P_{i-max}$. Thence, the stochastic variables E in Equation (10) can be depicted as:

$$E = \{y_{mi} \,|i = 1, 2, \cdots, N; m = 1, 2\} \tag{13}$$

In this paper, the objective function is described as:

$$\min f = \alpha_1 C_{DG} + \alpha_2 \cdot \text{ELIC} \tag{14}$$

where, α_1, α_2 are the weighting coefficients. C_{DG} is the costs of system operation planning associated with the DGs, which can be calculated by Equation (7). ELIC is the costs caused by EENS, which can be calculated by Equation (5).

4.3. Network Constraints

As shown in Equation (12), there are three kinds of network constraints, including equality constraints, inequality constraints, and chance constraints.

(1) The equality constraints include the well-known load-flow equations, which can be depicted in Equation (15):

$$\begin{cases} P_{is} = U_i \sum_{j\in i} U_j (G_{ij}\cos\theta_{ij} + B_{ij}\cos\theta_{ij}) \\ Q_{is} = U_i \sum_{j\in i} U_j (G_{ij}\cos\theta_{ij} - B_{ij}\cos\theta_{ij}) \end{cases} \tag{15}$$

where, P_{is} and Q_{is} are the total active, reactive output power of the generators at node i, U_i and U_j are the voltage amplitude at node i and j respectively. G_{ij} and B_{ij} are the conductance and susceptance between node i and j respectively. θ_{ij} is the voltage angle between node i and j.

For E, if $y_{1i} > 0$, $y_{2i} = 0$; if $y_{2i} > 0$, $y_{1i} = 0$. Namely, wind generating units and photovoltaic generating units cannot be installed in node i simultaneously. Therefore, an additional equality constraint is included in the equality constraints in this paper, which is shown in Equation (16):

$$y_{1i} \cdot y_{2i} = 0 \quad (i = 1, 2, \cdots, N) \tag{16}$$

(2) The inequality constraints include the given permitted penetration capacity of DGs in the distribution system and the upper limits of DGs capacities in candidate node i, which are shown in Equations (17) and (18) respectively:

$$\sum_{i=1}^{N} \sum_{m=1}^{2} y_{mi} \cdot P_{i-max} \leqslant P_{DG_{max}} \tag{17}$$

$$0 \leqslant y_{mi} \leqslant 1, \quad \forall i \in N \ ; m = 1, 2 \tag{18}$$

where, $P_{DG_{max}}$ is given permitted penetration capacity of DGs in the distribution system.

(3) The chance constraints in this paper mainly include node voltage constraints and branch transmission power constraints, which are shown in Equations (19) and (20) respectively:

$$P_r \left\{ U_i^{min} \leqslant U_i \leqslant U_i^{max} \right\} \geqslant \beta_u , \quad i = 1, 2, \cdots, N_b \tag{19}$$

$$P_r \left\{ |P_i| \leqslant P_i^{max} \right\} \geqslant \beta_p , \quad i = 1, 2, \cdots, N_a \tag{20}$$

where, $\{\cdot\}$ represents the probability of a certain event occurs. N_b is the total number of nodes in the distribution system, and N_a is the total number of branches in the distribution system. U_i^{min} and U_i^{max} are the upper and lower limits of the voltage U_i in node i respectively. β_u is the given confidence level of node voltage constraints, and β_p is the given confidence level of branch transmission power constraints. P_i is the transmission power in branch i, and P_i^{max} is upper limits of transmission power in branch i.

5. Solving Strategies

To solve the developed mathematical model of the CCP-based optimal siting and sizing of DGs in Equation (10), three steps are needed, as shown in Figure 1. In Step A, expected load interruption cost ELIC is calculated according to the output of wind generating units, photovoltaic generating units, and energy storage system. In Step B, total costs associated with the DGs C_{DG} are calculated according to the output of wind generating units, photovoltaic generating units, and energy storage system. In Step C, the objective function of optimal siting and sizing of DGs is firstly set up based on ELIC and C_{DG}. Then, with the network constraints, the developed mathematical model of the CCP-based optimal siting and sizing of DGs is solved, and the optimal solution E^* is obtained.

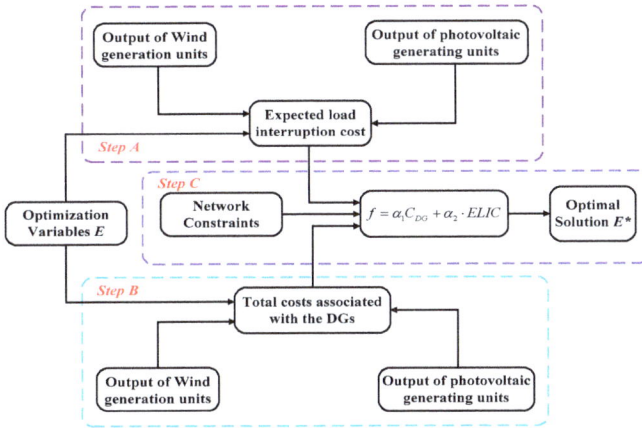

Figure 1. Flowchart of the developed chance constrained programming (CCP)-based method for optimal siting andsizing of distributed generators(DGs) in distribution systems.

5.1. Calculation of Expected Load Interruption Cost (ELIC)

In [21], a hierarchical risk assessment method based on discrete probability model of DGs, the enumeration method and Monte Carlosimulation was proposed to calculate EENS. This method is much effective, but is relatively complex. In [36], point estimate method (PEM) was used for transmission line overload risk assessment. This method is much effective, but is relatively complex. Compared with Monte Carlosimulation, PEM is slightly low in accuracy, but the computational cost of PEM is largely reduced. Therefore, point estimate method (PEM) is used to calculate the EENS parameterin this paper:

(1) The point estimate method (PEM), proposed by Hong in 1998, is used to calculate the estimation value $E(Z^j)$ of $F(X)$ ($Z = F(X) = F(X_1, X_2, \ldots, X_m)$) when the probability distribution functions of random variables $X = (X_1, X_2, \ldots, X_m)$ are known [37,38]. The mathematical principle of PEM can be depictedin Equation (21):

$$E(Z^j) \approx \sum_{l=1}^{m} \sum_{k=1}^{K} \omega_{l,k} F^j(\mu_{X1}, \cdots, x_{l,k}, \cdots, \mu_{Xm}) \tag{21}$$

where, $\omega_{l,k}$ is the weights of random variable X_i in $x_{l,k}$, and μ_{X_i} is the mathematical expectation of random variable X_i. In this paper, random variable X_i is the wind speed v or illumination intensity I_{PV}

of DG i. $X_i = \{X_{mi} \mid i = 1, 2, \ldots, N; m = 1, 2\}$, and X_{1i} is the wind speed v, X_{2i} is illumination intensity I_{PV}. $x_{l,k}$ can be got by Equation (22):

$$x_{i,k} = \mu_{X_i} + \xi_{i,k}\sigma_{X_i}, i = 1, 2, \ldots, m, k = 1, 2, \ldots, 2K-1 \tag{22}$$

where, σ_{X_i} is the mathematical variance of random variable X_i. The values of $\xi_{i,k}$ and $\omega_{l,k}$ can be obtained by Equation (23):

$$\begin{cases} \sum_{k=1}^{K} \omega_{i,k} \cdot \xi_{i,k}{}^j = \lambda_{i,j} , \quad j = 1, \cdots, 2K-1 \\ \lambda_{i,j} = E\left[(x_i - \mu_{X_i})^j\right]/\sigma_{X_i}{}^j \end{cases} \tag{23}$$

When $K = 3$, and $\xi_{i,3} = 0$ ($2m + 1$ PEM), Equation (23) can be solved as Equation (24):

$$\begin{aligned} \xi_{i,1} &= \frac{\lambda_{i,3}}{2} + \sqrt{\lambda_{i,4} - \frac{3}{4}\lambda_{i,3}{}^2} \\ \xi_{i,2} &= \frac{\lambda_{i,3}}{2} - \sqrt{\lambda_{i,4} - \frac{3}{4}\lambda_{i,3}{}^2} \\ \omega_{i,1} &= \frac{1}{\xi_{i,1}(\xi_{i,1} - \xi_{i,2})}, \omega_{i,2} = -\frac{1}{\xi_{i,2}(\xi_{i,1} - \xi_{i,2})} \\ \omega_{i,3} &= \frac{1}{m} - \frac{1}{\lambda_{i,4} - \lambda_{i,3}{}^2} \end{aligned} \tag{24}$$

(2) The procedure for computing the index of EENS is summarized in Figure 2. In Figure 2, the $2m + 1$ point estimate method (PEM) was used, namely $K=3$. $f(x_{1i,k})$ is calculated by Equation (2), in which $P_N = y_{1i} \cdot P_{i-max}$. $g(x_{2i,k}) = A \cdot \eta \cdot x_{2i,k}$, and $A \cdot \eta \cdot I_{PV}^{max} = y_{2i} \cdot P_{i-max}$. y_{1i} and y_{2i} satisfy the equality constraint Equation (16).

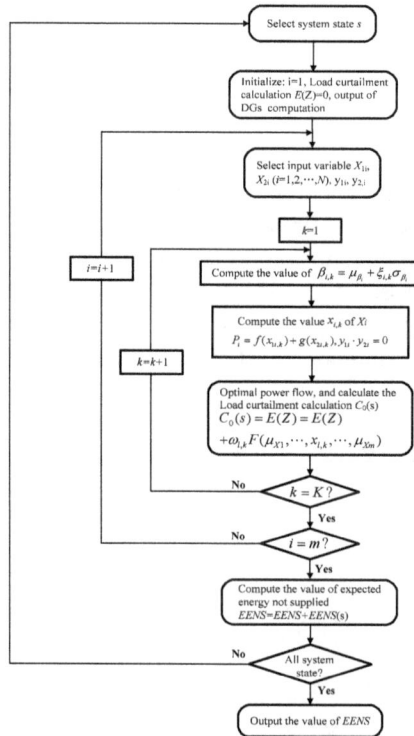

Figure 2. Flow chart of computing procedure for expected energy not supplied (EENS).

Also, optimal power flow was used to calculate the index of load curtailment calculation $C_0(s)$, which was described in [39]. Due to the limited space of this paper, the detailed process of optimal power flow is not covered.

(3) The UIC can be estimated by several methods such as a method based on customer damage functions or a method based on the revenue lost to a utility due to power outages [23]. In this paper, UIC is estimated by a method based on the revenue lost to a utility due to power outages, which can be referred in [27]. In light with EENS and UIC, the value of expected load interruption cost (ELIC) can be calculated by Equation (5).

5.2. Calculation of System Operation Planning Cost

As shown as Section 3.2, network loss cost C^L, investment cost C^I, maintenance cost C^M and the generation cost C^{DG} replaced by DGs can be calculated by Equations (8)–(11), respectively. The system operation planning cost C_{DG} can be calculated by Equation (7).

It should be noted that the output power of wind generating units and photovoltaic generating units in year are generated on an hourly basis. The unit electricity sale price C_e in Equation (8) and the unit on-grid electricity price C_{en}^m of DG m in Equation (11) are not constant, but can be approximately calculated by average prices of last year.

5.3. Solving Strategy of Chance Constrained Programming (CCP)-Based Optimization

In [40,41], a particle swarm optimization (PSO) algorithm is used to solve the CCP-based optimization described by Equation (12). In [18], a Monte Carlo simulation embedded genetic-algorithm approach is employed to solve the optimization problem described by Equation (12). The PSO algorithm requires a better set of initial values for the design variables and may easily fall into local optimum. GA can locate the solution in the whole domain, but it does not solve constraint problems easily, especially for exact constraints.

In order to overcome these drawbacks above, a hybrid genetic algorithm (HGA) [42], which combines the GA with traditional optimization methods, is used for solving the CCP-based optimization described by Equation (12) in this paper. In the first step of HGA, the GA is applied to provide a set of initial design variables, thereby avoiding the trial process. Thereafter, traditional algorithms are employed to determine the optimum results. The procedure for CCP-based optimization based on HGA is described as Figure 3.

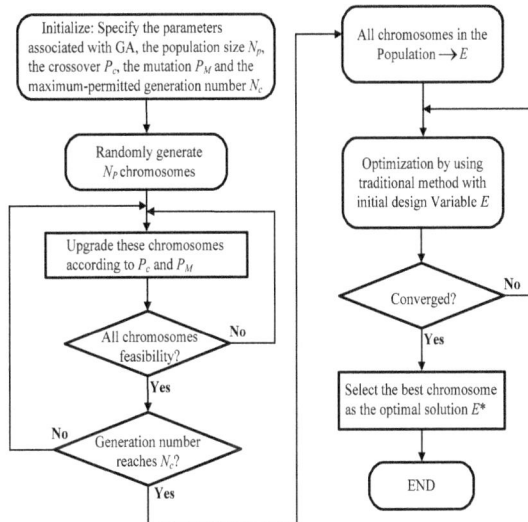

Figure 3. Flow chart of hybrid genetic algorithm (HGA) for CCP-based optimization procedure.

In Step C of Figure 1, objective function f is firstly got by system operation planning cost C_{DG} and expected load interruption cost ELIC. And then, chance constraints based on Equations (16) and (17) should be checked [43]. The detailed solving steps for CCP-based optimization based on HGA are as follows:

a   Specify the parameters associated with the GA, including the population size N_P, the crossover P_C, the mutation P_M and the maximum-permitted generation number N_C.

b   Randomly generate N_P chromosomes.

c   Update these chromosomes according to the crossover P_C and the mutation P_M.

d   Check the feasibility of all chromosomes until all the chromosomes are feasible. In this procedure, all network constraints including chance constraints should be checked for the chromosomes. In this paper, check of chance constraints is based on Monte Carlo simulation procedure, which was introduced in [18] in detail.

e   Repeat Steps b–e until the generation number reaches N_C.

f   All the chromosomes in e are regarded as E.

g   Calculate the objective function value of all chromosomes and the fitness value of each chromosome.

h   Select the best chromosome found in the above solving procedure as E^*.

6. Case Studies

To demonstrate the performance of the proposed model and method, it is applied to the modified IEEE 30-bus system shown in Figure 4. The parameters of wind generating units and photovoltaic generating units are given below, and other characteristic parameters are given in [44]. The base value is 100 MVA. For photovoltaic nodes, upper and lower voltage bounds are 1.1 and 0.95 p.u. and upper and lower voltage bounds are 1.06 and 0.94 p.u. for all other nodes. All case studies are carried out using MATLAB on an Advanced Micro Devices 64 Dual Core CPU 3.3 GHz PC (Lenovo, Wuhan, China).

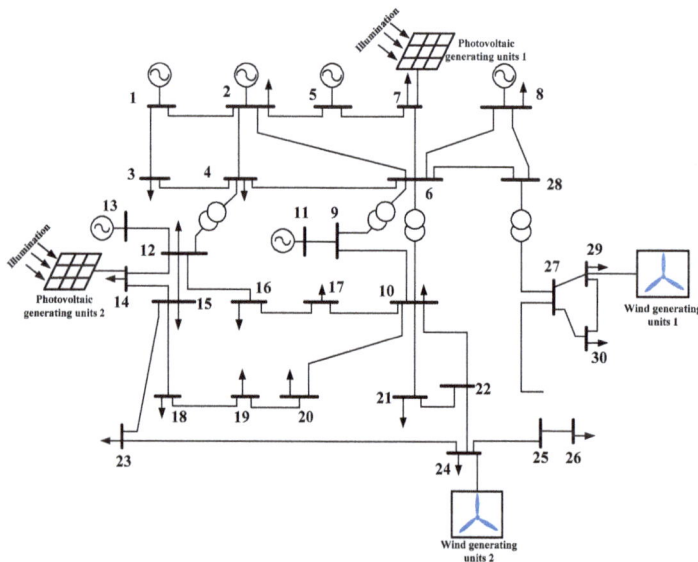

Figure 4. DGs connected to the IEEE 30-bus system.

The candidate notes for the species and maximum installed capacity of DGs are shown in Table 1. In Table 1, 1 and 2 represent wind generating units and photovoltaic generating units respectively. For different candidate notes, the weighting coefficients of investment cost $a_i = f(x_1, x_2, x_3)$ is different, which is also list in Table 1. In addition, the maximum total generation capacity of DGs in the IEEE 30-bus system accounts for 30% of total load.

Table 1. Candidate notes for the species and maximum capacity of DGs.

Candidate Notes	Species m	Maximum Installed Capacity of DG m/(MW)	a_i
3	1, 2	30, 20	1.01
7	1, 2	20, 20	1.03
14	1, 2	20, 20	1.05
19	2	20, 20	1.05
24	1, 2	20, 20	1.03
26	1, 2	30, 30	1.04
30	1, 2	20, 20	1.02

In all case studies, the per-unit capacity investment costs and per-unit capacity maintenance costs for wind generating units and photovoltaic generating units are shown in Table 2. To facilitate the calculation and simplify the model of Equation (12), electricity price is considered constant in one year, which can be estimated by the average value of last year. In this case, the unit electricity sale price C_e is 0.07 \$/kW· h. The unit on-grid electricity prices for wind generating units and photovoltaic generating units are also list in Table 2, in which government subsidies are considered.

Table 2. The per-unit capacity Cost of DGs.

m	Investment Costs /\$· kW^{-1}	Maintenance Costs /\$· (kW· h)$^{-1}$	On-Gridprice /\$· (kW· h)$^{-1}$
1	1200	0.04	0.13
2	1400	0.03	0.15

6.1. Calculation of ELIC

By the procedure for computing the index of EENS and the estimated value of UIC in Figure 2, the value of ELIC can be calculated. In this case, UIC is chosen as 0.32 \$/kW· h, which is also considered constant in one year.

For the wind generation units, v_N = 15 m/s, v_{in} = 4 m/s, v_{out} = 20 m/s. The shape index k = 6.25, the scale index c = 10.45, and the fitting coefficients a_1 = 0.0014848, a_2 = −0.041545, a_3 = 0.43333 and a_4 = −1.1636. For the photovoltaic generation units, the shape indexes α = 15.34, β = 4.2, A = 2.16 m, η = 13.44%.

For instance, wind generating units and photovoltaic generating units are installed in note 7, 14 respectively. Then, the value of ELIC was calculated by $2m$ + 1 point estimate method (PEM), which is shown in Figure 5.

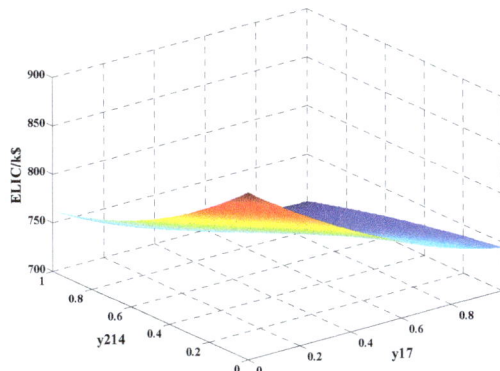

Figure 5. The value of ELIC impace with y_{17} and y_{214}.

As can be seen from Figure 5, ELIC decreases with the increment of DGs' capacity. When $y_{17} = 1$ and $y_{214} = 1$, the value of ELIC reach its minimum value, which is about 701.08 k\$. On the other hand, the locations of DGs also affect the value of ELIC. For instance, if wind generating units and photovoltaic generating units are installed in note 19, 30 respectively, the value of ELIC is shown in Figure 6.

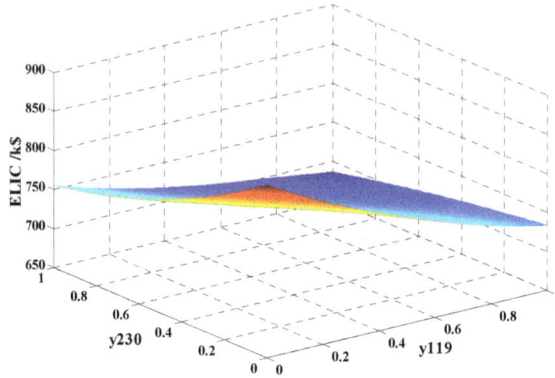

Figure 6. The value of ELIC impace with y_{119} and y_{230}.

Compared to Figure 5, although DGs' capacity has no changes, the value of ELIC changes when the locations of DGs change. For an example, ELIC = 701.08 k\$ when $y_{17} = 1$ and $y_{214} = 1$, but ELIC = 686.55 k\$ when $y_{119} = 1$ and $y_{230} = 1$. In addition, the distribution of DGs also affects the value of ELIC. Generally, ELIC decreases with the distribution of DGs. That is to say, the value of ELIC will be larger with more DGs connected to IEEE 30-bus system although the total capacity of DGs is constant.

In addition, it also indicates that wind generating units' power support is with better effectiveness than photovoltaic generating units, because the value of ELIC when $y_{17} = 0$ and $y_{214} = 1$ is much bigger than the value of ELIC when $y_{17} = 1$ and $y_{214} = 0$.

6.2. Calculation of System Operation Planning Cost

According to Section 5.2, the system operation planning cost C_{DG} can be calculated by Equation (7). The maximum hours of load loss in branch j $\tau_{j\max} = 3000$ h, the discount rate $r = 0.12$, and the lifetime of DGs $n_{DG} = 5$. The equivalent generation hours of DGs generation $T_{eq} = 3000$ h. For instance, wind generating units and photovoltaic generating units are installed in note 7, 14 respectively. Then, the value of cost C_{DG} can be shown in Figure 7, in which all network constraints are satisfied.

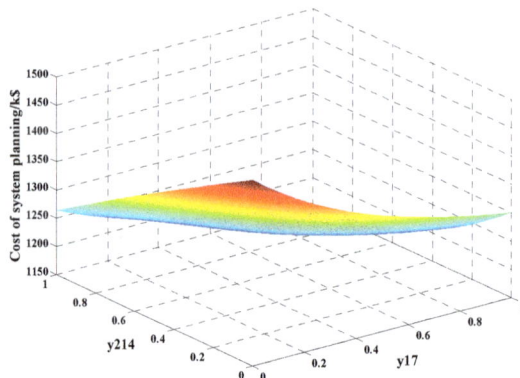

Figure 7. The value of C_{DG} impace with y_{17} and y_{214}.

As can be seen in Figure 7, the value of system operation planning cost C_{DG} = 1478.27 k\$ when y_{17} = 0 and y_{214} = 0, in which C^L = 1478.27 k\$, C^I = 0, C^M = 0, C^{DG} = 0. In pace with the increase of DGs'capacity y_{17} and y_{214}, system operation planning cost C_{DG} will cut down at a certain range. When y_{17} = 1 and y_{214} = 1, cost C_{DG} reach its minimum value 1186.56 k\$, in which C^L = 1203.63 k\$, C^I = 5487.65 k\$, C^M = 3741.25 k\$, C^{DG} = 9245.97 k\$.

This result shows that the installed capacities of DGs are not adequate. Namely, system operation planning cost C_{DG} can further cut down with the increase of DGs'capacities. In addition, it also shows that the locations of node 7 and 14 may not be the suitable locations for optimalsiting of DGs. Namely, the locations of DGs also affect the value of system operation planning cost C_{DG}. Anyway, the value of system operation planning cost C_{DG} can reach its minimum value when y_{mi} is at a certain fixed value for optimal sizing and siting of DGs.

6.3. Solving Strategy of CCP-Based Optimization

According to Section 5.3, an embedded genetic-algorithm approach is employed to solve the optimization problem described by Equation (12). In the objective function which is shown in Equation (14), the weighting coefficients α_1 and α_2 can greatly affect the optimization results of Equation (12). Therefore, the optimization problem described by Equation (12) is solved in four cases. In case 1, α_1 = 1 and α_2 = 0, which means that ELIC is not considered and this optimization of Equation (12) is the traditional method for optimal siting and sizing of distributed generators in distribution systems. In case 2, α_1 = 0.7 and α_2 = 0.3. In case 3, α_1 = 0.5 and α_2 = 0.5. In case 4, α_1 = 0 and α_2 = 1, which means that system operation planning cost C_{DG} is not considered. In all the cases, the parameters are specified as follows: N_C = 1000, N_P = 30, P_C = 0.3. Optimal sizing and siting results of DGs in cases 1–4 are shown in Table 3.

Table 3. Optimal sizing and siting of DGs in case n.

Notes	Species m of DGs	Optimal Sizing of DGs in Case n			
		1	2	3	4
3	1	18.56	17.33	17.19	15.28
	2	0	0	0	0
7	1	0	0	0	7.11
	2	12.68	0	0	0
14	1	0	0	0	13.26
	2	0	0	0	0
19	1	0	0	9.07	14.52
	2	0	16.91	0	0
24	1	0	0	0	5.81
	2	0	0	0	0
26	1	0	0	0	11.77
	2	0	0	18.35	0
30	1	17.27	18.44	15.67	17.23
	2	0	0	0	0

As can be seen from Table 3, the optimal installed capacities of DGs are different in different cases. In case 4, ELIC is only considered as the objective function in Equation (14). Because of the monotone decreasing characteristic of ELIC, ELIC reaches to its minimum value when installed capacities of DGs reach to its maximum value, which is restricted by network constraints. The results in case 4 also showed that ELIC decreases with the increment of dispersion degree. In addition, wind generating units' power support is with better effectiveness than photovoltaic generating units as the installed capacities of photovoltaic generating units y_{1n} = 0.

In case 1, system operation planning cost C_{DG} is only considered as the objective function in Equation (14). In this case, the optimal installed capacities of DGs are relative concentrated because dispersion degree of DGs will increase the investment cost and maintenance cost. In cases 2 and 3, the optimal installed capacities of DGs increase with the increase of α_2, which means that ELIC is given more consideration.

The total optimal installed capacities of DGs, the value of ELIC and system operation planning Cost C_{DG} in cases 1–4 are shown in Table 4. The values of ELIC and system operation planning Cost C_{DG} in cases 1–4 are also given in Table 4.

Table 4. The value indexes for optimal sizing and siting of DGs in case n.

Species m of DGs	Case n			
	1	2	3	4
total installed capacities/MW	48.51	52.68	60.28	84.98
C_{DG}/k\$	1032.64	1048.23	1056.31	1547.66
ELIC/k\$	642.03	592.78	573.82	560.67

In case 4, the value of system operation planning cost C_{DG} is much larger than any other cases. This is because of the inappropriate siting and sizing of DGs as ELIC is only considered. In case 1, the value of ELIC is much larger than any other cases. These two cases are not suitable for optimal sizing and siting of DGs because one index of C_{DG} and ELIC is much larger.

In cases 2 and 3, optimal sizing and siting of DGs are relatively more reasonable as the value of C_{DG} and ELIC is moderate. In case 3, the total costs of C_{DG} and ELIC are the lowest. With more consideration of ELIC is given, investment cost and maintenance cost will increase but network losses will be reduced. When $\alpha_2 > 0.5$, the increasing rate of C_{DG} will increase fast which is unwilling to occur for siting and sizing of DGs. Therefore, optimal sizing and siting of DGs are more reasonable when $\alpha_2 \leqslant 0.5$.

6.4. Comparison of Optimization Results

In case 1, ELIC is not considered and this optimization of Equation (12) is the traditional method for optimal siting and sizing of distributed generators in distribution systems. In cases 2 and 3, ELIC is considered with different degrees and this optimization of Equation (12) in these cases is the proposed method for optimal siting and sizing of distributed generators in distribution systems.

In Section 6.3, the simulation results have shown that the proposed method for optimal siting and sizing of distributed generators is more reasonable than the traditional method as the total costs of C_{DG} and ELIC in case 2 or 3 are less than the total costs in case 1. For further comparison, voltage variations, total harmonic distortion (THD) and voltage unbalance factor (VUF) at each node of the test feeder in cases 1–3 are analyzed.

From Figure 8, it could be observed that in cases 1–3, the voltage profile at each node of the test feeder has been greatly improved compared to the case of with no DGs installed in distribution systems. Also, the voltage profile at each node in case 2 or 3 is more excellent than the voltage profile in case 1 as the optimization results in case 2 or 3 is more reasonable than the optimization results in case 1. In addition, the voltage profile in case 3 is more excellent than the voltage profile in case 2 because more DGs are installed in case 3.

The value of total harmonic distortion (THD) in distribution system with DGs in cases 1–3 is also analyzed. For each installed DGs, the THD of grid-connected current I_{grid} is supposed as 5% and consist of only second harmonic. As can be seen from Figure 9, the value of THD at each node in case 2 or 3 is largely reduced compared to the value of THD in case 1. This can greatly illustrate that the optimization results in case 2 or 3 is more reasonable than the optimization results in case 1. Besides, the THD profile in case 3 is more excellent than the THD profile in case 2 because more operation risk cost ELIC is considered in case 3.

Voltage unbalance factor (VUF) profiles at each node in cases 1–3 are shown in Figure 10. The value of VUF at each node in case 2 or 3 is largely reduced compared to the value of VUF in case 1. In cases 2 and 3, the value of VUF at each node is less than 2%.

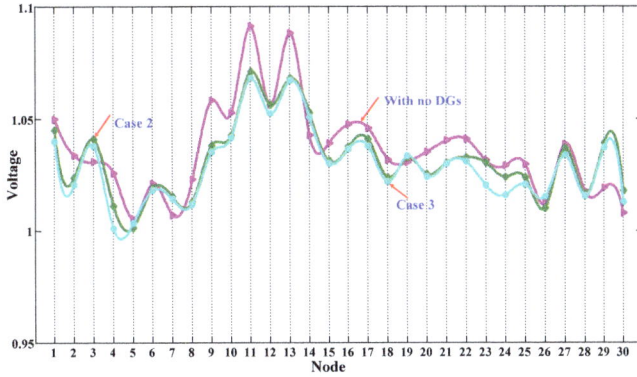

Figure 8. Voltage variations at each node with added DGs in case n.

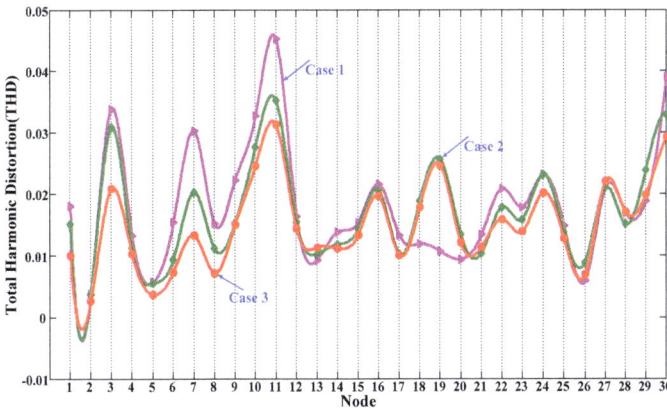

Figure 9. THD (total harmonic distortion) at each node with added DGs in case n.

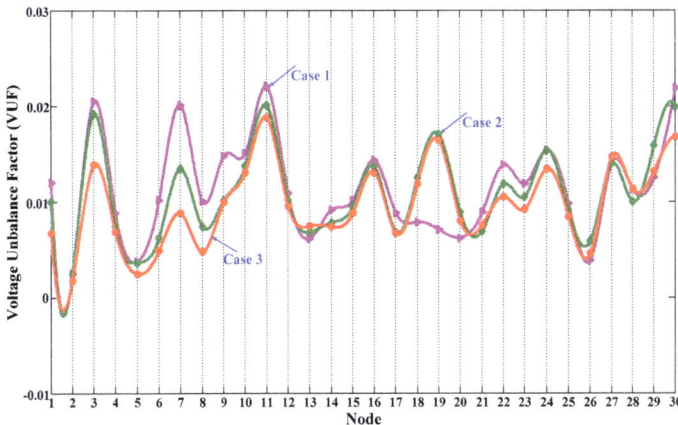

Figure 10. VUF (Voltage unbalance factor) at each node with added DGs in case n.

Generally, these simulation results demonstrate that the appropriate siting andsizing of DGs could improve voltage profiles and enhance power-supply reliability and the proposed method for optimal siting and sizing of distributed generators is more reasonable than the traditional method.

7. Conclusions

Expected load interruption cost is used as the assessment of operation risk indistribution systems, which is assessed by point estimate method (PEM). This proposed expected load interruption cost estimationmethod by PEM is slightly low in accuracy, but the computational cost of PEM is largely reduced compared with Monte Carlosimulation.

Combined the costs of system operation planning with expected load interruption cost (ELIC) in distribution systems, a novel mathematical model of chance constrained programming (CCP) framework for optimal siting and sizing of DGs in distribution systems is proposed considering the uncertainties of DGs. Then, a hybrid genetic algorithm (HGA), which combines the genetic algorithm(GA) with traditional optimization methods, is employed to solve the proposed CCP model. This HGA overcome the drawbacks of both GA and traditional optimization methods.

Test results have demonstrated that this proposed CCP model is more reasonable to determine the siting and sizing of DGs compared with traditional CCP model. Simulation results also demonstrate that the appropriate siting andsizing of DGs could lead to many positive effects on the distribution system concerned, such as the reduced total costs associated with DGs, reduced network losses, and improved voltageprofiles and enhanced power-supply reliability.

Acknowledgments: This work was supported by The National Key Technology R&D Program of China (2013BAA02B01) and Qinghai Province Key Laboratory of Photovoltaic Grid Connected Power Generation Technology (2014-Z-Y34A).

Author Contributions: Qingwu Gong proposed the concrete ideas of the proposed optimization method. Jiazhi Lei performed the simulations and wrote the manuscript. Jun Ye gave some useful suggestions to the manuscript. Both of the authors revised the manuscript.

Conflicts of Interest: The authors declare no conflict of interest.

References

1. Faria, P.; Vale, Z.; Baptista, J. Demand response programs design and use considering intensive penetration of distributed generation. *Energies* **2015**, *8*, 6230–6246. [CrossRef]
2. Arritt, R.F.; Dugan, R.C. Distribution system analysis and the future smart grid. *IEEE Trans. Power Ind. Appl.* **2011**, *47*, 2343–2350. [CrossRef]
3. Yu, W.; Liu, D.; Huang, Y. Operation optimization based on the power supply and storagecapacity of an active distribution network. *Energies* **2013**, *6*, 6423–6438. [CrossRef]
4. Huang, W.-T.; Yao, K.-C.; Wu, C.-C. Using the direct search method for optimal dispatch of distributed generation in a medium-voltage microgrid. *Energies* **2014**, *7*, 8355–8373. [CrossRef]
5. Gao, Y.; Liu, J.; Yang, J.; Liang, H.; Zhang, J. Multi-objectiveplanning of multi-type distributed generation considering timing characteristics and environmental benefits. *Energies* **2014**, *7*, 6242–6257. [CrossRef]
6. Li, R.; Ma, H.; Wang, F.; Wang, Y.; Liu, Y.; Li, Z. Game optimization theory and application in distribution system expansion planning, including distributed generation. *Energies* **2013**, *6*, 1101–1124. [CrossRef]
7. Soroudi, A.; Amraee, T. Decision making under uncertainty in energy systems: State of the art. *Renew. Sustain. Energy Rev.* **2013**, *28*, 376–384. [CrossRef]
8. Soroudi, A.; Ehsan, M. IGDT based robust decision making tool for dnos in load procurement under severe uncertainty. *IEEE Trans. Smart Grid* **2013**, *4*, 886–895. [CrossRef]
9. Charnes, A.; Cooper, W.W. Chance-constrained programming. *Manag. Sci.* **1959**, *6*, 73–79. [CrossRef]
10. Celli, G.; Pilo, F. Optimal distributed generation allocation in MV distribution networks. In Proceedings of the Transmission and Distribution Conference and Exposition, Sydeny, Australia, 20–24 May 2001; pp. 622–627.
11. Nara, K.; Hayashi, Y.; Ikeda, K.; Ashizawa, T. Application of tabu search to optimal placement of distributed generators. In Proceedings of the IEEE Power Engineering Society Winter Meeting, Columbus, OH, USA, 28 January–1 February 2001; pp. 918–923.

12. Lei, L.; Hai, B.; Hong, L. Siting and sizing of distributed generation based on the minimum transmission losses cost. In Proceeding of the Power Engineering and Automation Conference (PEAM), Whuhan, China, 8–9 September 2011; pp. 22–25.

13. Kim, J.O.; Park, S.K.; Park, K.W.; Singh, C. Dispersedgeneration planning using improved Hreford ranchalgorithm. *Electr. Power Syst. Res.* **1998**, 678–683. [CrossRef]

14. Gandomkar, M.; Vakilian, M.; Ehsan, M. A combinationof genetic algorithm and simulated annealing foroptimal DG allocation in distribution networks. In Proceeding of the IEEE Canadian Conference on Electrical and Computer Engineering, Saskatoon, SK, Canada, 1–4 May 2005; pp. 645–648.

15. El-Khattam, W.; Bhattacharya, K.; Hegazy, Y.; Salama, M.M.A. Optimal investment planning for distributed generation in a competitive electricity market. *IEEE Trans. Power Syst.* **2004**, *19*, 1674–1684. [CrossRef]

16. Celli, G.; Ghiani, E.; Mocci, S.; Pilo, F. A multiobjective evolutionary algorithm for the sizing and siting of distributed generation. *IEEE Trans. Power Syst.* **2005**, *20*, 750–757. [CrossRef]

17. Celli, G.; Ghiani, E.; Mocci, S.; Pilo, F. A multi-objective approach to maximize the penetration of distributed generation in distribution networks. In Proceedings ofthe International Conference on Probabilistic Methods Applied to Power Systems, Stockholm, Sweden, 11–15 June 2006; pp. 1–6.

18. Liu, Z.; Wen, F.; Ledwich, G. Optimal siting and sizing of distributedgenerators in distribution systems considering uncertainties. *IEEE Trans. Power Deliv.* **2011**, *26*, 2541–2551. [CrossRef]

19. Al Kaabi, S.S.; Zeineldin, H.H.; Khadkikar, V. Planning active distribution networks considering multi-DG configurations. *IEEE Trans. Power Syst.* **2014**, *29*, 785–793. [CrossRef]

20. Thota, P.; Kirthiga, M.V. Optimal siting and sizing of distributed generators in micro-grids. In Proceedings of the India Conference (INDICON), Kochi, India, 7–9 December 2012; pp. 731–735.

21. Jia, H.; Qi, W.; Liu, Z.; Wang, B.; Zeng, Y.; Xu, T. Hierarchical risk assessment of transmission system considering the influence of active distribution network. *IEEE Trans. Power Syst.* **2015**, *2*, 1084–1093. [CrossRef]

22. Feng, Y.; Wu, W.; Zhang, B.; Li, W. Power system operation risk assessment using credibility theory. *IEEE Trans. Power Syst.* **2008**, *23*, 1309–1318. [CrossRef]

23. Negnevitsky, M.; Nguyen, D.H.; Piekutowski, M. Risk assessment for power system operation planning with high wind power penetration. *IEEE Trans. Power Syst.* **2015**, *30*, 1359–1368. [CrossRef]

24. Billinton, R.; Allan, R. *Reliability Evaluation of Power Systems*; Plenum Publishing: New York, NY, USA, 1984.

25. Kirschen, D.S.; Bell, K.R.W.; Nedic, D.P.; Jayaweera, D.; Allan, R.N. Computing the value of security. *IEEE Proc. Gener. Transm. Distrib.* **2003**, *150*, 673–678. [CrossRef]

26. Celli, G.; Ghiani, E.; Pilo, F.; Soma, G.G. Active distribution network reliability assessment with a pseudo sequential Monte Carlo method. In Proceeding ofthe IEEE PowerTech, Trondheim, Norway, 19–23 June 2011; pp. 1–8.

27. Li, W. *Probabilistic Transmission System Planning*; Wiley-IEEE Press: Hoboken, NJ, USA, 2011.

28. Lin, Y.; Zhong, J.; Bollen, M.H. Evaluation of dip and interruption costs for a distribution system with distributed generations. In Proceeding ofthe 13th International Conference on Harmonics and Quality of Power, Wollongong, Australia, 28 September–1 October 2008; pp. 1–6.

29. Liu, C.; Chau, K.T.; Zhang, X. An efficientwind-photovoltaic hybrid generation system using doubly excitedpermanent-magnet brushless machine. *IEEE Trans. Ind. Electron.* **2010**, *57*, 831–839.

30. Prasad, A.R.; Natarajan, E. Optimization of integrated photovoltaic-wind power systems with battery storage. *Energy* **2006**, *31*, 1943–1954. [CrossRef]

31. Papavasiliou, A.; Oren, S.S.; O'Neill, R.P. Reserve requirements for wind power integration: A scenario-based stochastic programming framework. *IEEE Trans. Power Syst.* **2011**, *26*, 2197–2206. [CrossRef]

32. Li, J.; Wei, W.; Xiang, J. A simple sizing algorithm for stand-alone PV/wind/battery hybrid microgrids. *Energies* **2012**, *5*, 5307–5323. [CrossRef]

33. Yang, H.; Lu, L.; Zhou, W. A novel optimization sizing model for hybrid solar-wind powergeneration system. *Sol. Energy* **2007**, *81*, 76–84. [CrossRef]

34. Ramalakshmi, S.S. Optimal siting and sizing of distributed generation using fuzzy-EP. In Proceeding of the International Conference on Recent Advancements in Electrical, Electronics and Control Engineering, Sivakasi, India, 15–17 December 2011; pp. 470–477.

35. Naik, S.N.G.; Khatod, D.K.; Sharma, M.P. Analytical approach for optimal siting and sizing of distributed generation in radial distribution networks. *Gener. Transm. Distrib. IET* **2015**, *9*, 209–220. [CrossRef]

36. Li, X.; Zhang, X.; Wu, L.; Lu, P.; Zhang, S. Transmission line overload risk assessment for power systems with wind and load-power generation correlation. *IEEE Trans. Smart Grid* **2015**, *3*, 1233–1242. [CrossRef]

37. Morales, J.M.; Perez-Ruiz, J. Point estimate schemes to solve the probabilistic power flow. *IEEE Trans. Power Syst.* **2007**, *4*, 1594–1601. [CrossRef]

38. Hong, H.P. An efficient point estimate method for probabilistic analysis. *Reliab. Eng. Syst. Saf.* **1998**, *59*, 261–267. [CrossRef]

39. Meier, A. Power Flow Analysis. In *Electric Power Systems:A Conceptual Introduction*; Wiley-IEEE Computer Society Press: Hoboken, NJ, USA, 2006; pp. 195–228.

40. Hui, W.; Xu, F. Different load models based on particle swarm algorithm for the siting and sizing optimization problem for distributed power. In Proceeding of the IPEC, 2012 Conference on Power & Energy, Ho Chi Minh City, Vietnam, 12–14 December 2012; pp. 19–24.

41. Liu, K.Y.; He, K.; Sheng, W. Multiple-objective DG optimal sizing in distribution system using an improved PSO algorithm. In Proceeding of the 2nd IET Renewable Power Generation Conference (RPG 2013), Beijing, China, 9–11 September 2013; pp. 1–4.

42. Guo, P.; Wang, X.; Han, Y. The enhanced genetic algorithms for the optimization design. In Proceeding of the 3rd International Conference on Biomedical Engineering and Informatics (BMEI), Yantai, China, 16–18 October 2010; pp. 2990–2994.

43. Du Toit, N.E.; Burdick, J.W. Probabilisticcollision checking with chance constraints. *IEEE Trans. Robot.* **2011**, *27*, 809–815. [CrossRef]

44. Power System Test Case Archive. Available online: http://www.ee.washington.edu/research/pstca (accessed on 19 January 2016).

8

Reliability Analysis and Overload Capability Assessment of Oil-Immersed Power Transformers

Chen Wang [1], Jie Wu [2,*], Jianzhou Wang [3] and Weigang Zhao [4,5]

Academic Editor: Issouf Fofana

[1] School of Mathematics and Statistics, Lanzhou University, Lanzhou 730000, China; chenwang15@lzu.edu.cn
[2] School of Mathematics and Computer Science, Northwest University for Nationalities,
    Lanzhou 730030, China
[3] School of Statistics, Dongbei University of Finance and Economics, Dalian 116025, China; wjz@lzu.edu.cn
[4] Center for Energy and Environmental Policy Research, Beijing Institute of Technology, Beijing 100081, China;
    zwgstd@gmail.com
[5] School of Management and Economics, Beijing Institute of Technology, Beijing 100081, China
*   Correspondence: wuj19870903@gmail.com

Abstract: Smart grids have been constructed so as to guarantee the security and stability of the power grid in recent years. Power transformers are a most vital component in the complicated smart grid network. Any transformer failure can cause damage of the whole power system, within which the failures caused by overloading cannot be ignored. This research gives a new insight into overload capability assessment of transformers. The hot-spot temperature of the winding is the most critical factor in measuring the overload capacity of power transformers. Thus, the hot-spot temperature is calculated to obtain the duration running time of the power transformers under overloading conditions. Then the overloading probability is fitted with the mature and widely accepted Weibull probability density function. To guarantee the accuracy of this fitting, a new objective function is proposed to obtain the desired parameters in the Weibull distributions. In addition, ten different mutation scenarios are adopted in the differential evolutionary algorithm to optimize the parameter in the Weibull distribution. The final comprehensive overload capability of the power transformer is assessed by the duration running time as well as the overloading probability. Compared with the previous studies that take no account of the overloading probability, the assessment results obtained in this research are much more reliable.

Keywords: current measurement; losses; power transformers; reliability estimation; transformer windings

1. Introduction

The power grid is an important infrastructure for a nation's economic and social development, however, in recent years, the objective environment to guarantee the security and stability of the power grid is undergoing tremendous changes. Factors such as the rapid growth of the loads, the initial formation of the large area grid interconnection, as well as the influence of the global climate change all impact the electricity market and the effects on the power grid have become increasingly apparent, thus, guaranteeing the security and stability of the power grid represents a new challenge. To solve this problem, in recent years, smart grids have been constructed by comprehensively considering the market, safety, power quality and environmental factors. The term smart grid refers to a fully automated complicated power supply network, where each user and each node are monitored in real-time, to ensure a two-way flow of the current and information between the power plant and clients' appliances. The features of the smart grid can be summarized as: self-healing, compatibility, interaction, coordination, efficiency, quality, and integration.

Power transformers is one of the most vital pieces of equipment in the smart grid. In addition, it is a network equipment whose structure is the most complex and sophisticated. Any failure in transformers can cause damage to the power system, among which failures caused by overloading cannot be ignored. The consequences of overloaded operation of power transformers can be serious. As indicated, when the current flow in the windings exceeds the rated current stated on the nameplate, *i.e.*, the transformer operates under overload conditions, the load loss of transformers is proportional to the square of the current, conductor heating rises sharply, and the temperature of the windings and insulating oil surge accordingly. In this case, the transformer loss will increase due to the reason that power transformers are designed according to their rated capacity, so when the load of the transformer exceeds the rated capacity, the losses will increase. This will greatly affect the lifetime of the power transformer. In addition, transformers may fail due to the following two reasons: on the one hand, the transformer may be damaged since the overload operation would accelerate the cracking of insulating oil, generate bubbles, reduce the dielectric strength of the transformer, and cause an electrical breakdown. On the other hand, the excessive heat will reduce the mechanical strength of the windings, and when a short circuit occurs, coil deformation or mechanical instability will occur due to the external strong electric power. Therefore, overload capacity assessment is of particular importance in avoiding the catastrophic failure of power transformers and guaranteeing the normal operation of power grids.

Adequate and accurate assessment of power system reliability is a very challenging task that has been and still is under investigation. Previously developed power system reliability and security assessment models include the super components contingency model [1], the hybrid conditions-dependent outage model [2], and probability distribution based models such as the log-normal distribution [3] and the Weibull distribution [4]. As one of vital aspects in the power system reliability assessment, the overload capability of power transformers, has also been specifically surveyed by many researchers. For example, to make up the limitation of the American National Standards Instituteloading guide, which is only applicable to ambient temperatures above 0 °C, Aubin *et al.* [5] proposed a calculation method to assess the overload capacity of transformers for ambient temperatures below 0 °C. Tenbohlen *et al.* [6] developed on-line monitoring systems to assess the overload capacity of power transformers. Bosworth *et al.* [7] reported the development of electrochemical sensors for the measurement of phenol in transformer overloading evaluation. A stochastic differential equation was used by Edstrom *et al.* [8] to estimate the probability of transformer overloading. Estrada *et al.* [9] adopted magnetic flux entropy as a tool to predict transformer failures, and the overloading is just one aspect among the failures. Liu *et al.* [10] assessed the overload capacity of transformers through an online monitoring and overload factor calculated by a temperature reverse extrapolation approach. As known, when assessing network load capability, the hot-spot temperature is one the most significant factors. Thus, there are many studies devoted to hot-spot temperature forecasting such as the radial basis function network [11], a genetic algorithm based technique [12], and a local memory-based algorithm [13] provided by Galdi *et al.*, the Takagi-Sugeno-Kang fuzzy model presented by Siano [14], the optimal linear combination of artificial neural network approach used by Pradhan and Ramu [15], the grey-box model introduced by Domenico *et al.* [16], *etc.* Though these researches make tremendous contributions, efforts on overload capability assessments should not be stopped, and new overload capability measurement techniques with respect to power transformers still need to be developed and exploited to improve the accuracy of overload capability assessment and provide more techniques to prevent failure of transformers caused by emergency overloads.

This research gives a new insight into how to measure the overload probability of oil-immersed power transformers. As known, the hot-spot temperature is the most critical factor in measuring the overload capability of power transformers. Thus, the hot-spot temperature is first calculated to measure the duration of running time under overload conditions. Then, the overloading probability is fitted by a mature and attractive Weibull distribution. Finally, the comprehensive overload capability of the power transformer is assessed from both the duration of running time under the overload conditions and the overloading probability aspects. This research is innovative in the following aspects: (a) apart from the duration of running time under the overload conditions, the overload capability is also assessed according to the overloading probability of the power transformer, which is measured by the Weibull distribution in this paper; (b) though the Weibull distribution is a quite mature and attractive method for fitting the distribution of data series, this paper improves the fitting performance of the Weibull distribution by proposing a new objective function to obtain the parameters in the Weibull distribution; (c) different from other researches, the shape parameter in the Weibull distribution in this paper is determined according to the mean of the shape parameter values obtained under ten different mutation scenarios in the differential evolutionary (DE) algorithms, *i.e.*, the shape parameter is determined by taking results under different situations into account, this operation improves the accuracy of overload capability assessment further. The remainder of this paper is organized as follows: Section 2 introduces related techniques. Simulation results and discussions are presented in Section 3, while Section 4 concludes the whole research.

2. Related Techniques

2.1. Duration Running Time Calculation under Overloading Conditions

2.1.1. Steady-State Temperature Measurement

The final hot-spot temperature (θ_h) of the winding for power transformer is calculated by [17]:

$$\theta_h = \theta_a + \Delta\theta_{br}\left[\frac{1 + RK^2}{1 + R}\right]^x + 2\left[\Delta\theta_{imr} - \Delta\theta_{br}\right]K^y + Hg_rK^y \tag{1}$$

where θ_a is the air temperature (°C), $\Delta\theta_{br}$ is the temperature rise in bottom (K), $\Delta\theta_{imr}$ is the average winding temperature rise (K), R is the ratio between the load losses at the rated load and no-load losses, K is the load current per unit and y is the index of the winding.

For a forced-directed oil circulation and forced air circulation (ODAF) transformer, the oil flow in the windings is affected by the oil pump as well as the guide channel, the viscosity of the oil has little effect on the temperature change of the transformer, however, at this time, the temperature effect of the conductor resistance must be considered. Therefore, based on Equation (1), the final hot-spot temperature (θ'_h) of the winding for power transformer is corrected using [17]:

$$\theta'_h = \theta_h + 0.15(\theta_h - \theta_{hr}) \tag{2}$$

where θ_h is the final hot-spot temperature of the windings by not taking the effect of the conductor resistance into account and obtained by Equation (1), θ_{hr} is the hot-spot temperature under the rated operating conditions.

2.1.2. Transient Temperature Measurement

With the changes of the transformer load, the temperature of the transformer will change as well. It is found that the temperature rise stabilization time of the electric insulating oil, which is 1.5 h, is much longer than that of the conductor (usually 5–10 min). Thus the transient temperature is measured as follows:

$$\Delta\theta_{bt} = \Delta\theta_{bi} + (\Delta\theta_{bu} - \Delta\theta_{bi})(1 - e^{-t/\tau_0}) \tag{3}$$

where $\Delta\theta_{bi}$ is the initial bottom oil temperature rise, $\Delta\theta_{bu}$ is the bottom oil temperature rise of the applied load at the in the steady state, and τ_0 is the winding time constant.

Therefore, once the limit hot-spot temperature of the winding is determined, with the assistance of the thermal characterization parameters obtained in the factory test, and taking no account of the life lost, the overload capacity of the transformer can be calculated by Equations (1)–(3).

2.2. Overloading Probability Measurement

It is indicated that the relationship between the active power of the three-phase transformer and the current is as follows:

$$P = \sqrt{3}UI\cos\varphi \tag{4}$$

where P is the active power, U and I are the voltage and current respectively, and $\cos\varphi$ is called the power factor. Therefore, the probability value of the current located in the interval $[I_1, I_2)$ is as equal as that of the active power located in the interval $[\sqrt{3}UI_1\cos\varphi, \sqrt{3}UI_2\cos\varphi)$. This inspires us to carry out the overloading probability measurement by means of the active power probability fitting results, in the situation that the current values are unknown whereas the active power values are observed.

The Weibull distribution is one of the most commonly used the loss of life distributions in the reliability research of single samples. Its main feature is that the difference of shape parameters can reflect various failure mechanisms. Numerous experimental results demonstrate that the life of components, equipment, and systems that cause the global function to stop running owing to the failure or breakdown in certain parts obey the Weibull distribution [18]. Moreover, according to Reference [19], the life of liquid insulation obeys a Gumbel distribution, while the lifetime of solid insulation follows a two-parameter distribution or lognormal distribution. Therefore, this paper applies a two-parameter Weibull distribution to research the life distribution features of hot-spot absolute temperature insulation samples. The statistical analysis of Weibull life data is based on the following three assumptions [20]:

A1:   In each different stress level, the loss of life of hot-spot absolute temperature insulation samples all obeys the Weibull distribution. That is to say, the distribution type of life will not change with increasing stress level.

A2:   In each different stress level, the failure mechanism of hot-spot absolute temperature insulation samples mush keep consistent. However, owing to the randomness of experimental data, the shape parameters of Weibull distribution can be only approximately equal.

A3:   The life of hot-spot absolute temperature insulation samples that obeys the Weibull distribution should the function of trial voltage and temperature. If A1 and A2 are satisfied, the hot-spot absolute temperature insulation samples obey the Weibull distribution. Assume that the main aim of A3 is to realize the data extrapolation.

The three assumptions are built based on certain physics, and we can use professional knowledge and engineering experience to judge whether they are true. In the statistical analysis, both hypothesis testing and correlation coefficient test can be applied to confirm their existence.

The active power distribution of the transformer is surveyed with the assistance of the Weibull distribution in this paper. The probability density function of the Weibull distribution can be described by:

$$f(a) = (\frac{k}{c})(\frac{a}{c})^{k-1} \exp\left[-(\frac{a}{c})^k\right] \tag{5}$$

where a is the active power with the unit of kW, k is the dimensionless shape parameter and c is the scale parameter with the same unit of the active power.

2.3. Objective Function

To obtain the unknown shape and scale parameters, in this research, a new objective function is constructed and the results obtained by this new objective function are compared with those obtained by two other frequently used objective functions.

2.3.1. The New Proposed Objective Function

According to the Probability Density Function (PDF) of the Weibull distribution, the expected value ($E(a)$) and the variance ($Var(a)$) of the active power can be obtained by:

$$E(a) = c\Gamma(1 + \frac{1}{k}) \tag{6}$$

and:

$$Var(a) = c^2\Gamma(1 + \frac{2}{k}) - c^2\Gamma^2(1 + \frac{1}{k}) \tag{7}$$

The new objective function constructed in this paper benefits from the following idea. As known, the mean square error (MSE) defined as follows is always been used as the objective function:

$$MSE = \frac{1}{n}\sum_{i=1}^{n}(x_i - \hat{x}_i)^2 \tag{8}$$

where x_i and $\hat{x}_i$ are the observed and forecasted values, respectively. Let Y be a random variable and the possible values for Y are $y_1, y_2, \ldots, y_n$, where $y_i = x_i - \hat{x}_i$. Then Equation (8) can be written as:

$$MSE = \frac{1}{n}\sum_{i=1}^{n}y_i^2 \tag{9}$$

which can be seen as:

$$MSE = E(Y^2) \tag{10}$$

where $E(Y^2)$ represents the expected value of the variable Y^2. According to the following formula:

$$Var(Y) = E(Y^2) - [E(Y)]^2 \tag{11}$$

Equation (10) is equivalent to:

$$MSE = [E(Y)]^2 + Var(Y) \tag{12}$$

where $E(Y)$ and $Var(Y)$ denote the expected value and variance of the variable Y, respectively. Based on the calculation results obtained by Equations (6) and (7):

$$[E(a)]^2 + Var(a) = c^2\Gamma(1 + \frac{2}{k}) \tag{13}$$

However, there is always some error between the left side and the right side of the Equation (13). Thus, the residual value ε defined as below is used as the objective function:

$$\varepsilon_1 = [E(a)]^2 + Var(a) - c^2\Gamma(1 + \frac{2}{k}) \tag{14}$$

where $E(a)$ represents the mean value of the active power and $Var(a)$ denotes the variance of the active power. Then according to Equation (6), the scale parameter c can be obtained by:

$$c = \frac{E(a)}{\Gamma(1 + 1/k)} \tag{15}$$

So by substituting Equation (15) into Equation (14), the final objection function used to optimize the shape parameter k can be expressed as:

$$\varepsilon_1 = [E(a)]^2 + Var(a) - \frac{[E(a)]^2\Gamma(1 + 2/k)}{\Gamma^2(1 + 1/k)} \tag{16}$$

2.3.2. The First Comparison Objective Function

To verify the performance of the DE algorithm under different objective functions, the first objective function used to compare with the new one proposed in this paper is expressed as:

$$\varepsilon_2 = \frac{Var(a)}{[E(a)]^2} - \frac{\Gamma(1+2/k) - \Gamma^2(1+1/k)}{\Gamma^2(1+1/k)} \tag{17}$$

where $E(a)$ represents the mean value of the active power and $Var(a)$ denotes the variance of the active power. Similarly, Equation (17) is only used to optimize the shape parameter. The scale parameter in this comparison strategy is obtained by Equation (15) just as it did in the new proposed objective function. The construction of this objective function can be found in Appendix A.

2.3.3. The Second Comparison Objective Function

The second objective function, which used to compare with the new proposed one in this paper and is derived from the maximum likelihood estimation, can be expressed as:

$$\varepsilon_3 = k - \left[\frac{\sum_{i=1}^n a_i^k \ln a_i}{\sum_{i=1}^n a_i^k} - \frac{\sum_{i=1}^n \ln a_i}{n} \right]^{1/k} \tag{18}$$

where n is the active power sample number and $\{a_i\}_{i=1}^n$ is the active power series of the transformer. The construction of this objective function can be found in Appendix B. Once the value of the shape parameter k has been obtained, the scale parameter c is determined according to:

$$c = \left(\frac{1}{n} \sum_{i=1}^n a_i^k \right)^{1/k} \tag{19}$$

2.4. Intelligent Optimization Algorithms

To obtain the optimum shape and scale parameters, the differential evolution (DE) algorithm is used in this research. The usage of the DE algorithm is built on the basis of the three previous described objective functions. In general, the DE algorithm contains three procedures: mutation, crossover and selection [21].

Procedure 1 (mutation): In this step, ten different mutation scenarios are employed in this research to survey the performance of the three objective functions. Given a population with N parameter vectors X_i^G, $(i = 1, 2, 3, \dots, N$ for each generation G), these ten scenarios are expressed as follows:

$$\text{Scenario 1}: v_i^{G+1} = x_{r1}^G + F \times (x_{r2}^G - x_{r3}^G), r1 \neq r2 \neq r3 \neq i; \tag{20}$$

$$\text{Scenario 2}: v_i^{G+1} = x_i^G + F_1 \times (x_{best}^G - x_i^G) + F_2 \times (x_{r2}^G - x_{r3}^G); \tag{21}$$

$$\text{Scenario 3}: v_i^{G+1} = x_{best}^G + (x_{r1}^G - x_{r2}^G) \times ((1 - 0.9999) \times rand + F); \tag{22}$$

$$\text{Scenario 4}: v_i^{G+1} = x_{r1}^G + F_1 \times (x_{r2}^G - x_{r3}^G), F_1 = (1 - F) \times rand + F; \tag{23}$$

where the values of F_1 are the same for all of the parameters need to be estimated.

$$\text{Scenario 5}: v_i^{G+1} = x_{r1}^G + F_1 \times (x_{r2}^G - x_{r3}^G), F_1 = (1 - F) \times rand + F; \tag{24}$$

$$\text{Scenario 6}: v_i^{G+1} = x_i^G + F \times (x_{r2}^G - x_{r3}^G); \tag{25}$$

$$\text{Scenario 7}: v_i^{G+1} = x_{r1}^G + F \times (x_{r2}^G - x_{r3}^G + x_{r4}^G - x_{r5}^G); \tag{26}$$

$$\text{Scenario 8}: v_i^{G+1} = x_i^G + F \times (x_{r2}^G - x_{r3}^G + x_{r4}^G - x_{r5}^G); \tag{27}$$

$$\text{Scenario 9}: v_i^{G+1} = x_i^G + F \times (x_{best}^G - x_i^G) + 0.5 \times (x_{r2}^G - x_{r3}^G); \tag{28}$$

$$\text{Scenario } 10 : v_i^{G+1} = \begin{cases} x_{r1}^G + F \times (x_{r2}^G - x_{r3}^G), & \text{if } rand < 0.5 \\ x_{r1}^G + 0.5 \times (F+1) \times (x_{r1}^G + x_{r2}^G - 2 \times x_{r3}^G), & \text{if } rand \geqslant 0.5 \end{cases} \tag{29}$$

where $r1, r2, r3, r4, r5$ are integer numbers randomly selected from $\{1, 2, \dots, N\}$, F is the mutation factor chosen from the range $[0, 1]$, and x_i^G and x_{best}^G are the ith and the best individuals in generation G, respectively.

Procedure 2 (Crossover): The exponential crossover approach is employed in this step. Component update in the trial vector $U_i^{G+1} = (u_{1i}^{G+1}, u_{2i}^{G+1}, \dots, u_{Di}^{G+1})$ is described as:

$$u_{ji}^{G+1} = \begin{cases} v_{ji}^{G+1}, & \text{if } j \in \{k, \langle k+1 \rangle_n, \dots, \langle k+L-1 \rangle_n\} \\ x_{ji}^G, & \text{otherwise} \end{cases} \quad , j=1,2,\dots D \tag{30}$$

where k and L are random values selected from the set $\{1, 2, \dots, n\}$, and $\langle j \rangle_n$ is set to j in the case of $j \leqslant n$ while $j - n$ in the case of $j > n$.

Procedure 3 (Selection): This step is operated according to the following law:

$$X_i^{G+1} = \begin{cases} U_i^{G+1}, & \text{if } f(U_i^{G+1}) \leqslant f(X_i^G) \\ X_i^G, & \text{otherwise} \end{cases} \tag{31}$$

The DE algorithm is terminated in the case of the value of ε or the iteration number reaches the expected level.

2.5. New Proposed Overloading Probability Measurement Algorithm

Based on the above related techniques, a new proposed overloading probability measurement algorithm is proposed, the outline of this algorithm is shown in Algorithm 1.

Algorithm 1 New proposed overloading probability measurement algorithm

---

Input: Active power a—a sequence of sample data
Output: The probability density function of the active power
1. Initialize the shape parameter k
2. **WHILE** ($\varepsilon >$ predefined error level) **DO**
3. Update the shape parameter k with the DE algorithm
4. Calculate $\varepsilon = [E(a)]^2 + Var(a) - [E(a)]^2 \Gamma(1 + 2/k)/\Gamma^2(1 + 1/k)$ by using the new obtained k
5. **END WHILE**
6. Calculate $c = E(a)/\Gamma(1 + 1/k)$ by using the final value of k
7. $f(a) = (k/c)(a/c)^{k-1} \exp\left[-(a/c)^k\right]$
8. **RETURN** f

2.6. Fitting Performance Evaluation Criteria

In this paper, two error evaluation criteria named the Kolmogorov-Smirnov test error (KSE) [22] and the root mean square error (RMSE) [23], are applied to the further comparison among the new proposed and the comparison objective functions. The related definitions are as follows:

$$\text{KSE=max} \, |S(a) - O(a)| \tag{32}$$

$$\text{RMSE=} \left[\frac{1}{n} \sum_{i=1}^{n} (a_{oi} - a_{ci})^2 \right]^{1/2} \tag{33}$$

where $S(a)$ and $O(a)$ are the Cumulative Distribution Function (CDF)values of the active power not exceeding a obtained by the selected function and by the actual data, respectively, $\{a_{oi}\}_{i=1}^n$ and $\{a_{ci}\}_{i=1}^n$

are the probability data series obtained by the observed data and the selected probability density function respectively, n represents the number of the data.

3. Results and Discussion

In this paper, the overload capability of oil-immersed power transformers is assessed by the data sampled from three residential areas named Lake Neighborhood, North Neighborhood and Sunshine Mediterranean Neighborhood. The three-phase transformer used in the first residential area is a model S11-M-200/10 (HengAnYuan, Beijing, China), and those in the other two residential areas are both S11-M-400/10 units (HengAnYuan, Beijing, China), i.e., the rated capacity values of the transformers used in these three neighborhoods are 200, 400 and 400 kVA, respectively.

3.1. Overloading Probability Fitting Results

The DE algorithm is carried out and terminated when the objective function is no larger than 1×10^{-5} in this paper. Table 1 presents performance of the three objective functions by using different DE mutation scenarios in terms of the iteration number and the actual obtained objective function values when the termination condition is reached. For convenience, the new proposed objective function, the first comparison objective function and the second objective function are named the objective function 1, the objective function 2 and the objective function 3 in Table 1, respectively.

As seen from Table 1, when the new proposed objective function is applied to the shape parameter optimization, the iteration numbers of the DE algorithm needed to reach the objective function level are smaller than those obtained by the other two comparison objective functions under most of the mutation scenarios. For the Lake Neighborhood, the percentages by which the new proposed objective function outperforms the first comparison and the second comparison objective functions from the iteration numbers are 30% and 100%, respectively. For the North Neighborhood, these corresponding two values are 70% and 90%, respectively, while for the Sunshine Mediterranean Neighborhood, the values are both 100%. Note that in the case where the two objective functions have the same iteration numbers, the superior one is further selected by the actual obtained objective function values.

Furthermore, the shape parameter values obtained by the new proposed objective function is much closer to those obtained by the first comparison objective function. Since the iteration numbers need to reach the objective function level of the first comparison objective function are smaller than those obtained by the second comparison objective function, the first comparison objective function can be regarded as a better one as compared to the second comparison objective function from the iteration speed perspective. According to this, it can be concluded that the new proposed objective function is the best one among the three objective functions from the iteration speed perspective.

It can also be observed from Table 1 that the new proposed objective function is more sensitive to the change of the mutation scenarios as compared to the other two objective functions. This can be indicated by the ten shape parameter values under ten different mutation scenarios in the Sunshine Mediterranean Neighborhood, where those obtained by the new proposed objective function varied (though the variation is small) with the change of the mutation scenarios, while there are almost no change to the shape parameters obtained by the other two objective functions under different mutation scenarios). Thus, the new proposed objective function is better for its sensitivity.

As shown in Table 1, there is little difference among the shape parameter values obtained by the first objective function under the ten different mutation scenarios. Thus, to avoid the one-sidedness, the final shape parameter in this paper is determined by calculating the mean of the ten shape parameter values. As also seen from Table 1, the shape parameter values obtained by the new proposed and the first comparison objective functions are nearly equal. However, results obtained by the second comparison objective functions have larger difference as those gained by the new proposed objective functions. In the next section, this conclusion will be convinced by some statistics analysis and a test named the Moses Extreme Reactions (MER).

Table 1. Parameters obtained by three different objective functions.

	Lake Neighborhood				North Neighborhood					Sunshine Mediterranean Neighborhood				
Objective Function Type	Mutation Scenario	Iteration Number	Objective Function Value	k	Objective Function Type	Mutation Scenario	Iteration Number	Objective Function Value	k	Objective Function Type	Mutation Scenario	Iteration Number	Objective Function Value	k
1	1	18	6.9259×10^{-6}	1.7718	1	1	18	2.4220×10^{-6}	1.4420	1	1	13	3.8005×10^{-6}	1.5006
	2	13	9.2435×10^{-6}	1.7718		2	11	8.9312×10^{-6}	1.4420		2	4	6.9169×10^{-6}	1.5001
	3	10	7.4746×10^{-6}	1.7718		3	11	3.8947×10^{-6}	1.4420		3	6	5.8058×10^{-6}	1.5002
	4	16	6.9762×10^{-6}	1.7718		4	20	3.9435×10^{-6}	1.4420		4	11	6.5550×10^{-7}	1.5004
	5	20	5.6428×10^{-6}	1.7718		5	19	3.2357×10^{-6}	1.4420		5	14	9.6016×10^{-6}	1.5008
	6	19	9.4087×10^{-6}	1.7718		6	25	9.9202×10^{-6}	1.4420		6	23	4.2009×10^{-7}	1.5004
	7	17	4.6030×10^{-6}	1.7718		7	19	6.2421×10^{-6}	1.4420		7	9	4.0648×10^{-6}	1.5006
	8	34	4.9812×10^{-6}	1.7718		8	39	7.1700×10^{-6}	1.4420		8	7	2.6037×10^{-6}	1.5005
	9	13	7.4412×10^{-6}	1.7718		9	12	2.0244×10^{-6}	1.4420		9	7	6.3303×10^{-6}	1.5007
	10	13	9.7352×10^{-6}	1.7718		10	16	7.5910×10^{-7}	1.4420		10	12	5.6289×10^{-6}	1.5002
2	1	16	8.3027×10^{-7}	1.7718	2	1	18	8.4671×10^{-6}	1.4420	2	1	16	8.7020×10^{-8}	1.5004
	2	13	5.9090×10^{-6}	1.7718		2	14	7.8769×10^{-7}	1.4420		2	13	4.4437×10^{-6}	1.5004
	3	12	1.6624×10^{-6}	1.7718		3	12	8.9997×10^{-7}	1.4420		3	10	8.0433×10^{-6}	1.5004
	4	14	1.8828×10^{-6}	1.7718		4	19	3.1183×10^{-6}	1.4420		4	24	1.1569×10^{-6}	1.5004
	5	12	8.6969×10^{-6}	1.7718		5	18	5.0858×10^{-6}	1.4420		5	19	1.5519×10^{-6}	1.5004
	6	23	7.7303×10^{-6}	1.7718		6	38	6.8227×10^{-6}	1.4420		6	37	8.6291×10^{-6}	1.5004
	7	14	5.2478×10^{-6}	1.7718		7	19	5.2558×10^{-6}	1.4420		7	12	7.9567×10^{-6}	1.5004
	8	36	7.3341×10^{-6}	1.7718		8	40	3.7643×10^{-6}	1.4420		8	26	8.7335×10^{-6}	1.5004
	9	6	2.7901×10^{-6}	1.7718		9	17	5.2971×10^{-6}	1.4420		9	12	3.3013×10^{-6}	1.5004
	10	13	3.3655×10^{-6}	1.7718		10	18	1.6249×10^{-6}	1.4420		10	15	3.0078×10^{-6}	1.5004
3	1	21	5.5706×10^{-6}	1.8322	3	1	21	3.4492×10^{-6}	1.5010	3	1	21	9.1215×10^{-6}	1.5743
	2	16	6.0681×10^{-6}	1.8322		2	13	8.2479×10^{-6}	1.5010		2	13	9.8864×10^{-7}	1.5743
	3	15	2.6695×10^{-6}	1.8322		3	10	5.9725×10^{-6}	1.5010		3	13	7.4311×10^{-6}	1.5743
	4	19	8.4953×10^{-6}	1.8322		4	23	6.4584×10^{-6}	1.5010		4	23	5.0605×10^{-8}	1.5743
	5	23	2.9226×10^{-6}	1.8322		5	22	1.7365×10^{-6}	1.5010		5	19	1.0723×10^{-6}	1.5743
	6	50	2.9237×10^{-6}	1.8322		6	39	7.8088×10^{-6}	1.5010		6	42	4.8807×10^{-6}	1.5743
	7	26	1.4879×10^{-6}	1.8322		7	24	7.5942×10^{-6}	1.5010		7	18	3.6736×10^{-6}	1.5743
	8	43	4.7320×10^{-6}	1.8322		8	44	2.3022×10^{-6}	1.5010		8	50	8.4293×10^{-6}	1.5743
	9	16	9.1194×10^{-7}	1.8322		9	16	4.8699×10^{-7}	1.5010		9	15	9.2910×10^{-6}	1.5743
	10	18	8.8692×10^{-6}	1.8322		10	16	4.7118×10^{-6}	1.5010		10	24	3.0137×10^{-6}	1.5743

3.2. Three Objective Functions Comparison

In this section, the three objective functions are compared from the iteration number and the objective function value aspects. These three objective functions are firstly analyzed by comparing the corresponding results with regard to the three groups and two group pairs shown in Figure 1.

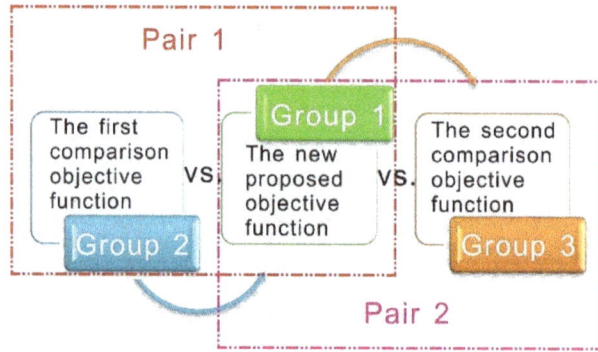

Figure 1. Groups and group pairs.

3.2.1. Analysis and Comparison Over the Three Groups

Boxplot Results Analysis

Figure 2(a1–c1) show the boxplots of the iteration number over the above defined three groups, where on each box, the central mark is the median, and the edges of the box are the lower quantile and the upper quantile, respectively. The lower quantile, the median and the upper quantile means the 0.25, 0.5 and 0.75 quantiles, respectively, where the f quantile corresponding to a datum $q(f)$ means that below this datum, approximately a decimal fraction f of the data can be found. It is calculated in this way: Sorting the data in a sequence $\{x_j\}_{j=1,2,...,n}$ in an ascending order. By this, the sorted data $\{x_{\langle i \rangle}\}_{i=1,2,...,n}$ have rank $i = 1, 2, \ldots, n$. Then the quantile value f_i for the datum $x_{\langle i \rangle}$ (equal to $q(f_i)$) is computed as:

$$f_i = \frac{i - 0.5}{n}, i = 1, 2, \ldots, n \tag{34}$$

While in the case of the desired quantile value f is equal to none of the f_i values shown in Equation (34), the f quantile $q(f)$ is found by linear interpolation, $i.e.,$:

$$q(f) = q(f_1) + \frac{f - f_1}{f_2 - f_1}[q(f_2) - q(f_1)] \tag{35}$$

where f_1 and f_2 are two unequal values selected from $\{0.5/n, 1.5/n, \ldots, (n - 0.5)/n\}$. Note that in the case of the probability value f is less than $0.5)/n$, the value $q(f)$ is assigned to the first value $x_{\langle 1 \rangle}$, while the value $q(f)$ is assigned to the last value $x_{\langle n \rangle}$, when the probability value f is greater than $(n - 0.5)/n$.

In addition, Figure 2(a1–c1) also show the outliers beyond the whiskers which are displayed using +. The whiskers in this paper are specified as 1.0 times the interquartile range, $i.e.,$ points larger than $q(0.75) + w[q(0.75) - q(0.25)]$ or smaller than $q(0.25) - w[q(0.75) - q(0.25)]$ are defined as outliers, where w is set to 1.0 in this paper.

(a1)

(b1)

(c1)

(a2) ANOVA Table

Source	SS	df	MS	F	Prob>F
Columns	38.871	2	19.4354	3.05	0.0641
Error	172.27	27	6.3804		
Total	211.141	29			

(b2) ANOVA Table

Source	SS	df	MS	F	Prob>F
Columns	3.785	2	1.89232	0.24	0.7876
Error	212.149	27	7.85738		
Total	215.934	29			

(c2) ANOVA Table

Source	SS	df	MS	F	Prob>F
Columns	0.226	2	0.1128	0.01	0.9896
Error	290.332	27	10.753		
Total	290.558	29			

(a3) No groups have means significantly different from Group 1

(b3) No groups have means significantly different from Group 1

(c3) No groups have means significantly different from Group 1

Lake Neighborhood → **North Neighborhood** → **Sunshine Mediterranean Neighborhood**

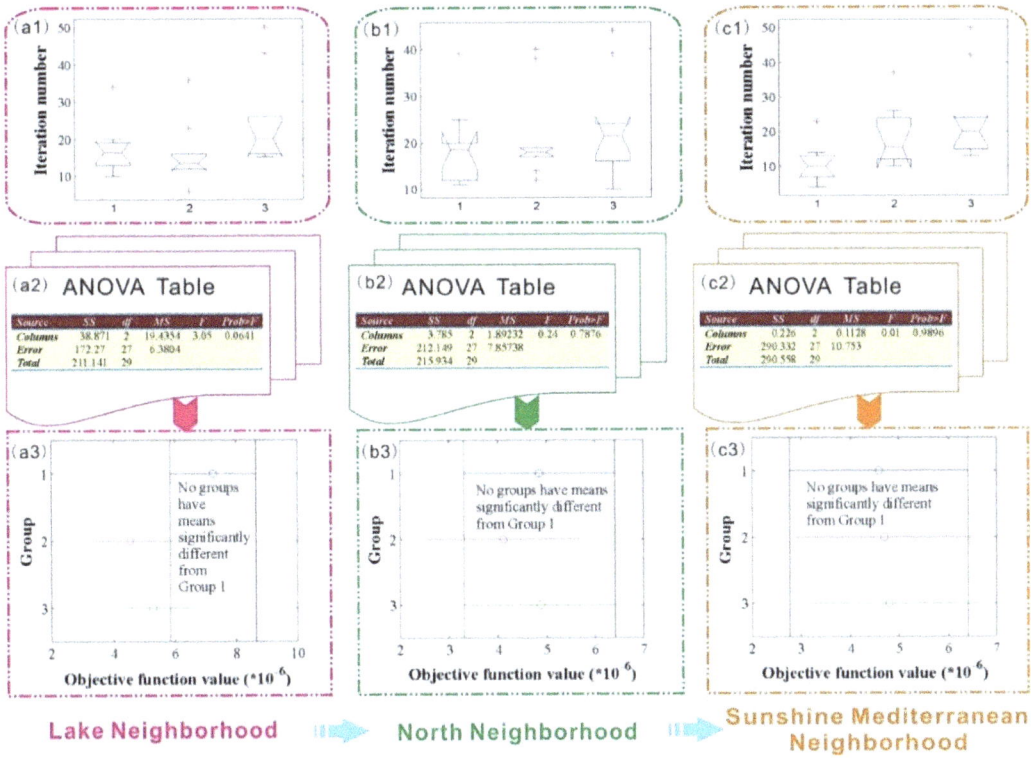

Figure 2. Boxplot and ANOVA comparison results of the three objective functions.

As seen from Figure 2(a1–c1), for the Lake Neighborhood, the number of outliers in the three groups are 1, 3, 2, respectively, while for the other two Neighborhoods, the number of outliers in Groups 1 and 3 remains the same, while values in Group 2 turns to 4 and 1 for the North Neighborhood and the Sunshine Mediterranean Neighborhood, respectively.

Analysis of Variance

Next one-way analysis of variance (ANOVA) was conducted to compare the objective function results of the three groups. Figure 2(a2–c2) provide the ANOVA results of the three neighborhoods, respectively, where SS, df, MS represent the sum of squares, degree of freedom and mean square, respectively, and $Columns$, $Error$ mean the feature between groups and feature within groups, respectively, and $Total$ indicates the sum of the $Columns$ and the $Error$. Specifically, we have the following definitions:

$$SS \text{ of the } Columns = \sum_{i=1}^{k} n_i (\overline{x}_i - \overline{x})^2 \tag{36}$$

$$SS \text{ of the } Error = \sum_{i=1}^{k} \sum_{j=1}^{n_i} (x_{ij} - \overline{x}_i)^2 \tag{37}$$

and

$$SS \text{ of the } Total = \sum_{i=1}^{k} \sum_{j=1}^{n_i} (x_{ij} - \overline{x})^2 \tag{38}$$

where k is the number of the groups, x_{ij} denotes the jth sample in the ith group, n_i represents the number of the samples in the i th group, and $\bar{x}_i$ and $\bar{x}$ indicates the mean of the samples in the ith group and the mean of the samples in all of the groups, respectively. The one-way ANOVA can be conducted according to the following four steps:

Step 1: Determine the null hypothesis. The null hypothesis of the one-way ANOVA is that samples in all of the groups are drawn from populations with the same mean.

Step 2: Select the test statistic. The test statistic of the one-way ANOVA used is the F statistic, which is defined as:

$$F = \frac{SS \text{ of the } Columns/(k-1)}{SS \text{ of the } Error/(n-k)} \tag{39}$$

where n is the total number of the samples, and k–1 and n–k are the degree of freedom of the SS of the $Columns$ and SS of the $Error$, respectively.

Step 3: Calculate the value of the test statistic as well as the corresponding probability value p.

Step 4: Make decisions according to the significance level α. In the case of $p < \alpha$, the null hypothesis should be rejected, and the decision that samples in all of the groups are not drawn from populations with the same mean is made; Otherwise, the null hypothesis should be accepted to demonstrate that samples in all of the groups are drawn from populations with the same mean.

As shown in Figure 2(a2–c2), all of the p values in the three neighborhoods are larger than the significance level α, which is set to 0.05 in this paper, this phenomenon indicates that for all of the three neighborhoods, the objective function samples in Groups 1–3 are drawn from populations with the same mean, Figure 2(a3–c3) display the mean comparison results of the three groups.

3.2.2. Test Over the Two Group Pairs

In this section, the three objective functions are analyzed by conducting the MER test over the two group pairs. The basic idea of the MER is that one group of the samples is regarded as the control group, while the other group of the samples is treated as the experimental group, then it is tested whether there are extreme reactions in the experimental group as compared to the control one. The conclusion of the MER is obtained by testing which one of the following hypothesis is accepted:

Null hypothesis: there is no significant difference between the distributions of samples in the control group and the experimental group; $vs.$ Alternative hypothesis: there is significant difference between the distributions of samples in the control group and the experimental group.

If the experimental group has extreme reactions, it is assumed that there is no significant difference between the distributions of the control group and the experimental group; instead, there is significant difference between the distributions of these two groups. The detailed analysis process is as follows:

1. First of all, samples in the two groups are mixed and ordered by ascending;
2. Calculate the minimum rank R_{min} and the maximum rank R_{max} of the control group, and obtain the span by:

$$S = R_{\max} - R_{\min} + 1$$

3. To eliminate the effect of the extreme values of the sample data on the analysis results, a proportional (usually this value is set to 5%) of the samples close to the left and the right ends are removed from the control group, and the span of the remaining samples which is named the trimmed span is calculated.

The MERs focus on the analysis of the span and the trimmed span. Obviously, if the values of the span or the trimmed span are small, the two sample groups cannot be mixed fully, and sample values in one group are greater than those in the other group, therefore, it can be regarded that as compared to the control group, the experimental group contains the extreme reactions, and thus the conclusion that there is significant difference between the distributions of these two groups can be obtained; otherwise, if the values of the span or the trimmed span are great, the two sample groups are mixed fully, and the phenomenon that sample values in one group are greater than those in the other group does not exist, therefore, it can be regarded that as compared to the control group, the experimental group does not contain extreme reactions, and thus the conclusion that there is no significant difference between the distributions of these two groups is reached. In general, the H statistics defined as below is used to evaluate the span or the trimmed span:

$$H = \sum_{i=1}^{m} (R_i - \overline{R})^2$$

where m is the number of the samples in the control group, R_i is the rank of the ith control sample in the mixed samples, $\overline{R}$ is the average rank of the control samples. It can be proved that for small samples, the H statistics obey the Hollander distribution, while for large samples, the H statistics approximately obey a normal distribution.

If the value of p is smaller than the given confidence level α, then the null hypothesis should be rejected, and it is regarded that there is significant difference between the distributions of samples in the control group and the experimental group; otherwise, the null hypothesis should be accepted, and the conclusion that there is no significant difference between the distributions of samples in the control group and the experimental group can be obtained. In this paper, the MER technique is used to compare the difference of the three objective functions furthermore. Section 3.2.1 analyzed the three objective function mainly from the shape parameter aspect, in this section, the three objective functions will be analyzed through the iteration number as well as the objective function value.

Table 2 lists the descriptive statistics of Pair 1 and Pair 2, where N denotes the number of the samples in the pair and the hth percentile is equivalent to the $h/100$ quantile. As seen from Table 2, apart from the objective function value of the Lake Neighborhood, the standard deviation of which in Pair 2 is smaller than the one in Pair 1, for other items, the standard deviation values in Pair 2 are all larger than the corresponding values in Pair 1, $i.e.$, when Groups 1 and 3 are mixed, their deviation is larger than the one obtained by mixing Groups 1 and 2. In addition, the difference between the maximum and the minimum present similar phenomenon: apart from the objective function value of the Lake Neighborhood, the difference values for other items in Pair 2 are all larger than those in Pair 1. Based on the descriptive statistics results in Table 2, Table 3 presents the MER test results. Note that in Table 3, the term Outliers trimmed means outliers trimmed from each end. It can be observed from Table 3 that there is only one probability value which is smaller than the predefined confidence level $\alpha = 0.05$, which appears in the iteration number of the Sunshine Mediterranean Neighborhood in Pair 2. This indicates that there is significant difference between the distributions of the iteration number in Groups 1 and 3, while no significant difference can be observed between the corresponding distributions in Groups 1 and 2.

In summary, it can be concluded from these analysis results that the iteration number and the objective function value of the new proposed objective function and the first comparison objective function can be regarded to have nearly no difference between each other. In addition, the shape parameter values obtained by the new proposed and the first comparison objective functions are nearly equal, however, the same conclusion cannot be concluded with regard to the new proposed objective function and the second comparison objective function. Therefore, in the following sections, only the error values obtained by the new proposed and the second objective functions will be compared.

Table 2. Descriptive statistics of the two pairs.

Neighborhood	Item	Pair 1								Pair 2							
		N	Mean	Standard Deviation	Minimum	Maximum	Percentiles			N	Mean	Standard Deviation	Minimum	Maximum	Percentiles		
							25th	50th	75th						25th	50th	75th
Lake Neighborhood	Iteration number	20	16.6000	7.3155	6.0000	36.0000	13.0000	14.0000	18.7500	20	21.0000	10.2341	10.0000	50.0000	15.2500	18.0000	22.5000
	Objective function	20	5.8941	2.6892	0.8303	9.7352	3.6749	6.4175	7.6664	20	6.2645	2.5157	1.4879	9.7352	4.6353	6.4970	8.7757
North Neighborhood	Iteration number	20	20.1500	8.8274	11.0000	40.0000	14.5000	18.0000	19.7500	20	20.9000	9.6404	10.0000	44.0000	13.7500	19.0000	23.7500
	Objective function	20	4.4833	2.7723	0.7591	9.9202	2.1238	3.9191	6.6776	20	4.8656	2.8425	0.4870	9.9202	2.3322	4.3277	7.4882
Sunshine Mediterranean Neighborhood	Iteration number	20	14.5000	7.9637	4.0000	37.0000	9.2500	12.5000	18.2500	20	17.2000	11.5421	4.0000	50.0000	9.5000	13.5000	22.5000
	Objective function	20	4.6370	3.0436	0.0870	9.6016	1.8149	4.2543	7.6968	20	4.6890	3.1493	0.0506	9.6016	1.4552	4.4728	7.3026

Table 3. The Moses Extreme Reactions (MER) test results of the two pairs.

Neighborhood	Item	Pair 1								Pair 2							
		Frequencies			Untrimmed		Trimmed		Outliers Trimmed	Frequencies			Untrimmed		Trimmed		Outliers Trimmed
		Control Sample	Experimental Sample	Total	Span	p	Trimmed Span	p		Control Sample	Experimental Sample	Total	Span	p	Trimmed Span	p	
Lake Neighborhood	Iteration number	10	10	20	18	0.500	11	0.089	1	10	10	20	18	0.500	12	0.185	1
	Objective function	10	10	20	15	0.070	13	0.325	1	10	10	20	16	0.152	13	0.325	1
North Neighborhood	Iteration number	10	10	20	19	0.763	17	0.957	1	10	10	20	17	0.291	16	0.848	1
	Objective function	10	10	20	20	1.000	15	0.686	1	10	10	20	19	0.763	16	0.848	1
Sunshine Mediterranean Neighborhood	Iteration number	10	10	20	17	0.291	12	0.185	1	10	10	20	17	0.291	10	0.035	1
	Objective function	10	10	20	19	0.763	13	0.325	1	10	10	20	19	0.763	13	0.325	1

3.2.3. Fitting Error Comparison

The error comparison analysis in this section is built on final shape parameter, which is determined by the mean of the ten shape parameter values. Since the shape parameter obtained by the new proposed and the first comparison objective functions are quite the same, this section only present the error results of the new proposed and the second comparison objective functions, for which the shape parameters are different.

Let the minimum and the maximum active power values of the transformer are MI and MA, respectively. Then each interval $[k, k + 1]$ can be divided into several subintervals with the same length, where k are the integers from $Floor(MI)$ to $Ceil(MA)$, and $Ceil(MA)$ denotes the integer larger than MA which has the minimum distance with MA, similarly, $Floor(MI)$ represents the integer smaller than or equal to MI which has the minimum distance with MI.

Figure 3 shows the PDF and CDF figures obtained by the new proposed and the second comparison objective functions where each unit interval $[k, k + 1]$ is divided into different subintervals: Figure 3(a–c, a1–c1, a2–c2) show the figures of the three neighborhoods where each unit interval is divided into five subintervals, respectively, Figure 3(a1–c1) are the corresponding figures where each unit interval is divided into two subintervals, respectively, and Figure 3(a2–c2) provide the results with no division to the unit interval. The corresponding error values are listed in Table 4.

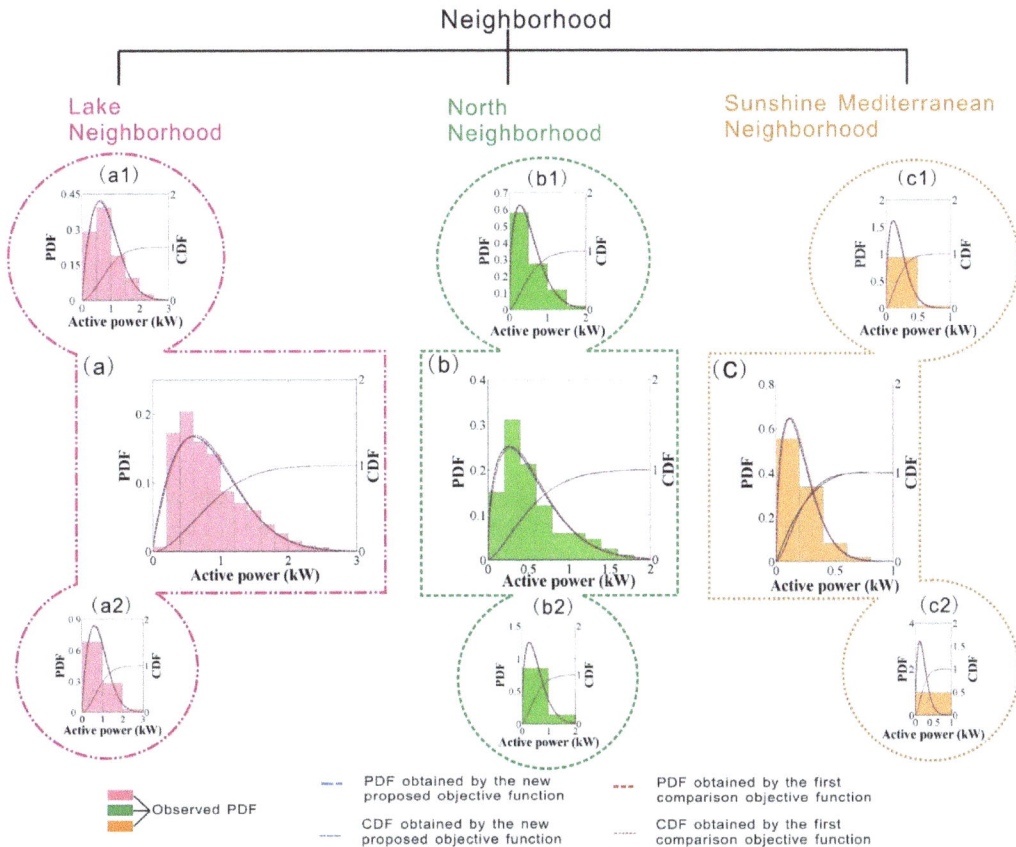

Figure 3. PDF and CDF results of the active power in the three neighborhoods by dividing the unit interval into different subintervals.

Table 4. Error values under different subinterval numbers. Kolmogorov-Smirnov test error (KSE); root mean square error (RMSE).

Neighborhood Name	Subinterval Numbers	The New Proposed Objective Function		The Second Comparison Objective Function	
		KSE	RMSE	KSE	RMSE
Lake Neighborhood	5 subintervals	0.05379	0.02199	0.04775	0.02236
	2 subintervals	0.02378	0.01916	0.03190	0.02414
	1 subinterval	0.02378	0.02190	0.02765	0.02646
North Neighborhood	5 subintervals	0.03878	0.02840	0.03896	0.02774
	2 subintervals	0.03739	0.03926	0.04639	0.04695
	1 subinterval	0.02687	0.02839	0.03079	0.03186
Sunshine Mediterranean Neighborhood	5 subintervals	0.00757	0.00518	0.01287	0.01247
	2 subintervals	0.01076	0.01082	0.01568	0.01571
	1 subinterval	0.00012	0.00012	0.00005	0.00005

3.3. Comprehensive Overload Capability Assessment Results

The comprehensive overload capability of power transformers is obtained based on the running time duration of the power transformers under overload conditions and the overloading probability calculation results: the running time duration of the power transformer is obtained according to the given ambient temperature and the rated load first, then the overloading probability is obtained from the probability of the current, which is derived from the probability of the corresponding active power. Overload capability measurement of power transformers based on the knowledge of overloading probability provides a more reliable assessment result.

The Weibull distribution can be used to evaluate transformer reliability. The scientific and reasonable assessment of reliability development trends is based on the research and mastery of a large amount of historical materials and accurate methods. On the basis of foregoing research, the reliability assessment of transformers can be performed by using the model of transportation load and test quantity. Therefore, the reliability assessment of transformers can be carried out in these two aspects. The valid assessment means the situation of transportation load and test quantity that can have an influence on the reliability of transformer so that we can obtain the future reliability assessment of the transformer.

The reliability model based on transportation overload is mainly based on the use of the hot-spot temperature of the transformer to evaluate the degree of thermal aging so that the fault probability of transformer can be obtained by analyzing the insulation aging damage. The hot-spot temperature is related to the operation load of the transformer and the environmental temperature; therefore, the key of the assessment is to evaluate the future load level and the environmental temperature. What largely affects the reliability change curve of a transformer is the increase of load level. Without great changes of the network structure, the assessment of future load increases can be conducted by evaluating the local load increases. If the load increase level in the assessment is fast, and the current transformer is burnt-in, one should consider adding new transformers in the future to reduce the load level of the current transformers and decrease the risk of accidents according to specific situations.

4. Conclusions

This paper measures the overload capability of oil-immersed power transformers, which is of particular importance in avoiding their catastrophic failure and guaranteeing the normal operation of power grids. The running time duration of the power transformers under overload conditions is calculated with the help of the hot-spot temperature. Then the overloading probability is fitted by the Weibull distribution, in which the desired parameters are computed according to a new proposed objective function. Compared with the previous two objective functions, the new proposed one acheived much better performance in terms of the convergence speed and the final objective function values. The integration of the running time duration and the overload probability provides a more comprehensive and reliable assessment results to the overload capability of power transformers.

Acknowledgments: The work was supported by the National Natural Science Foundation of China (Grant No. 71171102).

Author Contributions: Wang, C. and Wu, J. conceived and designed the experiments; Wang, C. and Zhao, W.G. performed the experiments; Wang, J.Z. and Zhao, W.G. analyzed the data; Wu, J. wrote the paper and Wang, C. checked the whole paper.

Conflicts of Interest: The authors declare no conflict of interest.

Appendix A

According to the PDF of the Weibull distribution, the mean ($\bar{a}$) and the standard deviation (σ) of the active power can be obtained by:

$$
\begin{aligned}
\bar{a} &= \int_0^{+\infty} a f(a)\,da \\
&= \int_0^{+\infty} a \left(\frac{k}{c}\right)\left(\frac{a}{c}\right)^{k-1} \exp\left[-\left(\frac{a}{c}\right)^k\right] da \\
&= \int_0^{+\infty} a \exp\left[-\left(\frac{a}{c}\right)^k\right] d\left(\frac{a}{c}\right)^k \\
&= \int_0^{+\infty} c\left[\left(\frac{a}{c}\right)^k\right]^{(1+1/k)-1} \exp\left[-\left(\frac{a}{c}\right)^k\right] d\left(\frac{a}{c}\right)^k \\
&= c \int_0^{+\infty} \left[\left(\frac{a}{c}\right)^k\right]^{(1+1/k)-1} \exp\left[-\left(\frac{a}{c}\right)^k\right] d\left(\frac{a}{c}\right)^k \\
&= c\,\Gamma\left(1+\frac{1}{k}\right)
\end{aligned}
\tag{A1}
$$

$$
\begin{aligned}
\sigma^2 &= E(a-\bar{a})^2 \\
&= Ea^2 - (Ea)^2 \\
&= \int_0^{+\infty} a^2 \left(\frac{k}{c}\right)\left(\frac{a}{c}\right)^{k-1} \exp\left[-\left(\frac{a}{c}\right)^k\right] da - c^2\Gamma^2\left(1+\frac{1}{k}\right) \\
&= \int_0^{+\infty} c^2 \left[\left(\frac{a}{c}\right)^k\right]^{(1+2/k)-1} \exp\left[-\left(\frac{a}{c}\right)^k\right] d\left(\frac{a}{c}\right)^k - c^2\Gamma^2\left(1+\frac{1}{k}\right) \\
&= c^2\Gamma\left(1+\frac{2}{k}\right) - c^2\Gamma^2\left(1+\frac{1}{k}\right)
\end{aligned}
\tag{A2}
$$

So:

$$
\frac{\sigma^2}{\bar{a}^2} = \frac{\Gamma(1+2/k) - \Gamma^2(1+1/k)}{\Gamma^2(1+1/k)}
\tag{A3}
$$

However, there is always some error between the left side and the right side of the Equation (A3). Thus, the residual value ε defined as below is used as the first objective function in this paper just as Liu *et al.* did in [16]:

$$
\varepsilon = \frac{\sigma^2}{\bar{a}^2} - \frac{\Gamma(1+2/k) - \Gamma^2(1+1/k)}{\Gamma^2(1+1/k)}
\tag{A4}
$$

Appendix B

Given the active power series $\{a_i\}_{i=1}^n$, the joint PDF of the Weibull distribution can be expressed as:

$$
\prod_{i=1}^n f(a_i; k, c) = \left(\frac{k}{c}\right)^n \left(\frac{a_1}{c}\cdot\frac{a_2}{c}\cdot\ldots\cdot\frac{a_n}{c}\right)^{k-1} \exp\left[-\left(\frac{a_1}{c}\right)^k - \left(\frac{a_2}{c}\right)^k - \cdots - \left(\frac{a_n}{c}\right)^k\right]
\tag{B1}
$$

Thus, according to the maximum likelihood approach, the parameters k and c can be calculated according to

$$
\begin{cases}
\dfrac{\partial \prod\limits_{i=1}^{n} f(a_i; k, c)}{\partial k} = 0 \\[4mm]
\dfrac{\partial \prod\limits_{i=1}^{n} f(a_i; k, c)}{\partial c} = 0
\end{cases}
\tag{B2}
$$

That is:

$$
c = \left(\frac{1}{n} \sum_{i=1}^{n} a_i^k \right)^{1/k}
\tag{B3}
$$

and

$$
k = \left[\frac{\sum_{i=1}^{n} a_i^k \ln a_i}{\sum_{i=1}^{n} a_i^k} - \frac{\sum_{i=1}^{n} \ln a_i}{n} \right]^{1/k}
\tag{B4}
$$

Generally, there will be an error between the right and the left side of Equation (B4). Therefore, the following equation has been set as the second objective function in this paper:

$$
\varepsilon = k - \left[\frac{\sum_{i=1}^{n} a_i^k \ln a_i}{\sum_{i=1}^{n} a_i^k} - \frac{\sum_{i=1}^{n} \ln a_i}{n} \right]^{1/k}
\tag{B5}
$$

References

1. Caro, M.A.; Rios, M.A. Super components contingency modeling for security assessment in power systems. *IEEE Lat. Am. Trans.* **2009**, *7*, 552–559. [CrossRef]
2. He, J.; Sun, Y.Z.; Wang, P.; Cheng, L. A hybrid conditions-dependent outage model of a transformer in reliability evaluation. *IEEE Trans. Power Deliv.* **2009**, *24*, 2025–2033. [CrossRef]
3. Zhu, Z.L.; Zhou, J.Y.; Yan, C.H.; Chen, L.J. Power system operation risk assessment based on a novel probability distribution of component repair time and utility theory. In Proceedings of the 2012 Asia-Pacific Power and Energy Engineering Conference (APPEEC), Shanghai, China, 27–29 March 2012.
4. Chen, Q.; Mili, L. Composite power system vulnerability evaluation to cascading failures using importance sampling and antithetic variates. *IEEE Trans. Power Syst.* **2013**, *28*, 2321–2330. [CrossRef]
5. Aubin, J.; Pierce, L.W.; Langhame, Y. Effect of oil viscosity on transformer loading capability at low ambient-temperatures. *IEEE Trans. Power Deliv.* **1992**, *7*, 516–524. [CrossRef]
6. Tenbohlen, S.; Stirl, T.; Stach, M. Assessment of overload capacity of power transformers by on-line monitoring systems. In Proceedings of the IEEE Power Engineering Society Winter Meeting, Columbus, OH, USA, 28 January–1 February 2001.
7. Bosworth, T.; Setford, S.; Heywood, R.; Saini, S. Electrochemical sensor for predicting transformer overload by phenol measurement. *Talanta* **2003**, *59*, 797–807. [CrossRef]
8. Edstrom, F.; Rosenlind, J.; Alvehag, K.; Hilber, P.; Soder, L. Influence of ambient temperature on transformer overloading during cold load pickup. *IEEE Trans. Power Deliv.* **2013**, *28*, 153–161. [CrossRef]
9. Estrada, J.H.; Ramírez, S.V.; Cortés, C.L.; Plata, E.A.C. Magnetic flux entropy as a tool to predict transformer's failures. *IEEE Trans. Magn.* **2013**, *49*, 4729–4732. [CrossRef]
10. Liu, W.J.; Wang, X.; Zheng, Y.H.; Li, L.X.; Xu, Q.S. The assessment of the overload capacity of transformer based on the temperature reverse extrapolation method. *Adv. Mater. Res.* **2014**, *860–863*, 2153–2156. [CrossRef]
11. Galdi, V.; Ippolito, L.; Piccolo, A.; Vaccaro, A. Neural diagnostic system for transformer thermal overload protection. *IEE Proc. Electr. Power Appl.* **2000**, *147*, 415–421. [CrossRef]
12. Galdi, V.; Ippolito, L.; Piccolo, A.; Vaccaro, A. Genetic Algorithm based parameters identification for power transformer thermal overload protection. In *Artificial Neural Nets and Genetic Algorithms*; Springer: Berlin, Germany, 2001; pp. 308–311.

13. Galdi, V.; Ippolito, L.; Piccolo, A.; Vaccaro, A. Application of local memory-based techniques for power transformer thermal overload protection. *IEE Proc. Electr. Power Appl.* **2001**, *148*, 163–170. [CrossRef]

14. Ippolito, L.; Siano, P. A power transformers' predictive overload system based on a Takagi-Sugeno-Kang fuzzy model. In Proceedings of the 12th IEEE Mediterranean Electrotechnical Conference, Dubrovnik, Croatia, 12–15 May 2004; pp. 301–306.

15. Pradhan, M.K.; Ramu, T.S. On-line monitoring of temperature in power transformers using optimal linear combination of ANNs. In Proceedings of the 2004 IEEE International Symposium on Electrical Insulation Conference, Indianapolis, IN, USA, 19–22 September 2004; pp. 70–73.

16. Villacci, D.; Bontempi, G.; Vaccaro, A.; Birattari, M. The role of learning methods in the dynamic assessment of power components loading capability. *IEEE Trans. Ind. Electron.* **2005**, *52*, 280–290. [CrossRef]

17. Jiang, Y.Q.; Lu, Z.H. Research on enhancement of overload capacity for 500 kV transformer. *East China Electr. Power* **2004**, *32*, 13–17.

18. Zhao, D.Y.; Fan, H.; Ren, Z.J. *Reliability Engineering and Applications*; National Defend Industry Press: Beijing, China, 2009.

19. Contin, A.; Montanari, G.C.; Ferraro, C. PD source recognition by weibull processing of pulse height distributions. *IEEE Trans. Dielectr. Electr. Insul.* **2000**, *7*, 48–58. [CrossRef]

20. Mao, S.; Wang, L. *Accelerated Life Testing*; Beijing Science Press: Beijing, China, 2000.

21. Wu, J.; Wang, J.Z.; Chi, D.Z. Wind energy potential assessment for the site of Inner Mongolia in China. *Renew. Sustain. Energy Rev.* **2013**, *21*, 215–228. [CrossRef]

22. Chang, T.P. Estimation of wind energy potential using different probability density functions. *Appl. Energy* **2011**, *88*, 1848–1856. [CrossRef]

23. Fyrippis, I.; Axaopoulos, P.J.; Panayiotou, G. Wind energy potential assessment in Naxos Island, Greece. *Appl. Energy* **2010**, *87*, 577–586. [CrossRef]

9

Decomposing Industrial Energy-Related CO_2 Emissions in Yunnan Province, China: Switching to Low-Carbon Economic Growth

Mingxiang Deng, Wei Li * and Yan Hu

Academic Editor: Robert Lundmark

State Key Laboratory of Water Environment Simulation, School of Environment, Beijing Normal University, Beijing 100875, China; dmx221@163.com (M.D.); isabella9210@hotmail.com (Y.H.)
* Correspondence: weili@bnu.edu.cn

Abstract: As a less-developed province that has been chosen to be part of a low-carbon pilot project, Yunnan faces the challenge of maintaining rapid economic growth while reducing CO_2 emissions. Understanding the drivers behind CO_2 emission changes can help decouple economic growth from CO_2 emissions. However, previous studies on the drivers of CO_2 emissions in less-developed regions that focus on both production and final demand have been seldom conducted. In this study, a structural decomposition analysis-logarithmic mean Divisia index (SDA-LMDI) model was developed to find the drivers behind the CO_2 emission changes during 1997–2012 in Yunnan, based on times series energy consumption and input-output data. The results demonstrated that the sharp rise in exports of high-carbon products from the metal processing and electricity sectors increased CO_2 emissions, during 2002–2007. Although increased investments in the construction sector also increased CO_2 emissions, during 2007–2012, the carbon intensity of Yunnan's economy decreased substantially because the province vigorously developed hydropower and improved energy efficiency in energy-intensive sectors. Construction investments not only carbonized the GDP composition, but also formed a carbon-intensive production structure because of high-carbon supply chains. To further mitigate CO_2 emissions in Yunnan, measures should promote the development and application of clean energy and the formation of consumption-based economic growth.

Keywords: CO_2 emissions; drivers; structural decomposition analysis; less-developed regions; low-carbon pilot; Yunnan province

1. Introduction

As the top CO_2 emitter in the world, China's annual CO_2 emissions in 2013 were more than the sum of the amount from the United States and the European Union (EU) [1]. In addition, its per capita CO_2 emissions exceeded the EU for the first time [1]. Facing severe international pressure to mitigate its CO_2 emissions, China has pushed its provinces to reduce their CO_2 emissions. As a mitigating action, China's central government has chosen provinces to conduct pilot work in low-carbon development in alignment with the development stage and characteristics of the different regions. Yunnan province was chosen as a pilot province and tasked with finding ways to achieve low-carbon transition by utilizing its underdevelopment to its advantage [2]. Yunnan is located in southwest China and is famous for its abundant hydropower, tourism and biological and cultural diversity [3] (see Figure S1 in the Supplementary Materials (SM)). The per capita gross domestic product (GDP) in Yunnan is approximately 60% of the national average [4]. In 2012, approximately 17% of the population (approximately eight million) was labeled as living in poverty [5]. Thus, Yunnan faces challenges to develop its economy and reduce poverty.

Since the 12th Five Year Plan (FYP) period (2010–2015), the central government has set a compulsory target to reduce the carbon intensity (CO_2 emissions per unit of GDP) by 17% and has disaggregated the target to its provinces [6]. As one of China's low-carbon pilot provinces, the carbon intensity reduction target in Yunnan is higher than the national government's standard for the country as a whole [2]. In 2014, the central government further committed to prevent CO_2 emissions from increasing after 2030 [7] instead of only reducing the carbon intensity. Under these circumstances, in the near future, Yunnan will face an even stricter target for controlling the volume of its CO_2 emissions. To achieve these targets, Yunnan faces large challenges in its transition to a low-carbon future.

From 1997–2012, Yunnan's total energy consumption more than tripled, rising from 34.3 million tons (Mt) of coal equivalent in 1997 to 104.3 in 2012 (Figure 1). The total energy-related CO_2 emissions grew 3.5-times during the same period, from 56.6 Mt–197.1 Mt (Figure 1). The CO_2 emissions are predicted to reach 216.6 Mt in 2015 and 278.3 Mt in 2020 [8]. Industrial CO_2 emissions in Yunnan increased by 209.6% during the period 1997–2007, a figure that is 102.4% higher than the national growth rate over the same period [9]. This growth rate is 4.6-, 2.6- and 1.5-times greater than that of Beijing municipality [10], Liaoning province [11] and Jiangsu province [12], respectively. These facts raise a question: what are the driving forces behind the rapid growth of CO_2 emissions in a less-developed region with abundant low-carbon resources? From 1997–2012, the carbonization process of Yunnan's economy experienced three different stages (Figure 1). From 1997–2002, low CO_2 emission growth was accompanied by low economic growth. During this period, the average annual CO_2 emission elasticity ratio (CEER) (industrial CO_2 emissions' growth rate/GDP growth rate) was one. From 2002–2007, a high CO_2 emissions increase was accompanied by rapid economic growth. The average annual CEER in this period was 1.6. From 2007–2012, a low CO_2 emissions increase was accompanied by rapid economic growth. During this period, the average annual CEER was 0.4. It can be seen that from 2007–2012, Yunnan experienced relatively low-carbon economic growth. The factors driving the changes in the economic growth pattern in Yunnan have attracted our attention because the question of how to achieve rapid economic growth in less-developed regions while maintaining relatively low CO_2 emissions is important and is a challenge for policy makers [13].

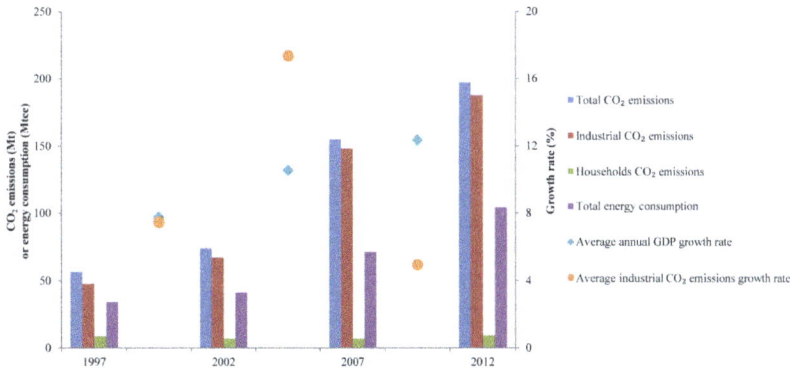

Figure 1. Annual GDP growth rate, total energy consumption and related CO_2 emissions in Yunnan from 1997–2012.

Currently, research on CO_2 emissions in Yunnan focuses on CO_2 inventories [14], and prediction analysis is based on the realization of carbon intensity reduction targets [8]. These studies have provided some data resources and analytical approaches for low-carbon development in Yunnan. However, empirical studies on the drivers of Yunnan's CO_2 emissions that focus on the entire economic system from both production and final demand perspectives have been insufficient, lacking a relevant basis for formulating a low-carbon economic growth policy. Thus, an in-depth analysis to identify the drivers of CO_2 emission changes is helpful for finding a way to decouple the economic growth from CO_2 emissions in Yunnan.

There are two main decomposition techniques for identifying the drivers: structural decomposition analysis (SDA) and index decomposition analysis (IDA) [15]. The latest comparisons between IDA and SDA are shown in [16]. IDA has been applied to analyze the drivers of CO_2 emissions in China [17–19]. However, it can only analyze direct CO_2 emissions and provides fewer details about economic sectors than SDA. [20]. Compared to IDA, SDA is based on environmental input-output analysis [21], which is a widely-accepted method for quantifying sectoral CO_2 emissions considering both direct effects from the production process and indirect effects from the entire supply chain [16]. For the regional economy in Yunnan, SDA can analyze more comprehensive issues and is helpful for the low-carbon management of the entire economic chain from production to final demand. Because of the advantages of SDA, it was widely applied to analyze the drivers of the increase of CO_2 emissions in China as a whole [9,20,22] and in specific regions, such as Beijing [10,23], Jiangsu [12] and Liaoning [11]. These studies contributed to understanding the determinants of changes in CO_2 emissions and to forming policy implications that are suitable for local conditions. However, all of these studies focus on the national level or on developed regions in eastern China. Studies for less-developed regions (such as Yunnan province) in central or western China are not sufficient. Low-carbon development in less-developed regions is vital to achieve China's national carbon reduction target, because emissions in those regions are projected to increase substantially in the coming decades. In the time scale, most of these studies analyzed drivers from 1997–2007, but the impact of the 2008 Global Financial Crisis on changes in regional CO_2 emission drivers has seldom been analyzed. Moreover, these studies did not analyze the impact of changes in the GDP composition on CO_2 emissions. For the less-developed regions, GDP composition greatly changed over time; therefore, it is very important to study how GDP composition changes affect regional CO_2 emissions.

There are different decomposition forms in SDA, such as additive and multiplicative SDA [16]. The specific decomposing methods in SDA are also abundant, such as the method developed by Dietzenbacher and Los [24] (D&L) and the logarithmic mean Divisia index (LMDI) method [16]. Su and Ang [16] reviewed the new development of SDA in energy and emission fields and provided a guide on method selection in additive SDA for empirical studies. According to the review of the literature from 2000–2010 by Su and Ang [16], most energy and emission SDA research applied an additive SDA form and D&L method. However, recent SDA research on changes in the CO_2 emissions embodied in exports [25] and carbon intensity [26] in China applied the multiplicative SDA form. Due to the advantages of the simple decomposition process and the easy results representation [16], the additive SDA form is adopted in this study. The wide applications of additive SDA in China have proven its applicability and practicality. For methods selection in additive SDA, Su and Ang [16] have suggested that the LMDI method should be selected when the factors are more than four. However, all of the SDA studies in China's regions applied the D&L method [24]. The SDA-D&L model has non-unique decomposition problems, and when the drivers increase, the calculation will increase greatly [16]. Compared to the SDA-D&L model, the SDA-LMDI model is characterized by a simple expression, and its computation requirements are minimal [27]. The advantages of the SDA-LMDI model are more obvious for studies with four or more drivers [16]. Since the number of factors is seven in this study and the LMDI method is easy to use and has a simple computation in this situation, the LMDI method was selected in the decomposition analysis following the suggestions by Su and Ang [16]. At present, the SDA-LMDI model has been widely used in different countries and regions to analyze the driving forces behind changes in the economic systems [27], energy consumption and energy intensity [28,29] and greenhouse gas (GHG) emissions [30]. All of the previous research indicated that the SDA-LMDI model is applicable to this study.

This paper has two aims. The first is to understand the factors that drove the rapid CO_2 emissions' growth in Yunnan from 1997–2012. The second is to determine how the drivers contributed to the relatively low-carbon economic growth pattern in Yunnan from 2007–2012. We focused on the low-carbon development of the less-developed regions, of which Yunnan is the best case. Based on newly-released data, the time scale was extended to 2012. Seven drivers were analyzed in our decomposition analysis, including changes in the GDP composition. We focused on analyzing energy-related CO_2 because of the availability of data and the large share of energy-related CO_2 in the total emissions. According to China's GHG inventories [31], energy-related CO_2 accounted for more than 90% of the total CO_2 emissions. Following other SDA research, we focused on the emissions created by the production of goods and services for consumers and did not consider the direct fuel use of households [20]. Direct CO_2 emissions from household energy consumption accounted for 4%–16% of the total emissions in Yunnan and continued to decline with the development of the economy (Figure 1).

The contents of the next sections are as follows: Section 2 is the description of the SDA-LMDI model employed in this paper, along with information on the data sources and the required preparation processes for using the data. Section 3 presents the results from the analysis, an in-deep discussion of them and a number of policy recommendations. Section 4 offers conclusions.

2. Method and Data

2.1. Method

Before conducting SDA, some assumptions in the input-output analysis framework should be made. The most important assumptions are "processing and normal exports" [32] and "competitive *versus* non-competitive imports" [33]. Uniform exports are assumed in this study, which means that processing and normal exports have the same input structures, since the processing exports in Yunnan are less and the data of processing and normal exports in the interregional exports for a region in China are unavailable. Imports are assumed as a competitive import assumption. Since Yunnan's primary input-output tables (IOTs) are compiled based on the competitive import assumption, excluding imports from these IOTs will cause the information on regional imports to be missed and will introduce uncertainties into the final results (see more discussions on the technology assumption about imports in the SM).

Changes in regional CO_2 emissions are affected by many factors [34]. The SDA-LMDI model can be used to analyze each factor's contribution to CO_2 emission changes from the perspective of the entire economic system. The model can be expressed mathematically as Equation (1):

$$e = F(I-A)^{-1}y = FLy \tag{1}$$

where $e = (e_i)$ is the $n \times 1$ vector of CO_2 emissions from n sectors in an industrial system and e_i is the direct CO_2 emissions from the i-th sector's production. This means the CO_2 emission responsibilities are allocated to the producers under Equation (1). In addition, $F = (f_{ij})$ is an $n \times n$ diagonal matrix, where the diagonal element f_{ii} is the ratio of the CO_2 emissions from sector i to the sector's total output; i.e., the direct carbon intensity (DCI) of sector i. I is the $n \times n$ identity matrix. The $n \times n$ matrix $A = (a_{ij})$ is the direct requirement coefficient matrix, where a_{ij} is the value of the intermediate products sector i required to bring about a unit increase in the output of sector j. $L = (I-A)^{-1} = (l_{ij})$ is the $n \times n$ Leontief inverse matrix, where l_{ij} represents the output increase of sector i by a unit increase in final demand in sector j. $y = (y_i)$ is an $n \times 1$ vector, where y_i represents the final demand of sector i.

The vector y can be further divided into four components: final demand structure, W; GDP composition, v; per capita GDP, g; and population, p . Each element, w_{ik}, in the $n \times m$ matrix, W, represents the contribution rates of the products demanded by sector i to the sum of the k-th final demand column, where k indicates the final demand categories, such as household consumption, government consumption, capital investment, exports and imports and m indicates the number of final demand categories. The element v_k in the $m \times 1$ vector v indicates the total of k-th final demand conlumn divided by GDP. Similarly, F can be further divided into three components: CO_2 emission coefficient, c; energy structure, J; and energy intensity, Q. The element c_d in the $1 \times z$ vector c is the CO_2 emission coefficient of energy d, and z indicates number of energy types. Each element, j_{di}, in the $z \times n$ matrix, J, represents the proportion of the amount of energy d in the total energy consumption by sector i. $Q = (q_{ij})$ is an $n \times n$ diagonal matrix, where the diagonal element q_{ii} is the ratio of the energy consumption from sector i to the sector's total output; *i.e.*, the energy intensity of sector i. As a result, CO_2 emissions (Equation (2)) can be specified as follows:

$$e = cJQLWvgp \tag{2}$$

As CO_2 emission coefficient c is constant in different years, changes in the technology (or energy efficiency) of the economic system are captured by two drivers, which are the changes in the energy structure and changes in the energy intensity. Therefore, seven CO_2 emission drivers are finally included in this study. The purpose of decomposition is to demonstrate changes in CO_2 emissions, Δe, in a particular period through the variation of each factor in Equation (2) as follows (Equation (3)):

$$\Delta e = \Delta J + \Delta Q + \Delta L + \Delta W + \Delta v + \Delta g + \Delta p \tag{3}$$

where ΔJ indicates changes in the energy structure; ΔQ represents changes in the energy intensity; ΔL means changes in the Leontief inverse matrix (*i.e.*, changes in the production structure); Δw is changes in the final demand structure; Δv is changes in the GDP composition; Δg is changes in the per capita GDP; and Δp is changes in the population.

In this study, changes in the CO_2 emissions have been decomposed into seven factors. De Boer [27] has developed the detailed model on which our work is based by utilizing two factors as an example. By extending De Boer's development in a similar manner to seven factors, we obtain the following: $e = cJQLwvgp$, where $e_i = \sum_{d=1}^{z} \sum_{j=1}^{n} \sum_{k=1}^{m} c_d j_{di} q_{ii} l_{ij} w_{jk} v_k g p$. Taking 0 and 1 to represent the base year and end year, respectively, and setting $e_{dijk} = c_d j_{di} q_{ii} l_{ij} w_{jk} v_k g p$, the following equations can be deduced (Equations (3a)–(3g)):

$$\Delta J = \sum_{d=1}^{z} \sum_{j=1}^{n} \sum_{k=1}^{m} \frac{e_{dijk}(1) - e_{dijk}(0)}{ln\left[e_{dijk}(1)/e_{dijk}(0)\right]} ln\left[j_{di}(1)/j_{di}(0)\right] \tag{3a}$$

$$\Delta Q = \sum_{d=1}^{z} \sum_{j=1}^{n} \sum_{k=1}^{m} \frac{e_{dijk}(1) - e_{dijk}(0)}{ln\left[e_{dijk}(1)/e_{dijk}(0)\right]} ln\left[q_{ii}(1)/q_{ii}(0)\right] \tag{3b}$$

$$\Delta L = \sum_{d=1}^{z} \sum_{j=1}^{n} \sum_{k=1}^{m} \frac{e_{dijk}(1) - e_{dijk}(0)}{ln\left[e_{dijk}(1)/e_{dijk}(0)\right]} ln\left[l_{ij}(1)/l_{ij}(0)\right] \tag{3c}$$

$$\Delta W = \sum_{d=1}^{z} \sum_{j=1}^{n} \sum_{k=1}^{m} \frac{e_{dijk}(1) - e_{dijk}(0)}{ln\left[e_{dijk}(1)/e_{dijk}(0)\right]} ln\left[w_{jk}(1)/w_{jk}(0)\right] \tag{3d}$$

$$\Delta v = \sum_{d=1}^{z} \sum_{j=1}^{n} \sum_{k=1}^{m} \frac{e_{dijk}(1) - e_{dijk}(0)}{ln\left[e_{dijk}(1)/e_{dijk}(0)\right]} ln\left[v_k(1)/v_k(0)\right] \tag{3e}$$

$$\Delta g = \sum_{d=1}^{z} \sum_{j=1}^{n} \sum_{k=1}^{m} \frac{e_{dijk}(1) - e_{dijk}(0)}{ln\left[e_{dijk}(1)/e_{dijk}(0)\right]} ln\left[g(1)/g(0)\right] \tag{3f}$$

$$\Delta p = \sum_{d=1}^{z} \sum_{j=1}^{n} \sum_{k=1}^{m} \frac{e_{dijk}(1) - e_{dijk}(0)}{ln\left[e_{dijk}(1)/e_{dijk}(0)\right]} ln\left[p(1)/p(0)\right] \tag{3g}$$

The model is characterized by simple expressions, and the calculations are straight-forward. However, difficulty in applying the model arises when there is a zero or negative number in the dataset. To solve this problem, we follow the practice of De Boer [27] and replace the zero value in the preliminary dataset with 10^{-29}. For negative numbers, we follow the processing method of Ang and Liu [35].

From the consumption perspective, CO_2 emissions can be allocated to the sectors whose products were consumed by the final demand categories [36,37], which can be expressed mathematically as (Equation (4)):

$$e^f = F^f(I-A)^{-1}y^f = Ty^f \tag{4}$$

where $e^f = (e_i^f)$ is the $1 \times n$ vector of CO_2 emissions caused by the n sectors' final demand consumption and e_i^f is the CO_2 emissions from the consumption of i-th sector's products. This means that the CO_2 emission responsibilities are allocated to the consumers under Equation (4). In addition, $F^f = (f_i^f)$ is a $1 \times n$ vector, where the element f_i^f is the DCI of sector i. $T = F^f(I-A)^{-1} = (t_i)$ is a $1 \times n$ vector, where the element t_i is the total carbon intensity (TCI) of sector i, which represents CO_2 emissions from sector i's supply chains (including direct and indirect emissions) by the final demand unit products' value from sector i. Therefore, indirect carbon intensity (ICI) can be obtained by TCI minus DCI. $y^f = (y_{ij}^f)$ is an $n \times n$ diagonal matrix, where the diagonal element y_{ii}^f represents the final demand of sector i.

2.2. Data Sources and Processing

Two datasets were employed in our analysis: the input-output tables (IOTs) in time series and the sectoral energy use and CO_2 emission factor statistics. In China, provincial IOTs are published every five years, and the most recent IOT is from 2012. Yunnan's IOTs for 1997, 2002, 2007 and 2012 were obtained from the Yunnan Statistical Bureau. The 1997 Table contained 40 sectors, whereas other tables contained 42 sectors. Nineteen types of energy sources were included in the calculations: raw coal, cleaned coal, other washed coal, briquettes, coke, coke oven gas, other gas, crude oil, gasoline, kerosene, diesel, fuel oil, liquefied petroleum gas, natural gas, other petroleum products, other coke products, heat, electricity and other energy. The energy data for agriculture, industry, construction, transport, storage and post, wholesale, retail trade and hotels, restaurants and other services originated from Yunnan's energy balance sheets from 1997, 2002, 2007 and 2012, which can be found in the China Energy Statistical Yearbooks for the various years [38]. Energy data for the industrial sub-sectors for the years 2002, 2007 and 2012 came from the Yunnan Energy Statistical Yearbook [39]. Energy data for the industrial sub-sectors in 1997 were missing. To compensate, these data were estimated based on the total industrial energy data that are provided in available energy balance sheets and the energy proportion of industrial sub-sectors in 2000.

The energy and CO_2 emission statistics were processed following the method proposed by Peters *et al.* [40] and the National Development and Reform Commission of China [31]. Because coal usage in China is usually inefficient [41], instead of the default values of the Intergovernmental Panel on Climate Change (IPCC), we employed the specific CO_2 emission factors (including carbon content, net calorific value and carbon oxidation factors), which are more suitable for China [31]. The final demand columns of Yunnan's 2007 and 2012 IOTs consisted of rural residential consumption, urban residential consumption, government consumption, fixed capital formation, stock changes, interprovincial and international exports and interprovincial and international imports. One of the final demand columns in the Yunnan 1997 IOT is called "others", which represents the margin of error among different data sources [20]. In our analysis, each sector's total output in 1997 was calculated as the sum of intermediate deliveries and final demand; thus, the "others" column was excluded [12]. Because Yunnan's 1997 and 2002 IOTs did not differentiate interprovincial and international trade, they were combined into two columns titled exports and imports. Since we do not know the detailed sources of imports, we assumed that all imports are produced under the same conditions in Yunnan; therefore, the emissions embodied in these imports were designated as the "emissions avoided in Yunnan through imports" [20]. By synthesizing sectors in the IOTs and sectors with available energy data, all of the sectors were combined into a total of 29 sectors according to "classification of national economic industries (GB/T 4754-2011, GB/T represents the recommended national standards and it is the abbreviation of Guo Biao/Tui in Chinese.)" [42] (Table 1). To avoid the influence of price changes, we utilized the double deflation method [43] and transformed all of the IOTs into constant prices in 2002. Price index data were obtained from the Yunnan Survey Yearbook 2014 [5], and population data were taken from the Yunnan Statistical Yearbook [4] (see more discussions on the double deflation method in the SM).

After comprehensively considering the availability of sectoral energy data and the consistency of sector classification in IOTs from different years, 29 sectors were finally adopted in this study. However, this would cause sector aggregation issues mentioned by Su *et al.* [44]. Through comparing the results of CO_2 emissions embodied in China and Singapore's exports under different sector aggregation situations, Su *et al.* [44] suggested a sector aggregation of more than 40 sectors. If so, the 29 sector aggregation in this study might cause some uncertainties to the results. In addition, the extended IOTs in 2000, 2005 and 2010 were not compiled in Yunnan, instead being done at the national scale or for some developed regions, such as Beijing. Therefore, the time intervals of IOTs were determined as 5 years in this study. This would cause the temporal aggregation issues mentioned by Su and Ang [45], and also cause some uncertainties to the results. Furthermore, the trend of changes in the drivers would be missed and regarded as some kind of average values within the sub-periods 1997–2002, 2002–2007 and 2007–2012, respectively. Missing some peculiarities of the drivers would increase the uncertainties when the results were discussed. As these uncertainties existed, the results in this study were regarded as approximate values. However, these uncertainties were unlikely to extremely affect the scale and trend of the results in any one direction [46]. Keeping these uncertainties in mind, detailed and in-depth analysis and discussions were conducted by combining our results with other statistics and materials in both official publications and literature, before policy suggestions were made. The results and discussions could be improved when more detailed data are available.

Table 1. Sectoral carbon intensity in 1997, 2002, 2007 and 2012 in Yunnan (unit: tons/10^4 Yuan of 2002 constant price).

	Sectors	Direct Carbon Intensity				Indirect Carbon Intensity				Total Carbon Intensity			
		1997	2002	2007	2012	1997	2002	2007	2012	1997	2002	2007	2012
1-AGR	Agriculture	0.3	0.4	0.4	0.3	1.2	2.0	1.4	1.4	1.5	2.4	1.8	1.7
2-MWC	Mining and washing of coal	8.2	4.1	2.0	0.2	1.4	7.5	5.6	4.1	9.6	11.6	7.6	4.3
3-EPN	Extraction of petroleum and natural gas	0.0	0.0	0.0	0.0	0.0	5.4	0.0	0.0	0.0	5.4	0.0	0.0
4-MPM	Mining and processing of metal ores	1.4	1.6	1.7	1.1	2.8	7.2	6.5	5.5	4.2	8.8	8.2	6.6
5-MPN	Mining and processing of nonmetal ores	1.1	1.1	3.8	1.6	3.5	6.5	6.8	5.5	4.6	7.6	10.6	7.1
6-MFT	Processing and manufacture of food and tobacco	0.2	0.1	0.2	0.1	0.7	1.8	1.0	0.8	0.9	1.9	1.2	0.9
7-MOT	Manufacture of textiles	0.4	0.5	2.7	1.0	2.1	4.4	3.6	2.3	2.5	4.9	6.3	3.3
8-MCL	Manufacture of clothes, leather and related products	0.3	0.3	0.5	0.1	2.3	3.5	3.1	1.7	2.6	3.8	3.6	1.8
9-MWF	Manufacture of wood products and furniture	0.2	0.5	0.6	0.1	4.5	5.0	3.0	1.3	4.7	5.5	3.6	1.4
10-PMP	Paper making, printing and articles manufacture	0.5	0.4	0.8	0.5	3.5	4.0	2.1	2.0	4.0	4.4	2.9	2.5
11-PPC	Petroleum processing and coking	6.3	10.6	1.5	1.2	4.5	7.4	3.6	2.8	10.8	18.0	5.1	4.0
12-CHE	Chemistry	3.9	4.0	1.9	1.9	5.9	6.7	5.9	3.7	9.8	10.7	7.8	5.6
13-MNM	Manufacture of non-metallic mineral products	6.0	4.6	15.5	7.0	5.2	6.9	7.7	5.3	11.2	11.5	23.2	12.3
14-SPM	Smelting and pressing of metals	5.8	5.2	2.8	2.8	6.9	8.1	4.2	3.6	12.7	13.3	7.0	6.4
15-MMP	Manufacture of metal products	0.2	0.3	0.2	0.1	8.0	8.5	4.3	3.3	8.2	8.8	4.5	3.4
16-MGS	Manufacture of general and special purpose machinery	0.3	0.3	0.2	0.1	3.8	6.8	3.5	3.2	4.1	7.1	3.7	3.3

Table 1. *Cont.*

Sectors		Direct Carbon Intensity				Indirect Carbon Intensity				Total Carbon Intensity			
		1997	2002	2007	2012	1997	2002	2007	2012	1997	2002	2007	2012
17-MTE	Manufacture of transportation equipment	0.3	0.2	0.1	0.2	4.1	5.9	2.7	2.6	4.4	6.1	2.8	2.8
18-MEM	Manufacture of electrical machinery and equipment	0.0	0.1	0.0	0.0	5.9	6.7	4.6	3.8	5.9	6.8	4.6	3.8
19-MCE	Manufacture of communication and electronic equipment	0.9	0.5	0.0	0.0	3.8	4.5	2.5	1.4	4.7	5.0	2.5	1.4
20-IMC	Manufacture of measuring instruments and machinery for cultural activity and office work	0.0	0.0	0.0	0.0	4.4	5.6	2.8	1.9	4.4	5.6	2.8	1.9
21-MAO	Manufacture of artwork and other manufacturing	0.0	0.1	0.3	0.7	4.2	6.1	3.2	2.4	4.2	6.2	3.5	3.1
22-RDW	Recycling and disposal of waste	0.1	0.1	0.1	0.0	0.0	0.0	0.0	0.3	0.1	0.1	0.1	0.3
23-PSE	Production and supply of electric and heat power	30.2	11.1	12.9	6.8	3.9	5.2	8.7	4.5	34.1	16.3	21.6	11.3
24-PSG	Production and supply of gas	5.3	2.8	4.6	0.1	5.5	7.0	3.7	1.4	10.8	9.8	8.3	1.5
25-PSW	Production and supply of water	0.1	0.0	0.0	0.0	2.9	5.1	1.6	2.5	3.0	5.1	1.6	2.5
26-CON	Construction	0.1	0.1	0.1	0.1	5.4	7.2	5.8	4.5	5.5	7.3	5.9	4.6
27-TSP	Transport, storage and post	1.5	3.3	3.7	4.1	2.8	6.3	2.3	2.2	4.3	9.6	6.0	6.3
28-WRH	Wholesale, retail trade and hotel, restaurants	0.1	0.1	0.1	0.2	1.6	2.1	1.4	1.2	1.7	2.2	1.5	1.4
29-OSE	Other services	0.1	0.1	0.0	0.1	1.5	2.7	1.7	1.5	1.6	2.8	1.7	1.6

3. Results and Discussion

3.1. Overall Trend of CO_2 Emission Variations in Yunnan from 1997–2012

From 1997–2012, Yunnan's industrial CO_2 emissions grew from 47.8 Mt–187.6 Mt, a four-fold increase. Most of this total growth of 139.8 Mt was concentrated in the period 2002–2007. In this period, CO_2 emissions grew by 80.9 Mt, which constitutes 57.9% of the growth that occurred in the 1997–2012 period, with the remaining 28.2% (39.5 Mt) in 2007–2012 and 13.9% (19.4 Mt) in 1997–2002.

From a production perspective, 23-PSE (production and supply of electric and heat power) (33.7%), 14-SPM (smelting and pressing of metals) (29.6%), 27-TSP (transport, storage and post) (12.3%) and 13-MNM (manufacture of non-metallic mineral products) (10.6%) contributed 86.2% (120.5 Mt) of the total CO_2 emissions increase during 1997–2012 (see Table S1 in SM). These sectors have a high DCI in their production processes (Table 1) and their contributions to the CO_2 emissions increase were much larger than their contributions to GDP growth (Table S1). From a final demand perspective, 26-CON (construction) (64.8%), 14-SPM (35.7%), 29-OSE (other services) (19.6%) and 23-PSE (10.7%) contributed 130.9% (183.0 Mt) of the total CO_2 emissions increase (Table S2 in SM). 14-SPM and 23-PSE not only have a high DCI, but also a high TCI (Table 1). The DCI for 26-CON and 29-OSE is extremely low, whereas their TCI is much higher than their DCI (Table 1). The expansion of the final demand volume of products in these sectors will boost their carbon-intensive supply chains. For example, the final demand sector 26-CON increased its CO_2 emissions by 90.7 Mt during 1997–2012 with 28.2% contributed by 13-MNM, 25.9% by 23-PSE, 23% by 14-SPM and 10.4% by 27-TSP in the construction supply chains (Figure 2). Compared to most other sectors, 29-OSE has lower carbon intensity in both DCI and TCI (Table 1). The increments of expenditure on products from 29-OSE have a 33.7% share in GDP, while contributing only 19.6% to the CO_2 emissions increase (Table S2) (see more analysis about sectoral carbon intensity changes in the SM).

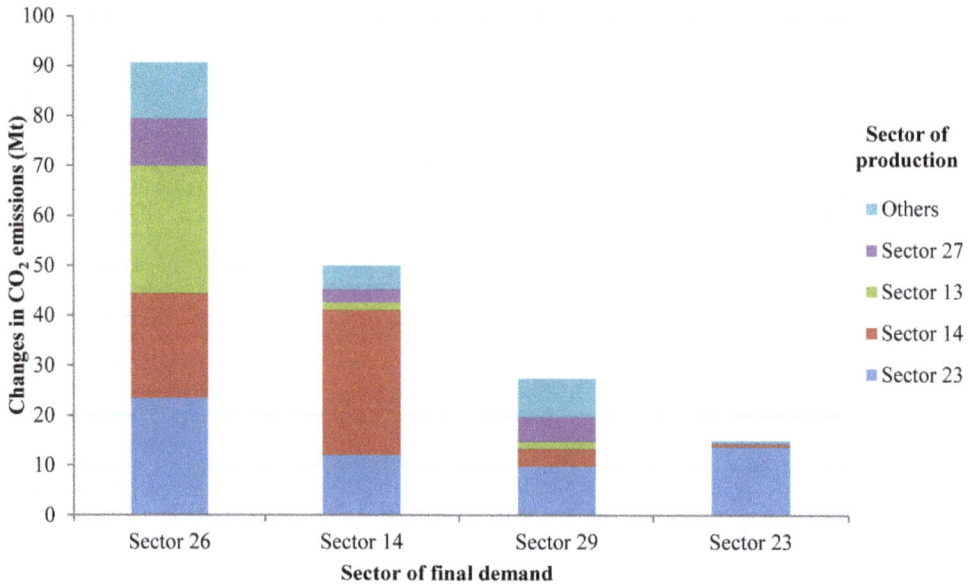

Figure 2. Major sectors contributing to the CO_2 emissions increase from the final demand perspective during 1997–2012 and their carbon-intensive supply chains.

3.2. Drivers from Decomposition Analysis

The largest driver of changes in CO_2 emissions was the growth of per capita GDP, leading to a 172.5-Mt CO_2 emissions increase from 1997–2012 (Figure 3). In the previous decades, Yunnan had been struggling to develop its economy and reduce poverty. The per capita GDP of Yunnan in 2012 was more than 3.5-times that of 1997 [4]. Yunnan's per capita GDP is expected to continue increasing to catch up with other developed regions in China, which will signify its contribution to CO_2 emission increments in the following decades.

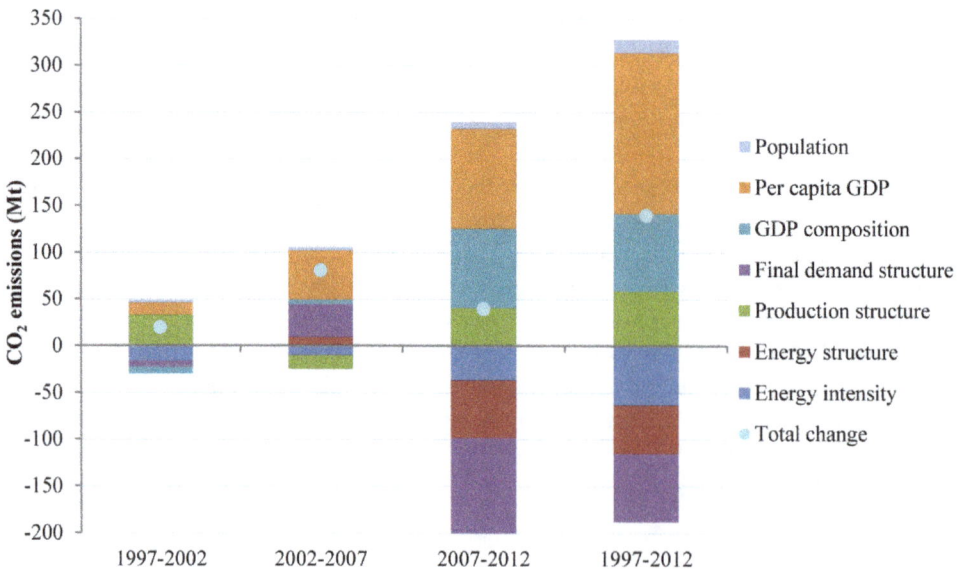

Figure 3. Drivers contributing to changes in Yunnan's industrial CO_2 emissions in different time periods from a decomposition analysis.

The second largest driver was GDP composition change, causing an 82.7-Mt increase in CO_2 emissions from 1997–2012 (Figure 3). The sub-periods data showed a positive correlation between changes in the fraction of capital investment in GDP and CO_2 emissions (see Figure 3 and Table 2). The Yunnan government was aware of the problem and aimed to decrease capital investment in its economy since the 11th FYP period (2006–2010) [47], but this policy did not achieve the desired effect because of the Global Financial Crisis. Under the impact of the four-trillion-Yuan stimulus plan in China [48], the fraction of capital investment in GDP increased by 38.9% during 2007–2012 in Yunnan (Table 2). This GDP composition change caused an 84.9-Mt increase in CO_2 emissions during this period (Figure 3).

Table 2. Changes in the GDP composition in Yunnan from 1997–2012 (unit: %).

Year	Fraction of Investment in GDP	Fraction of Consumption in GDP	Fraction of Exports in GDP
1997	42.8	59.8	33.3
2002	39.8	68.4	38.1
2007	44.3	61.9	63.8
2012	83.2	61.2	51.5

Production structure modifications caused a 58.6-Mt increase in the total CO_2 emissions during 1997–2012 (Figure 3). This production structure change reflects the expansion of the heavy industry in Yunnan as it transitioned from an economy dominated by the tobacco industry [49]; the proportion of heavy industry among total industry increased from 47.8% in 1997 to 69.5% in 2012 [4]. During 2002–2007, changes in the production structure had a positive effect decreasing CO_2 emissions by 14.9 Mt (Figure 3). This seemed to be the policy dividend of Yunnan's promotion of development in the green economy since 2000 [50]. Relatively low-carbon industries, such as biological development and tourism, developed rapidly during this period [47], but these benefits were stopped by the 2008 Global Financial Crisis. Under China's four-trillion-Yuan stimulus plan [48], capital investments (most in the construction sector) in Yunnan increased rapidly (Table 2), and the carbon-intensive supply chains of the construction sector increased sharply during 2007–2012. Carbonization of the production structure increased CO_2 emissions by 40.9 Mt during this period (Figure 3).

The driving effect of population growth was small; its contribution to the growth in CO_2 emissions was only 14.0 Mt from 1997–2012 (Figure 3). This was primarily due to the strict implementation of the "one-child policy" in Yunnan. In each FYP, a target is set to control the population growth rate. It is expected that Yunnan will continue following the population planning policy formulated by the central government to control the population growth rate. The contribution of changes in population to CO_2 emissions will remain small (see more information about the "one-child policy" in China in the SM).

From 1997–2012, the largest driver leading to a reduction in industrial CO_2 emissions was changes in the carbon intensity (the sum of changes in the energy intensity and energy structure; Figure 3). Of the 115.1-Mt CO_2 emissions offset by improvements in carbon intensity during this period, 45.1% was from 23-PSE, 37.1% was from 13-MNM and 12.0% was from 14-SPM (Table S3 in SM). However, a DCI increase in 27-TSP led to 5.9 Mt in carbon intensity-related CO_2 increments (Table 1 and Table S3). Figure 4 shows the changes in energy intensity and energy structure in these four energy-intensive sectors. It highlights the expansion of renewable electricity (mostly from hydropower) share in total electricity generation during 2007–2012. Hydropower generation in Yunnan grew seven-fold, from 16.8 terawatt hours in 1997 to 123.8 terawatt hours in 2012, and the share of hydropower in total electricity generation was 71% in 2012 [39]. The application of non-fossil energy in sectors 13-MNM and 14-SPM also increased through a "combination of mineral industry and hydropower" policy in Yunnan [47]. Changes in the energy structure led to a reduction of CO_2 emissions as much as 61.8 Mt during 2007–2012 (Figure 3 and Table S4). From a sector perspective, 14-SPM, 13-MNM and 23-PSE contributed to 26.2-, 16.2- and 13.4-Mt energy structure-related CO_2 emissions decreases during this period, respectively (Table S4). Although the process was lengthy and complex, the implementation of key energy-saving projects and the elimination of backward production capacity under a mandatory target set by the central government to encourage the provincial government to reduce energy intensity since the 11th FYP period (2006–2010) [47] resulted in significant improvements in energy intensity in key energy consumption sectors, such as 13-MNM, 14-SPM and thermal power. From 1997–2012, changes in the energy intensity reduced CO_2 emissions constantly by 63.1 Mt, with 58.0% of this reduction in 2007–2012, 25.8% in 1997–2002 and only 16.2% in 2002–2007 (Figure 3 and Table S4). From a sector perspective, 23-PSE and 13-MNM contributed 65.6% (−41.4 Mt) and 45.4% (−28.7 Mt) to energy intensity-related CO_2 emissions reduction during 1997–2012, respectively (Table S4). However, coal and petroleum still dominated energy use in the 13-MNM, 14-SPM and 27-TSP sectors. Meanwhile, energy intensity and structure improvements were mostly found in industrial sectors. For transportation (27-TSP), both the energy intensity and structure deteriorated. Road transportation composes more than 90% of total transportation in Yunnan, and most roads in Yunnan are labeled as low grade [4].

Optimization of the final demand structure was another driving force offsetting the increase of CO_2 emissions by 72.9 Mt between 1997 and 2012 (Figure 3). However, it has not always played a positive role in CO_2 mitigation. During 2002–2007, changes in the final demand structure increased CO_2 emissions by 35.3 Mt (Figure 3). The next section presents a detailed analysis showing how changes in the final demand structure affect CO_2 emissions in separate final demand categories.

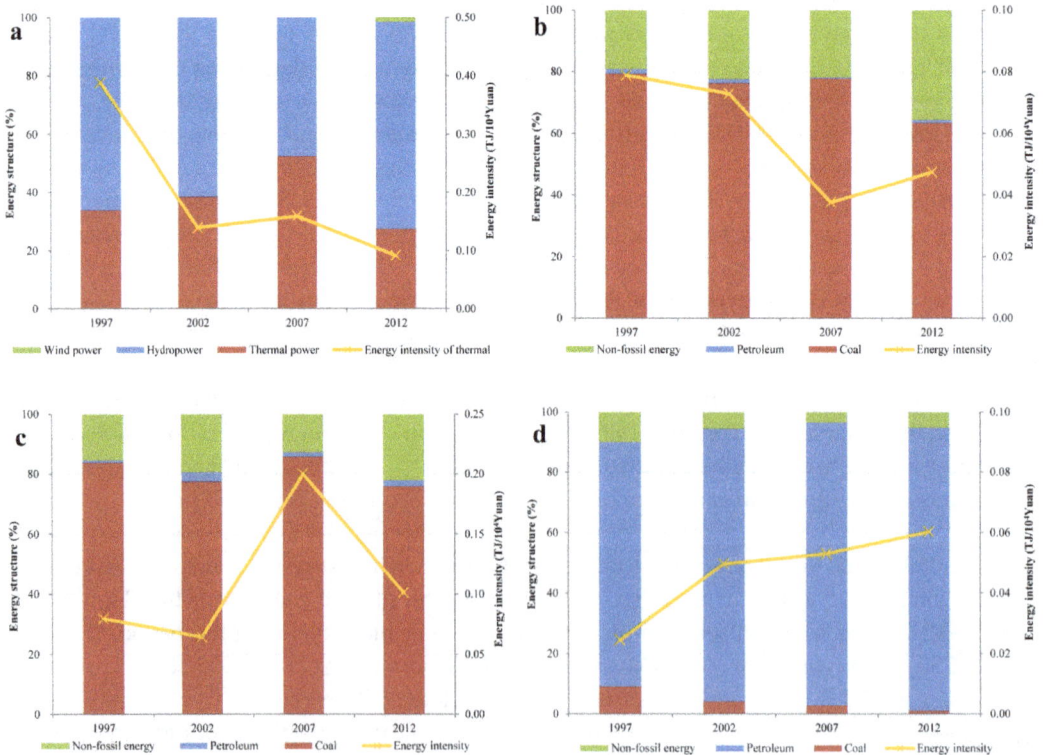

Figure 4. Energy intensity and structure change in major sectors from a production perspective from 1997–2012. (**a**) 23-PSE; (**b**) 14-SPM; (**c**) 13-MNM; and (**d**) 27-TSP.

3.3. Drivers from a Final Demand Perspective

From a final demand perspective, capital investment (the sum of fixed capital formation and stock changes) was the largest driving force between 1997 and 2012. It drove a CO_2 emissions increase of 195.3 Mt, 69.5% of which occurred in the period 2007–2012 (Figure 5). To reduce the impact of the Global Financial Crisis on the economy, the Chinese government adopted the four-trillion-Yuan stimulus plan, of which 45% was invested in constructing infrastructure, such as roads, airports and buildings [6]. This plan caused the growth of CO_2 emissions in construction investments in Yunnan to explode (Figures 6a and 7a).

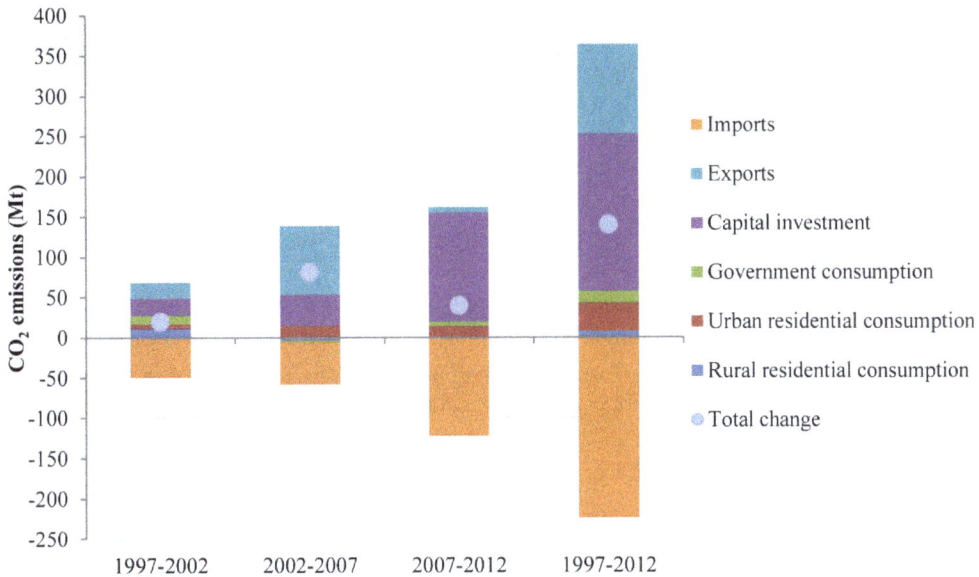

Figure 5. Drivers contributing to changes in Yunnan's industrial CO_2 emissions in different time periods when analyzed from a final demand perspective.

Exports were the second largest driver of the CO_2 emissions increase in Yunnan during 1997–2012. Of the 111.2-Mt increase in export-related CO_2 emissions during this period, 76.8% occurred in 2002–2007, 17.2% occurred in 1997–2002 and only 6.0% in 2007–2012 (Figure 5). Figures 6b and 7b demonstrate the main decomposition drivers and sectors contributing to the increase in export-related CO_2 emissions. They indicate that from 2002–2007, the export-related CO_2 emissions increase was caused primarily by growth in the exports of carbon-intensive products from 14-SPM and 23-PSE. During this period, exports from these two sectors increased by 1106.2% and 269.6%, respectively (Table S5). This change also caused carbonization of the exports' structure, which led to an increase of 40.5 Mt in export-related CO_2 during this period (Figure 6b). The share of carbon-intensive products from 14-SPM and 23-PSE in total exports increased from 9.4% and 2.7% in 2002 to 38.7% and 3.4% in 2007, respectively (Table S5). During 2007–2012, the situation was different. Although carbon-intensive product exports from 14-SPM and 23-PSE increased by 0.3% and 3.2%, respectively, their shares in total exports decreased to 26.6% and 2.4% in 2012 (Table S5). Low-carbon products from the agriculture and service sectors increased their share in total exports during the same period (Table S5). Changes in the exports' structure led to a decrease of 16.8 Mt CO_2 during this period (Figure 6b). Most of Yunnan's exports were raw materials to support the development of developed coastal regions in China. For example, Yunnan was important to China's "West to East Transmission Scheme", increasing deliveries of electricity to developed eastern areas (e.g., Guangdong Province) [3] from 0.2 terawatt hours in 1997 to 42.6 in 2012 [39].

Urban residential consumption was another important driver of the growth of CO_2 emissions. This was especially the case between 2002 and 2007, when urban residential consumption exceeded both rural residential consumption and government consumption and became the third largest driving force (Figure 5). The urban population of Yunnan grew from 23.4% of the total population in 2000 to 39.3% in 2012, increasing by 8.4 million in these twelve years [4]. The increase in urban population strengthened the driving effect of urban residential consumption, which increased CO_2 emissions by 34.5 Mt during 1997–2012 (Figure 5). Growth in consumption level was the primary driver of the 43.2-Mt increase in urban-related CO_2 emissions during 1997–2012 (Figure 6c). From a sectoral perspective, increased consumption of life-cycle carbon-intensive products, such as electricity, petroleum and cars, contributed to an urban-related CO_2 emissions increase in recent years (Figures 6c and 7c). The urbanization rate in Yunnan is far below the national average. In 2012, 39.3% of Yunnan's population was classified as urban, whereas the figure was 52.6% for the country as a whole [4]. With increasing urbanization, it is foreseeable that urban residential consumption in Yunnan will continue to strengthen as a driving force.

Rural residential consumption and government consumption had lesser driving effects during 1997–2012, increasing CO_2 emissions by only 7.9 Mt and 14.8 Mt, respectively (Figure 5). Meanwhile, for the period 1997–2012, increasing imports of carbon-intensive products from construction and heavy manufacturing sectors avoided a 224.0-Mt CO_2 emissions increase if these products were produced in Yunnan (see Figures 5, 6d and 7d). Nonetheless, it will increase the carbon footprint (CO_2 emissions caused by final demand within a region) of Yunnan. Furthermore, it will intensify the debate regarding the responsibility for the redistribution of CO_2 emissions [37]. Thus, fundamentally decreasing local carbon footprint should become the focus of policy makers.

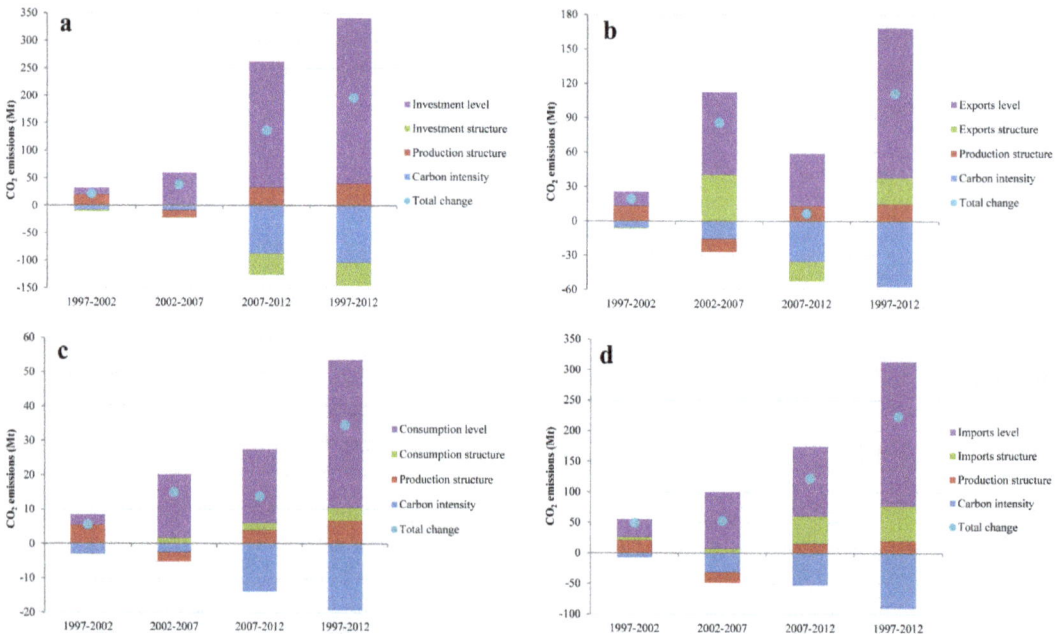

Figure 6. Main decomposition drivers contributing to changes in CO_2 emissions by different final demand categories from 1997–2012. (**a**) Capital investment; (**b**) exports; (**c**) urban residential consumption; and (**d**) imports.

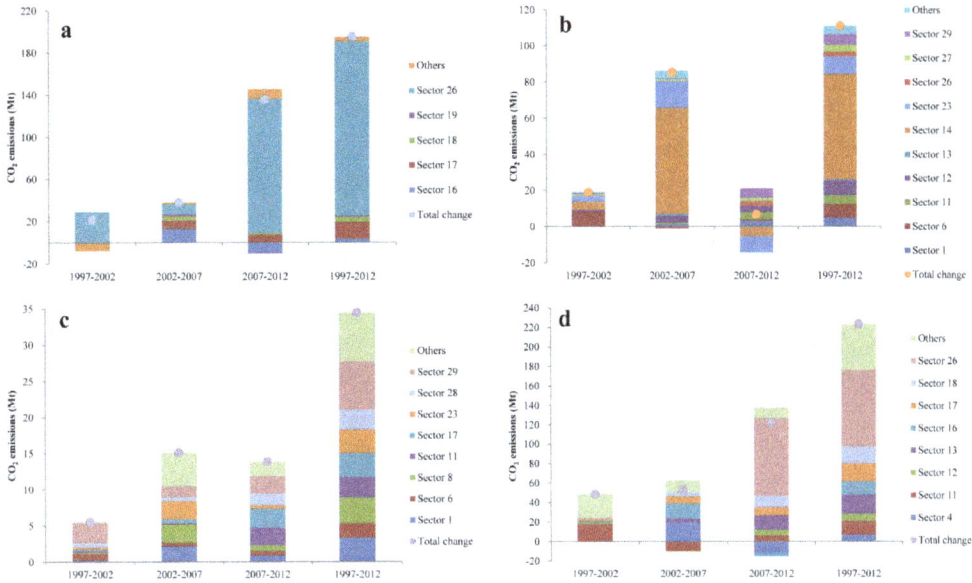

Figure 7. Main sectors contributing to changes in CO_2 emissions by different final demand categories from 1997–2012. (**a**) Capital investment; (**b**) exports; (**c**) urban residential consumption; and (**d**) imports.

3.4. Policy Recommendations for Further Mitigating CO_2 Emissions in Yunnan

Based on the above analysis, policy recommendations are suggested to further mitigate CO_2 emissions in Yunnan from the perspective of both production and final demand.

3.4.1. Policy Recommendations from a Production Perspective

Promote the Development and Application of Clean Energy

Developing hydropower substantially decarbonized Yunnan's energy structure in recent years, but applying clean energy in the production sectors is not enough. Thus, policy makers should formulate policies supporting the development of clean energy, such as hydropower, wind, solar and biomass, according to local conditions. However, some details should be considered in this process: (1) the cooperation among the Yunnan government, China's central government and power grid company; (2) the ecological environment problems coupled with clean energy development; and (3) the coordination of the layout of the clean energy and industrial sectors. In addition, building a distributed clean energy supply network can increase the use of clean energy and a greener energy structure in production sectors. Increasing the use of clean energy, such as hydropower, hydrogen and natural gas, in the transportation system can also reduce a sector's dependence on carbon-intensive petroleum products.

Further Improve the Energy Efficiency in Energy-Intensive Sectors

According to the front analysis, the decreased energy intensity in energy-intensive sectors was greatly due to the administrative measures, such as mandatory reduction targets. To further reduce sectoral energy intensity, more economic measures should be taken. If economic instruments (for example, financial subsidies and tax rebates) are integrated into sectoral low-carbon policies, the effects of technology improvements in reducing CO_2 emissions are expected to be critical. For less-developed regions, such as Yunnan, technology and funds for research and development (R&D) are limited. Therefore, the forming of mechanism to promote the transfer of funds and low-carbon technologies from developed regions is worth the consideration for policy makers.

The above analysis indicated that sectors 13-MNM, 14-SPM, 23-PSE and 27-TSP were the most energy-intensive sectors, and their production processes contributed over 80% of the CO_2 emissions increase during 1997–2012 in Yunnan. Therefore, strengthening the measures to improve energy efficiency in these sectors will play a more effective role in reducing CO_2 emissions. For the cement sector (13-MNM), typical measures, including upgrading the existing equipment (such as the rotary kilns, mills and dryers), developing new energy-saving technologies (such as heat recovery for power generation in rotary kilns, raw meal blending systems in dry process and process control systems in all kilns) and eliminating backward production techniques (such as wet kilns and dry hollow kilns), can be applied [51]. For the metal processing sector (14-SPM), the blast furnace top gas recovery turbine (TRT) unit, converter negative energy steelmaking technology, blast furnace gas and converter gas should be well applied and used in the iron and steel making process; oxygen-rich flash and oxygen-enriched bath smelting, large pre-baked electrolytic cells, oxygen bottom blown technology and wet processing technology should be applied in copper, aluminum, lead and zinc smelting, respectively. For the electricity sector (23-PSE), encouraging the construction of large coal-fired units over 600 megawatts (MW) and large-capacity gas combined cycle plants to replace smaller and less efficient units is essential. Actively developing a smart and ultra-high voltage power grid and phasing out the low voltage electricity distribution network can reduce the electricity loss in the transmission process. For Yunnan's carbon-intensive transportation system (27-TSP), strengthening the relatively low-carbon railway transportation system can reduce the dependence on high-carbon road transportation. Meanwhile, upgrading the road transportation system (such as building more high-grade roads), improving the fuel economy of vehicles and promoting intelligent traffic management systems are also important measures to improve the energy efficiency in the transportation sector [8].

Adjust the Industrial Structure According to the Total Carbon Intensity of Sectors and Local Advantageous Resources

It is very important to construct a low-carbon industrial system for Yunnan's low-carbon development. Both sides, including the carbon intensity of the entire industrial supply chain (*i.e.*, total carbon intensity of a sector) (Table S6) and the combination with local advantageous resources, should be considered in this process. Thus, policy measures should continue to promote low-carbon industries, such as plateau characteristic agriculture (1-AGR), clean power, bio-resources (6-MFT (processing and manufacture of food and tobacco)), tourism (29-OSE) and ethnic culture (29-OSE). Vigorously developing the productive service industry (29-OSE) not only increases the economic share of the low-carbon service industry, but also promotes the development of technology-intensive industries. In addition, the policy encouraging the transfer of labor-intensive (7-MOT (manufacture of textiles), 8-MCL (manufacture of clothes, leather and related products), 10-PMP (paper making, printing and articles manufacture) and 28-WRH (wholesale, retail trade and hotel, restaurants)) and technology-intensive industries (19-MCE (manufacture of communication and electronic equipment) and 22-RDW (recycling and disposal of waste)) to Yunnan from developed regions can gradually replace resource-intensive industries and create a lighter production structure.

3.4.2. Policy Recommendations from a Final Demand Perspective

Reduce the Dependence of Economic Growth on Investments and Exports and Encourage a Consumption-Based Economic Growth

Capital investments in infrastructure construction and exports of raw materials not only carbonized the GDP composition, but also formed a carbon-intensive production structure due to the high-carbon supply chains. Compared to capital investments and exports, consumption is more dependent on the relatively low-carbon service and agriculture industries, which contribute less to CO_2 emissions' growth (Table S7). Therefore, policy makers should vigorously promote policies that are conducive to consumption-based economic growth. For example, improving the minimum wage standards and establishing a sound social security system can effectively promote the growth of consumer spending.

Optimize the Life-Cycle Supply Chains for Final Demand Products

For the carbon-intensive supply chain in the construction, metal processing and electricity sectors, strengthening the recycling of resources in the production process can reduce new production of high-carbon raw materials, such as coal, cement, steel and electricity. Thus, improving the standards for clean production and developing circular economy from a life-cycle perspective in a sector's supply chain is important.

Adjust the Final Demand Structure

The adjustment of the final demand structure is very important for Yunnan's low-carbon development. Policy makers should focus on adjusting the export structure, the consumption structure of urban residents and the infrastructure investment direction.

Taking effective measures to reduce the exports of high-carbon raw materials is very important. China's central government had canceled export tax rebates and levied export tariffs for international exports of products with high resource consumption and high emissions. For interprovincial exports, there are no export tariffs, because the central government promotes free trade among all regions of the country. On this background, policies that subsidize the exports of low-carbon products from the service and agriculture industries and strengthen cooperation among trade partners are necessary. Considering that Yunnan exports a great deal of electricity to Guangdong, a compensatory mechanism should be formed to encourage Guangdong to provide renewable energy technology to Yunnan to reduce the carbon intensity of Yunnan's exports.

Optimizing the consumption structure for urban residents is essential. For example, the implementation of new established policies by the Chinese government, such as a ladder price for electricity, higher taxes on oil and a ban on disposable plastic bags, will encourage a low-carbon consumption structure in Yunnan. For urban transportation, it is imperative to increase investments in public transportation and to improve low-carbon transportation, such as walking and bicycling, to reduce the dominance of cars.

The direction of infrastructure investments will play a vital role in decreasing Yunnan's CO_2 emissions. For example, investments in mass transit or renewable energy production will benefit future low-carbon development. Thus, a low-carbon investment plan should be developed to guide long-term infrastructure investments.

4. Conclusions

In this study, a structural decomposition analysis-logarithmic mean Divisia index (SDA-LMDI) model was developed to find the drivers behind the CO_2 emission changes during 1997–2012 in Yunnan, a low-carbon pilot province in China. The potential relationship between economic growth and CO_2 emissions in a less-developed region was explored in order to promote the low-carbon economic development strategy.

As a less-developed region undergoing rapid development of industrialization and urbanization, the growth of per capita GDP driving by the export and capital investment increase was the primary driver of CO_2 emissions' growth in Yunnan. However, changes in the export and investment structure have become more important driving forces for CO_2 emissions' growth in Yunnan in recent years. Therefore, adjusting the economic structure to a low-carbon direction should have more attention paid to it. Vigorously developing hydropower and improving energy efficiency in energy-intensive sectors have become the backbone of CO_2 emissions' reduction. Thus, enhancing the application of clean energy and energy efficient technologies in production sectors can be essential measures to decease the CO_2 emissions.

Maintaining economic growth and eliminating poverty are still important tasks for Yunnan. However, the pillar industries of Yunnan's economy are high-carbon product sectors depending more on imports from other regions, which are not conducive to the low-carbon development in Yunnan from a long-term perspective. Hence, economic structure adjustment and technological improvements should be encouraged in the future. Based on this analysis, the further policy recommendations from both production and final demand perspectives to mitigate CO_2 emissions in Yunnan are proposed. The most important measures should focus on promoting the development and application of clean energy and the formatting of consumption-based economic growth.

This study provided a theoretical basis for developing a low-carbon policy in Yunnan by revealing driving forces behind changes in local CO_2 emissions, which also offered guidance for a low-carbon development mode in other less-developed regions undergoing rapid development of industrialization and urbanization. However, key economic entities in Yunnan's CO_2 emissions, such as major cities and industrial parks, were not analyzed due to the lack of relevant input-output tables in China. The ecological network analysis procedures might be helpful in compiling the input-output tables for the economic entities; however, more studies are needed in the future. Furthermore, as for the SDA-LMDI model, some improvements should be made in the future. For example, the use of more than two matrices or logarithms in the model will complicate the calculations; thus, a strengthened programming of this model is needed in the future.

Acknowledgments: We are grateful for the contributions of emeritus Prof. Robert B. Wenger from the University of Wisconsin-Green Bay, for his kind help in providing language and structural improvements to the manuscript. Thanks for the support provided by the Grant-Funded Projects of the China Clean Development Mechanism Fund (fund-key: 1213075) supervised by the National Development and Reform Commission, China. Additionally, we also want to thank the three anonymous reviewers for their valuable comments on this paper.

Author Contributions: Mingxiang Deng and Wei Li designed the framework of this research. Mingxiang Deng and Yan Hu collected the initial data and processed the data. Mingxiang Deng and Wei Li discussed the results and formed the policy recommendations and conclusions. All the authors wrote the paper.

Conflicts of Interest: The authors declare no conflict of interest.

References

1. Friedlingstein, P.; Andrew, R.M.; Rogelj, J.; Peters, G.P.; Canadell, J.G.; Knutti, R.; Luderer, G.; Raupach, M.R.; Schaeffer, M.; van Vuuren, D.P.; *et al.* Persistent growth of CO_2 emissions and implications for reaching climate targets. *Nat. Geosci.* **2014**, *7*, 709–715. [CrossRef]

2. Qi, Y. *Annual Review of Low-Carbon Development in China (2013)*; Social Science Academic Press: Beijing, China, 2013. (In Chinese)

3. Hennig, T.; Wang, W.L.; Feng, Y.; Ou, X.K.; He, D.M. Review of Yunnan's hydropower development. Comparing small and large hydropower projects regarding their environmental implications and socio-economic consequences. *Renew. Sustain. Energy Rev.* **2013**, *27*, 585–595. [CrossRef]

4. Statistical Bureau of Yunnan Province (SBYP). *Yunnan Statistical Yearbook 1998–2013*; China Statistics Press: Beijing, China, 2013.

5. National Bureau of Statistics Survey Office in Yunnan (NBSSOY). *Yunnan Survey Yearbook 2014*; China Statistics Press: Beijing, China, 2014.

6. Guan, D.B.; Klasen, S.; Hubacek, K.; Feng, K.S.; Liu, Z.; He, K.B.; Geng, Y.; Zhang, Q. Determinants of stagnating carbon intensity in China. *Nat. Clim. Chang.* **2014**, *4*, 1017–1023. [CrossRef]

7. Tollefson, J. US-China climate deal raises hopes for Lima talks. *Nature* **2014**, *515*, 473–474. [CrossRef] [PubMed]

8. Yunnan Development and Reform Commission (YDRC). Low-carbon development plan in Yunnan (2011–2020). 2011. Available online: http://www.yn.gov.cn/yn_zwlanmu/yn_srmzfgb/201108/t20110804_537.html (accessed on 13 November 2014). (In Chinese).

9. Minx, J.C.; Baiocchi, G.; Peters, G.P.; Weber, C.L.; Guan, D.B.; Hubacek, K. A "Carbonizing Dragon": China's fast growing CO_2 emissions revisited. *Environ. Sci. Technol.* **2011**, *45*, 9144–9153. [CrossRef] [PubMed]

10. Wang, Y.F.; Zhao, H.Y.; Li, L.Y.; Liu, Z.; Liang, S. Carbon dioxide emission drivers for a typical metropolis using input-output structural decomposition analysis. *Energy Policy* **2013**, *58*, 312–318. [CrossRef]

11. Geng, Y.; Zhao, H.Y.; Liu, Z.; Xue, B.; Fujita, T.; Xi, F.M. Exploring driving factors of energy-related CO_2 emissions in Chinese provinces: A case of Liaoning. *Energy Policy* **2013**, *60*, 820–826. [CrossRef]

12. Liang, S.; Zhang, T.Z. What is driving CO_2 emissions in a typical manufacturing center of South China? The case of Jiangsu Province. *Energy Policy* **2011**, *39*, 7078–7083. [CrossRef]

13. Guan, D.B.; Barker, T. Low-carbon development in the least developed region: a case study of Guangyuan, Sichuan province, southwest China. *Nat. Hazards* **2012**, *62*, 243–254. [CrossRef]

14. Tang, X.L.; Zhang, Y.Y.; Yi, H.H.; Ma, J.Y.; Pu, L. Development a detailed inventory framework for estimating major pollutants emissions inventory for Yunnan Province, China. *Atmos. Environ.* **2012**, *57*, 116–125. [CrossRef]

15. Hoekstra, R.; van der Bergh, J. Comparing structural and index decomposition analysis. *Energy Econ.* **2003**, *25*, 39–64. [CrossRef]

16. Su, B.; Ang, B.W. Structural decomposition analysis applied to energy and emissions: Some methodological developments. *Energy Econ.* **2012**, *34*, 177–188. [CrossRef]

17. Chen, L.; Yang, Z.F.; Chen, B. Decomposition analysis of energy-related industrial CO_2 emissions in China. *Energies* **2013**, *6*, 2319–2337. [CrossRef]

18. Lu, J.C.; Fan, W.G.; Meng, M. Empirical research on China's carbon productivity decomposition model based on multi-dimensional factors. *Energies* **2015**, *8*, 3093–3117. [CrossRef]

19. Pan, L.Y.; Guo, Z.; Liu, P.; Ma, L.W.; Li, Z. Comparison and analysis of macro energy scenarios in China and a decomposition-based approach to quantifying the impacts of economic and social development. *Energies* **2013**, *6*, 3444–3465. [CrossRef]

20. Peters, G.P.; Weber, C.L.; Guan, D.B.; Hubacek, K. China's growing CO_2 emissions-A race between increasing consumption and efficiency gains. *Environ. Sci. Technol.* **2007**, *41*, 5939–5944. [CrossRef] [PubMed]

21. Leontief, W. *Input-Output Economics*; Oxford University Press: New York, NY, USA, 1986.

22. Guan, D.B.; Hubacek, K.; Weber, C.L.; Peters, G.P.; Reiner, D.M. The drivers of Chinese CO_2 emissions from 1980 to 2030. *Glob. Environ. Change* **2008**, *18*, 626–634. [CrossRef]

23. Tian, X.; Chang, M.; Tanikawa, H.; Shi, F.; Imura, H. Structural decomposition analysis of the carbonization process in Beijing: A regional explanation of rapid increasing carbon dioxide emission in China. *Energy Policy* **2013**, *53*, 279–286. [CrossRef]

24. Dietzenbacher, E.; Los, B. Structural decomposition techniques: sense and sensitivity. *Econ. Syst. Res.* **1998**, *10*, 307–324. [CrossRef]

25. Su, B.; Ang, B.W. Attribution of changes in the generalized Fisher index with application to embodied emission studies. *Energy* **2014**, *69*, 778–786. [CrossRef]

26. Su, B.; Ang, B.W. Multiplicative decomposition of aggregate carbon intensity change using input–output analysis. *Appl. Energy* **2015**, *154*, 13–20. [CrossRef]

27. De Boer, P. Additive structural decomposition analysis and index number theory: an empirical application of the Montgomery decomposition. *Econ. Syst. Res.* **2008**, *20*, 97–109. [CrossRef]

28. Chai, J.; Guo, J.E.; Wang, S.Y.; Lai, K.K. Why does energy intensity fluctuate in China? *Energy Policy* **2009**, *37*, 5717–5731. [CrossRef]

29. Wachsmann, U.; Wood, R.; Lenzen, M.; Schaeffer, R. Structural decomposition of energy use in Brazil from 1970 to 1996. *Appl. Energy* **2009**, *86*, 578–587. [CrossRef]

30. Wood, R. Structural decomposition analysis of Australia's greenhouse gas emissions. *Energy Policy* **2009**, *37*, 4943–4948. [CrossRef]

31. National Development and Reform Commission (NDRC). *China Greenhouse Gas Inventory Study 2005*; China Environmental Press: Beijing, China, 2014. (In Chinese).

32. Su, B.; Ang, B.W.; Low, M. Input-output analysis of CO_2 emissions embodied in trade and the driving forces: Processing and normal exports. *Ecol. Econ.* **2013**, *88*, 119–125. [CrossRef]

33. Su, B.; Ang, B.W. Input-output analysis of CO_2 emissions embodied in trade: Competitive *versus* non-competitive imports. *Energy Policy* **2013**, *56*, 83–87. [CrossRef]

34. Hubacek, K.; Feng, K.S.; Chen, B. Changing lifestyles towards a low carbon economy: an IPAT analysis for China. *Energies* **2012**, *5*, 22–31. [CrossRef]

35. Ang, B.W.; Liu, N. Negative-value problems of the logarithmic mean Divisia index decomposition approach. *Energy Policy* **2007**, *35*, 739–742. [CrossRef]

36. Davis, S.J.; Caldeira, K. Consumption-based accounting of CO_2 emissions. *Proc. Natl. Acad. Sci. USA* **2010**, *107*, 5687–5692. [CrossRef] [PubMed]

37. Peters, G.P. From production-based to consumption-based national emission inventories. *Ecol. Econ.* **2008**, *65*, 13–23. [CrossRef]

38. National Bureau of Statistics (NBS). *China Energy Statistical Yearbook 1998–2013*; China Statistics Press: Beijing, China, 2013.

39. Statistical Bureau of Yunnan Province (SBYP). *Yunnan Energy Statistical Yearbook 2000–2013*; Yunnan Science and Technology Press: Kunming, China, 2013.

40. Peters, G.P.; Weber, C.L.; Liu, J.R. *Construction of Chinese Energy and Emissions Inventory*; Norwegian University of Science and Technology: Trondheim, Norway, 2006.

41. Guan, D.B.; Liu, Z.; Geng, Y.; Lindner, S.; Hubacek, K. The gigatonne gap in China's carbon dioxide inventories. *Nat. Clim. Chang.* **2012**, *2*, 672–675. [CrossRef]

42. National Bureau of Statistics (NBS). Classification of national economic industries (GB/T 4754–2011). 2011. Available online: http://www.stats.gov.cn/tjsj/tjbz/hyflbz/ (accessed on 13 November 2014) . (In Chinese).

43. United Nations Department for Economic and Social Affairs Statistics Division (UNDESASD). *Hand Book of Input-Output Table Compilation and Analysis*; United Nations: New York, NY, USA, 1999.

44. Su, B.; Huang, H.C.; Ang, B.W.; Zhou, P. Input-output analysis of CO_2 emissions embodied in trade: The effects of sector aggregation. *Energy Econ.* **2010**, *32*, 166–175. [CrossRef]

45. Su, B.; Ang, B.W. Structural decomposition analysis applied to energy and emissions: aggregation issues. *Econ. Syst. Res.* **2012**, *24*, 299–317. [CrossRef]

46. Weber, C.L.; Peters, G.P.; Guan, D.B.; Hubacek, K. The contribution of Chinese exports to climate change. *Energy Policy.* **2008**, *36*, 3572–3577. [CrossRef]

47. People's Government of Yunnan Province (PGYP). The Eleventh Five Year Plan for economic and social development in Yunnan province. 2006. Available online: http://xxgk.yn.gov.cn/bgt_Model1/newsview.aspx?id=130772 (accessed on 13 November 2014). (In Chinese).

48. Yuan, C.Q.; Liu, S.F.; Xie, N.M. The impact on Chinese economic growth and energy consumption of the Global Financial Crisis: An input-output analysis. *Energy* **2010**, *35*, 1805–1812. [CrossRef]

49. Eng, I. Agglomeration and the local state: The tobacco economy of Yunnan, China. *Trans. Inst. Br. Geogr.* **1999**, *24*, 315–329. [CrossRef]

50. People's Government of Yunnan Province (PGYP). The Tenth Five Year Plan for economic and social development in Yunnan province. 2001. Available online: http://www.bjpc.gov.cn/fzgh_1/guihua/10_5/gj_ws/200710/P020071024539269370799.pdf (accessed on 13 November 2014). (In Chinese).

51. Wang, Y.; Zhu, Q.H.; Geng, Y. Trajectory and driving factors for GHG emissions in the Chinese cement industry. *J. Clean. Prod.* **2013**, *53*, 252–260. [CrossRef]

10

A New Fault Location Approach for Acoustic Emission Techniques in Wind Turbines

Carlos Quiterio Gómez Muñoz * and Fausto Pedro García Márquez

Academic Editor: Frede Blaabjerg

Ingenium Research Group, Department of Business Management, University of Castilla-La Mancha, Ciudad Real 13071, Spain; FaustoPedro.Garcia@uclm.es
* Correspondence: carlosquiterio.gomez@uclm.es

Abstract: The renewable energy industry is undergoing continuous improvement and development worldwide, wind energy being one of the most relevant renewable energies. This industry requires high levels of reliability, availability, maintainability and safety (RAMS) for wind turbines. The blades are critical components in wind turbines. The objective of this research work is focused on the fault detection and diagnosis (FDD) of the wind turbine blades. The FDD approach is composed of a robust condition monitoring system (CMS) and a novel signal processing method. CMS collects and analyses the data from different non-destructive tests based on acoustic emission. The acoustic emission signals are collected applying macro-fiber composite (MFC) sensors to detect and locate cracks on the surface of the blades. Three MFC sensors are set in a section of a wind turbine blade. The acoustic emission signals are generated by breaking a pencil lead in the blade surface. This method is used to simulate the acoustic emission due to a breakdown of the composite fibers. The breakdown generates a set of mechanical waves that are collected by the MFC sensors. A graphical method is employed to obtain a system of non-linear equations that will be used for locating the emission source. This work demonstrates that a fiber breakage in the wind turbine blade can be detected and located by using only three low cost sensors. It allows the detection of potential failures at an early stages, and it can also reduce corrective maintenance tasks and downtimes and increase the RAMS of the wind turbine.

Keywords: acoustic emission; wind turbine; fault detection and diagnosis; macro-fiber composite; non-destructive testing

1. Introduction

The renewable energy industry is undergoing continuous improvement to cover the current demands of electricity, wind energy being one of the most important. The new technologies, communication systems and advances in mathematical models for signal processing aid in achieving that goal [1]. The complexity of these devices causes a reduction of the reliability, availability, maintainability and safety of the system (RAMS) and increases the maintenance costs due to the occurrence of non-monitored failures [2–4].

Nowadays, fault detection and diagnosis (FDD) by non-destructive testing (NDT) is employed in maintenance management [5–7], for example in structural health monitoring (SHM) [8]. SHM enables identifying and diagnosing the fault and its location by detecting changes in the static and dynamic features of the structure [9,10]. SHM can be remotely managed, reducing the costs of manual inspections and the time between the fault occurrences, and this has been noted [11,12]. This will lead to an increase in the productivity, reducing the potential downtimes for the wind farms and increasing the RAMS of the wind turbine [13–15].

The purpose of this paper is to design an FDD model for the SHM of a wind turbine blade [16–18]. The case study proposes a novel localization method using signals from macro-fiber composite (MFC) sensors. Three MFC sensors are strategically located along a blade section to detect incipient breakages in the structure [19,20]. The case study involves some considerations, e.g., the appearance of the scattering phenomena, the orientation of the sensors when the excitation is received, *etc*. However, it will be demonstrated that the proposed method can set the location with high accuracy. The analysis identifies a single point obtained from a graphical method that is analytically set by nonlinear equations.

The accuracy of this method depends on the transducer sensitivity, the type of composite material, irregularities in the material, the environmental noise, *etc*. The localization precision of the emission source will be affected by the type of composite material, the sensitivity of the materials, environmental noise, false positives due to impacts on the piece, *etc*. Moreover, in real working conditions, considering environmental conditions, e.g., rain or hail, or impacts on the blade, it can cause false alarms.

In working conditions, it would be possible to distinguish between the frequencies associated with the vibration of the blade (low frequencies) and the frequencies associated with the acoustic emission of the fiber breakage (frequencies within the audible range and the ultrasonic range) [21]. It is possible to filter the frequencies associated with the vibration from the collected signal. The authors demonstrated this in [22].

2. Experiments

The experiments are done in a section of the wind turbine blade. The fragment, shown in Figure 1, is made of glass fiber-reinforced polymer (GFRP), with dimensions of 100×79.5 cm. The section is composed of a honeycomb central layer embedded between two fiberglass layers made of polyester resin. This type of material has good structural properties, resistance to fatigue and other advantages. The attenuation of the acoustic emission in the blade is high, and it depends on the material, wave frequency and travelling distance between the failure source and the sensor location [23].

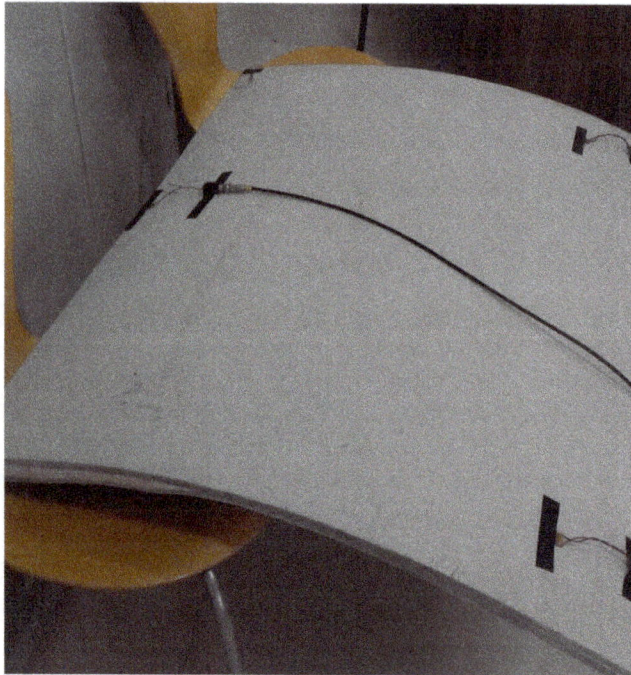

Figure 1. Wind turbine section with sensors for acoustic emission location.

The waves with the same velocity form a circular wave front when they propagate through an isotropic material. The velocity generally does not depend on the direction of propagation, but in anisotropic materials, e.g., the composite materials of the wind turbines, the velocity depends on the direction of propagation. A slowness factor could be introduced in order to consider the propagation direction, e.g., it has been observed that the configuration of layers (+45/−45) for a composite has a strong dependency of the direction of propagation. However, it has been demonstrated that the direction of propagation does not affect the velocity in the blade section studied in this paper. Therefore, the slowness factor has not been introduced in these experiments.

3. Location of the Fiber Breakage by the Triangulation Methodology

The SHM on wind turbine blades is employed to detect the defect online and to locate it with accuracy [24]. The wind turbine blades are becoming larger and more complex, and this requires setting the exact location of a fiber breakage to reduce the maintenance cost and the productivity.

The glass fiber breakages of a wind turbine blade have been simulated in the laboratory on a real blade. A novel location method by triangulation has been developed. The aim of the paper is to locate the acoustic emission source in four different points on the blade section. The acoustic emission produced by the division of the glass fibers is simulated by breaking the tip of the lead from a mechanical pencil [25–27]. Three MFC transducers (A, B and C) were used to detect the acoustic emissions. The three transducers are used as sensors that collect the wave front of the mechanical wave produced by the acoustic emission. These signals received by the sensors present a low amplitude, and therefore, they need to be pre-amplified before being acquired by the oscilloscope [28].

In working conditions, there are many factors that could influence the configuration of the arrangement of the sensors on a blade, for example the length of the blades, the intensity of the acoustic emission, the accuracy of the sensors, the background noise, attenuation, *etc.* Depending on these factors, many groups of three sensors would be established, as they are required to cover the entire blade.

The propagation velocity of the acoustic emission (see Figure 2) has been experimentally calculated by breaking a pencil lead and measuring the delays in the excitement of the sensors S1 and S2 (Figure 3).

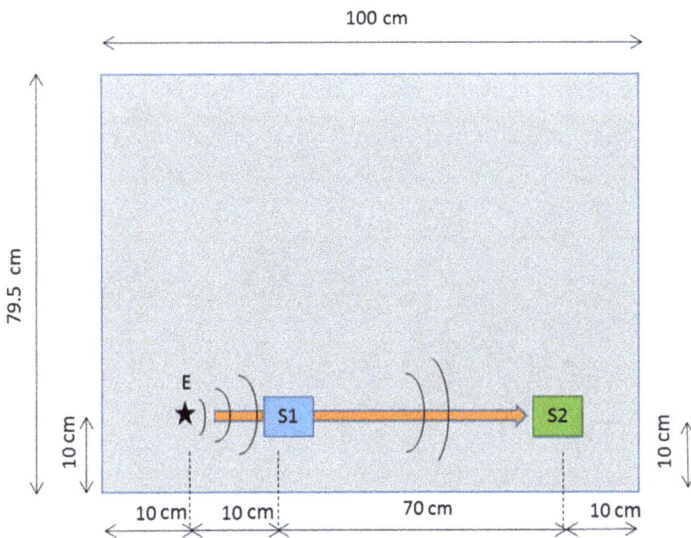

Figure 2. Measuring the experimental propagation velocity in the composite material.

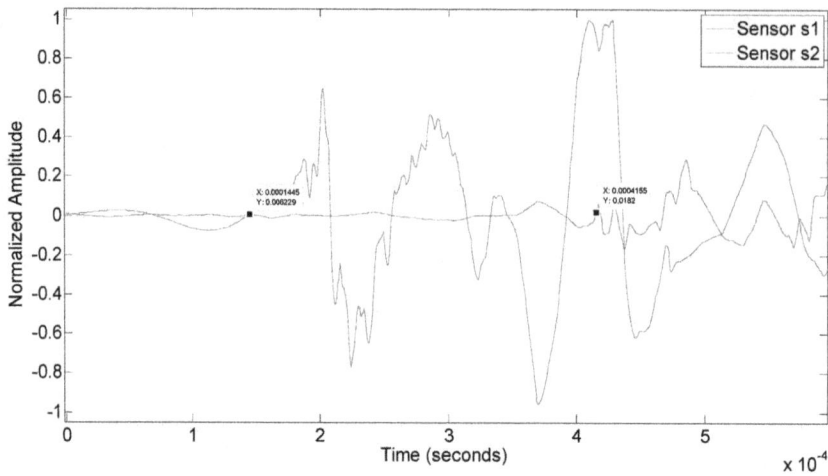

Figure 3. Peak detection of the acoustical emission collected by Sensor 1 (**blue**) and Sensor 2 (**green**) to obtain the experimental propagation velocity in the composite material.

The delay between Signals 1 and 2 is 271 μs (542 samples), and the propagation velocity for the composite material is 2583 m/s.

Four experiments have been conducted at four different locations of the acoustic emission. Twelve tests have been done applying the same force, angle of inclination and length (1 mm approximately). The main objective is to get similar signals for all of the case studies. The data are also filtered for the signal processing, where undesired frequencies are filtered [29]. The peak detection algorithm identifies the wave front of each signal. This process is complex because the waves are compounded by a large number of frequencies. Moreover, there are multiple elements in the blade that could affect the scattering of the acoustic signal, such as the edges of the geometry, the junction with the beam, adhesives, *etc*.

The signal processing consists of a pass band filter that eliminates low and high frequencies and carries out a comparison of the peaks of the wave front in the same frequency range of Signals A, B and C, generated by the above-mentioned MFC sensors, A, B and C (Figure 4).

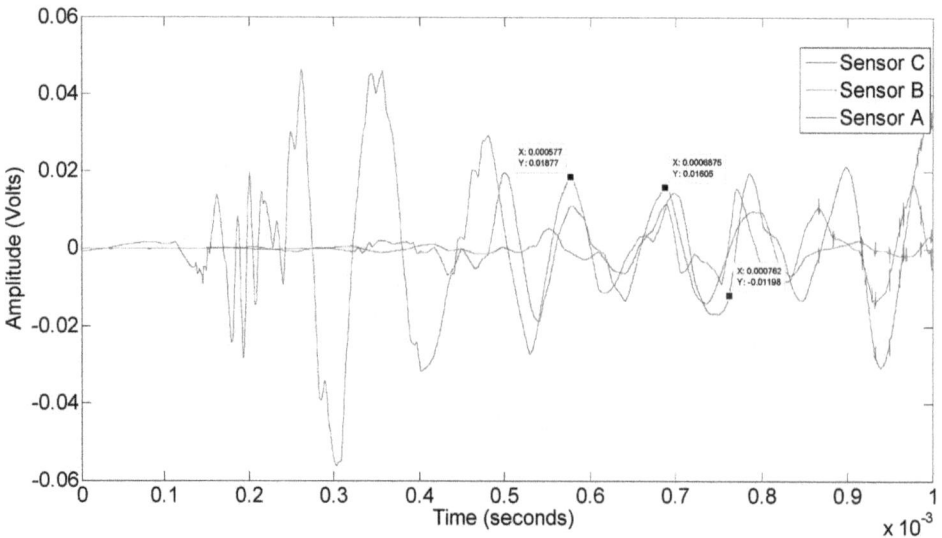

Figure 4. Pre-processing of the signal. Wave front collected by Sensors C (**blue**), B (**green**) and A (**red**).

The MFC Sensors A, B and C are placed as an equilateral triangle (see Figure 5). D is the location of the emission source by the breaking, D (D is known in this experiment).

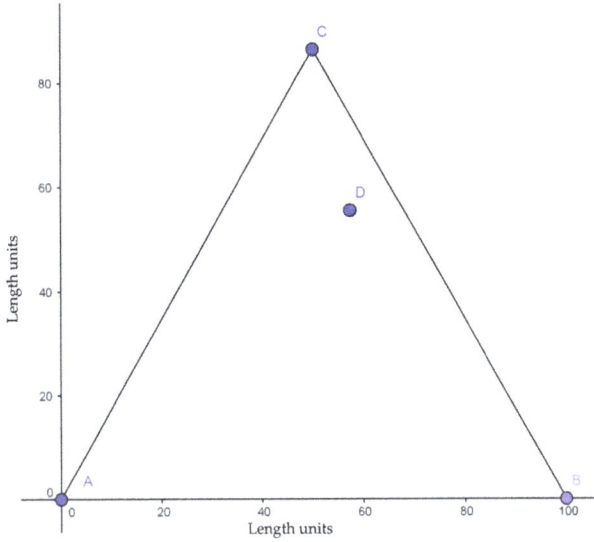

Figure 5. Location of Vertices A, B and C and the, defect D.

The nearest Sensor C is the first to be excited due to the wave front coming from the acoustic emission (Figure 6).

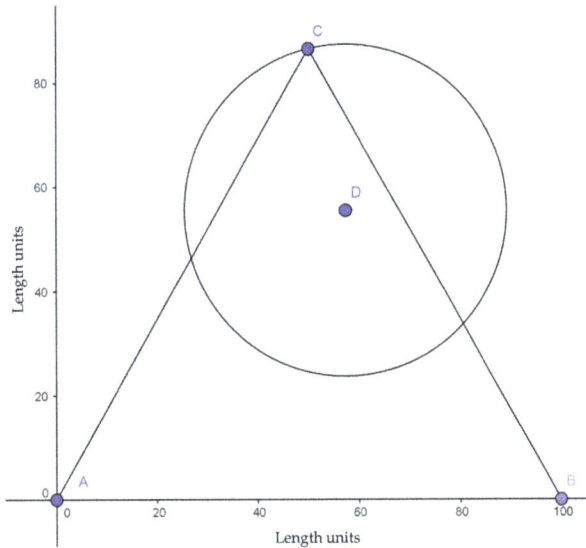

Figure 6. Wave front of the acoustic emission collected by the nearest Sensor C.

The delay between the excitation of the first Sensor C and the second closest Sensor B to the defect, D, is given by the distance from E to B (Figure 7). The delay time and the speed of the wave propagation on the blade is calculated by Equation (1):

$$D_{EB} = v \times t_{CB} \tag{1}$$

where D_{EB} is the distance between E and B, v is the propagation velocity of the wave (obtained experimentally) and t_{CB} is the time delay between the excitation of Sensors C and B.

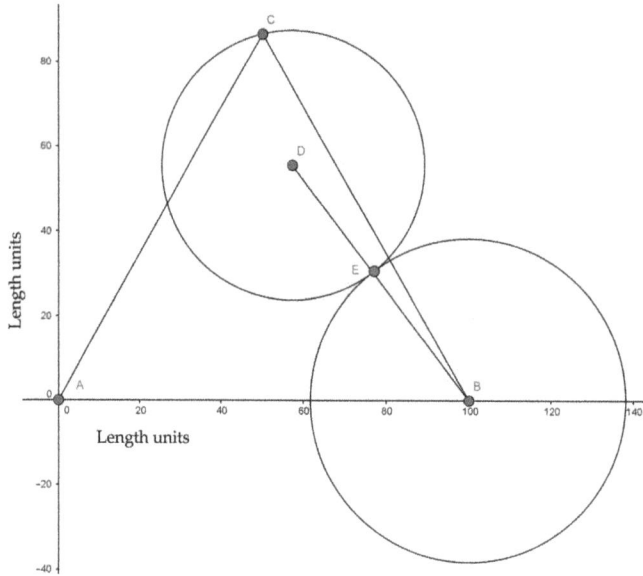

Figure 7. Location of Point E, set by the delay between the excitation time in Sensors C and B.

The delay between Sensor C and Sensor A, the farthest one from Defect D, is given by the distance from F to A D_{FA} in Equation (2).

$$D_{FA} = v \times t_{CA} \tag{2}$$

where t_{CA} is the time delay between the excitation of Sensor C and Sensor A. Figure 8 shows the scheme of the triangulation approach, the delay being represented by a circle.

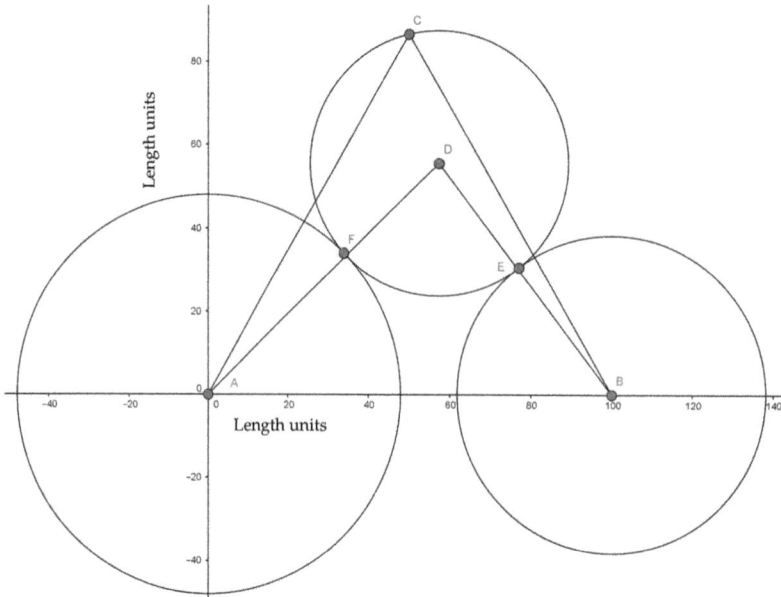

Figure 8. Scheme of the acoustic emission delays for locating the source.

In a real case study, Point D is unknown regarding the time and location, and the delays between the different sensors can be calculated. This condition is shown in Figure 9, where the circumferences represent the delays of the signal that comes to each sensor with respect to the first sensor (C).

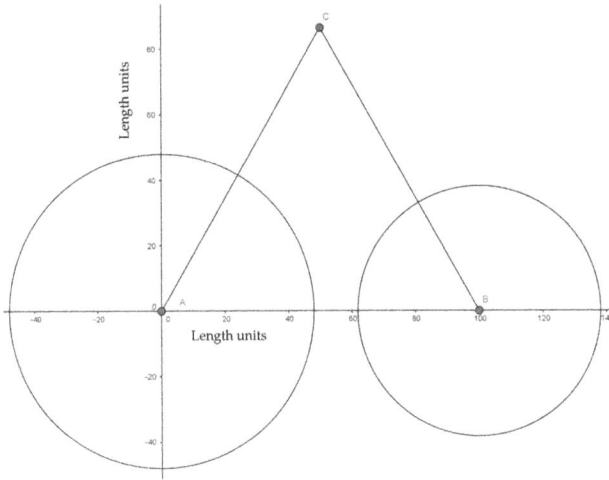

Figure 9. Initial conditions to locate the source of the acoustic emission.

The objective is to find the source of the acoustic emission D mentioned above. This point is the center of a circle that is tangential to two given circles and passes through Point C (see Figure 8). The solution is obtained in this paper employing a graphical method and an analytical method using a system of seven nonlinear equations.

4. Triangulation Equations System

The seven nonlinear equations to solve this problem are given by Equations (3) to (9), considering the scheme shown in Figure 4, where the MFC sensors are located at Points A, B and C, and the defect is at Point D. The coordinates and radius are:

- x_c: x-coordinate at the top of the triangle.
- y_c: y-coordinate at the top of the triangle.
- x_a: x-coordinate at the left lower corner of the triangle.
- y_a: y-coordinate at the left lower corner of the triangle.
- x_b: x-coordinate at the right lower corner of the triangle.
- y_b: y-coordinate at the right lower corner of the triangle.
- r_a: radius of the circle originated from A (delay of Sensor A).
- r_b: radius of the circle originated from B (delay of Sensor B).

The data mentioned above are known. The unknown variables are x_1, x_2, x_3, x_4, x_5, x_6 and x_7, being:

- x_1 and x_2 the coordinates of the emission Source D.
- x_3 and x_4 the coordinates of the tangency of Point F.
- x_5 and x_6 the coordinates of the tangency of Point E.
- x_7 is the radius of the circumference with the center D.

The following equations define the method analytically.
Equation (3) considers a circle with the center at D and passing through C:

$$F\,(1) = (x_c - x_1)^2 + (y_c - x_2)^2 - (x_7)^2 \tag{3}$$

Equation (4) represents a circle with the center at D and passing through F:

$$F\,(2) = (x_3 - x_1)^2 + (x_4 - x_2)^2 - (x_7)^2 \qquad (4)$$

Equation (5) sets a circle with the center at D and passing through E:

$$F\,(3) = (x_5 - x_1)^2 + (x_6 - x_2)^2 - (x_7)^2 \qquad (5)$$

Equation (6) represents a circle with the center at A and passing through F:

$$F\,(4) = (x_1 - x_a)^2 + (x_4 - y_a)^2 - r_a^{\ 2} \qquad (6)$$

Equation (7) considers a circle with the center at B and passing through E:

$$F\,(5) = (x_5 - x_b)^2 + (x_6 - y_b)^2 - r_b^{\ 2} \qquad (7)$$

Equation (8) provides the straight line passing through Points A and F:

$$F\,(6) = \frac{(x_4 - y_a)}{(x_3 - x_a)} \times x_1 + \left(y_a - \frac{(x_4 - y_a)}{(x_3 - x_a)} \times x_a\right) - x_2 \qquad (8)$$

Equation (9) sets the straight line passing through Points B and E:

$$F\,(7) = \frac{(y_b - x_6)}{(x_b - x_5)} \times x_1 + \left((y_b - \frac{(y_b - x_6)}{(x_b - x_5)})) \times x_b\right) - x_2 \qquad (9)$$

5. Experimental Procedure and Results

The time of flight and distances are set in this section for Sensors B and A regarding C, C being the first sensor to receive the acoustic signal of the breakage. The experiments are repeated four times to take into account the deviations of the results. The algorithm gives the exact location of the defect, as well as a graphic outline, knowing the radius of the circles with centers at B and C. The dimensions of the blade section, the distribution of the sensors and the emission source (star) in the wind turbine blade are shown in Figure 10. The mathematical results obtained with the algorithm are given in Tables 1–8.

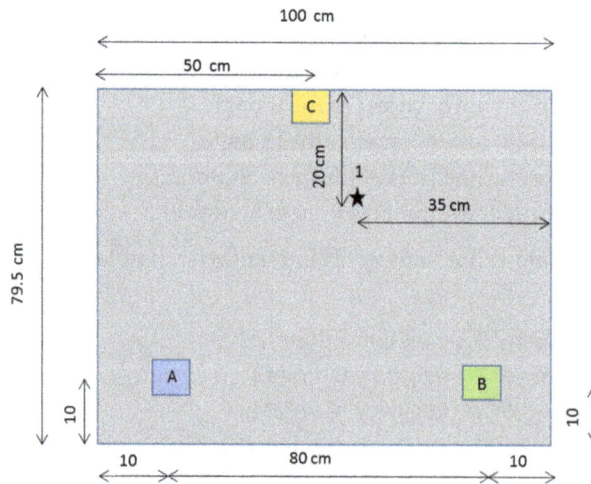

Figure 10. First experiment. Case Study 1.

Table 1. First case study: detection time; delay with C; delay; theoretical distance; experimental distance.

Sensors	Detection Time (Samples)	Delay with C (Samples)	Delay (s)	Theoretical Distance (m)	Experimental Distance (m)
C	1152	-	-	-	-
B	1381	229	1.15×10^{-4}	0.30	0.30
A	1528	376	1.88×10^{-4}	0.49	0.49

Table 2. Initial data of the first case study.

Locations	x-Coordinate (m)	y-Coordinate (m)	Radius (m)
A	0	0	0.49
B	0.8	0	0.30
C	0.4	0.69	-
1	0.55	0.495	-

Table 3. Second case study: detection time; delay with C; delay; theoretical distance; experimental distance.

Sensors	Detection Time (Samples)	Delay with C (Samples)	Delay (s)	Theoretical Distance (m)	Experimental Distance (m)
C	912	-	-	-	-
B	1063	151	7.55×10^{-5}	0.20	0.20
A	1296	384	1.92×10^{-4}	0.50	0.50

Table 4. Initial data of the second case study.

Locations	x-Coordinate (m)	y-Coordinate (m)	Radius (m)
A	0	0	0.50
B	0.8	0	0.20
C	0.4	0.69	-
2	0.65	0.495	-

Table 5. Third case study: detection time; delay with C; delay; theoretical distance; experimental distance.

Sensors	Detection Time (Samples)	Delay with C (Samples)	Delay (s)	Theoretical Distance (m)	Experimental Distance (m)
C	962	-	-	-	-
B	1298	336	1.68×10^{-4}	0.43	0.43
A	1087	125	6.25×10^{-5}	0.17	0.16

Table 6. Initial data of the third case study.

Locations	x-Coordinate (m)	y-Coordinate (m)	Radius (m)
A	0	0	0.16
B	0.8	0	0.43
C	0.4	0.69	-
3	0.2	0.445	-

Table 7. Fourth case study: detection time; delay with C; delay; theoretical distance; experimental distance.

Sensors	Detection Time (Samples)	Delay with C (Samples)	Delay (s)	Theoretical Distance (m)	Experimental Distance (m)
C	1155	-	-	-	-
B	1650	495	2.48×10^{-4}	0.64	0.64
A	1385	230	1.15×10^{-4}	0.29	0.30

Table 8. Initial data of the fourth case study.

Locations	x-Coordinate (m)	y-Coordinate (m)	Radius (m)
A	0	0	0.30
B	0.8	0	0.64
C	0.4	0.69	/
4	0.05	0.645	/

5.1. Case Study 1

The breaking of the lead is made in the following coordinates from Sensor A at Point 1 (star); see Figure 10. Sensor A is the coordinate origin.

- Coordinate x: 0.55.
- Coordinate y: 0.495.

The location of the source employing the algorithm is: Point 1: (x: 0.5533, y: 0.4920). The error in the location is: coordinate x: 3.3 mm; coordinate y: 30 mm.

5.2. Case Study 2

In this case, the emission source was generated at Point 2 (star), shown in Figure 11:

- Coordinate x: 0.65.
- Coordinate y: 0.495.

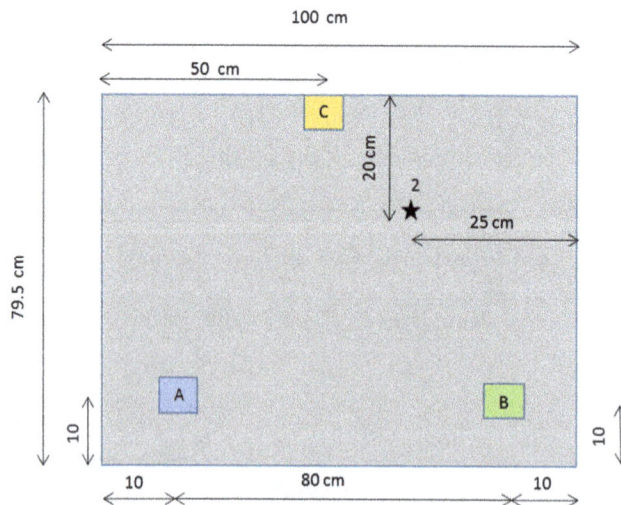

Figure 11. Scheme for Case Study 2.

The location of the source employing the algorithm is: Point 2: (x: 0.6502, y: 0.4950). The errors in the location are: coordinate x: 0.2 mm; coordinate y: 0.00 mm.

5.3. Case Study 3

In this case, the emission source was generated at Point 3 (star); see Figure 12:

-   Coordinate x: 0.20.
-   Coordinate y: 0.445.

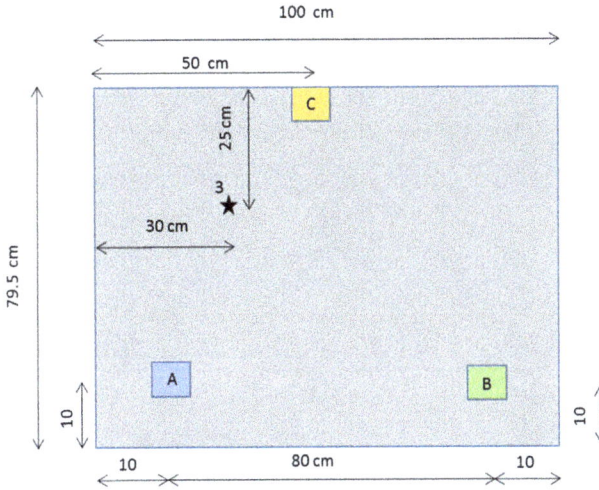

Figure 12. Scheme for Case Study 3.

The location of the source employing the algorithm is: Point 3 (x: 0.1914, y: 0.4434). The errors in the location are: coordinate x: 8.6 mm; coordinate y: 1.6 mm.

5.4. Case Study 4

In this case, the emission source was generated at Point 4 (star), and it is shown in Figure 13:

-   Coordinate x: 0.05.
-   Coordinate y: 0.645.

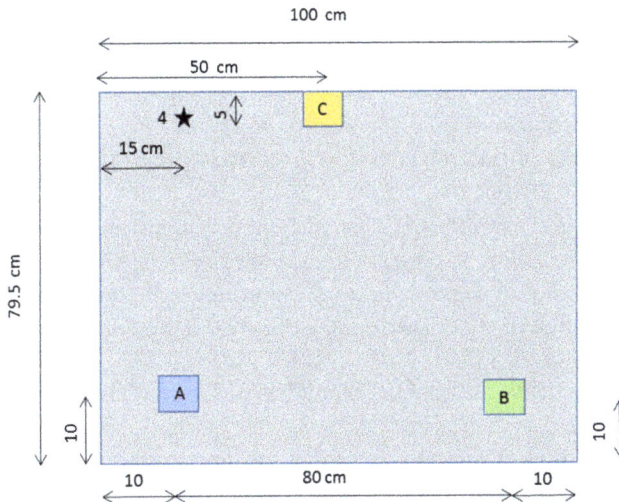

Figure 13. Scheme for Case Study 4.

The location of the source employing the algorithm is: Point 4: (x: 0.050, y: 0.6495). The errors in the location are: coordinate x: 0 mm; coordinate y: 4.5 mm.

Different waves with different speeds appear as a result of the scattering phenomena when a large number of frequencies are excited by the breakage. This makes the identification of peaks to measure the delays of the signals complicated. The orientation of the sensors, when they receive the excitation, can affect the shape of the signal collected.

It is observed that the algorithm provides correct and coherent results. It detects the location of the acoustic emission with an accuracy of two decimals (millimeters). The maximum error registered was 9 mm.

Finally, the algorithm shows the position of the acoustic emission point with the real dimensions of the blade. Figure 14 shows the location of the acoustic emission for the first case study.

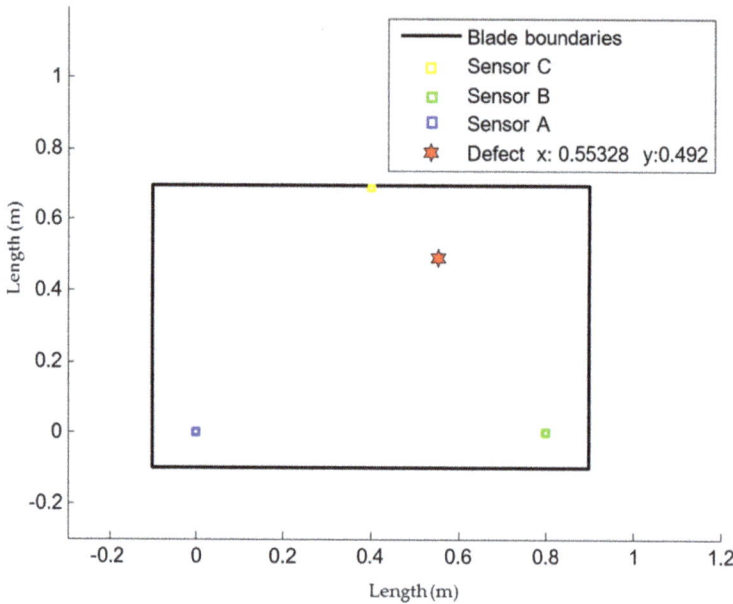

Figure 14. Scheme of the location of the acoustic emission for the first case study.

6. Conclusions

The development of a localization approach presented in this paper is set using macro-fiber composites to detect cracks in blades in an SHM system. This approach, based on NDT, automatically identifies and locates an acoustic emission source coming from a fiber's breakage in a wind turbine blade section by a novel signal processing method. It can be extrapolated to other similar structures, e.g., airplane wings.

Three sensors are strategically located in the blade. It is demonstrated that the approach is able to detect the location of the simulated defect accurately employing acoustic emissions signals. The signal processing is based on a graphical method of triangulation and seven nonlinear equations. The signals are previously filtered. Different experiments are performed to demonstrate the effectiveness of the proposed method.

The approach detects the location of the acoustic emission with high accuracy, 9 mm being the maximum error registered.

There are conditions that affect the accuracy of the emission source location, e.g., the type of composite material, the sensitivity of the transducers, environmental noise, false positives due to impacts on the piece, *etc.* The method shows the position of the acoustic emission point with the real dimensions of the blade.

Acknowledgments: This project is partly funded by the EC under the FP7 framework program (Ref. 322430), OPTIMUS and the MINECO project WindSeaEnergy (Ref. DPI2012-31579), where OPTIMUS is supporting to the new condition Monitoring System, and WindSeaEnergy the novel signal processing approach presented in this research paper.

Author Contributions: Carlos Quiterio Gómez Muñoz and Fausto Pedro García Marquez conceived and designed the experiments; Carlos Quiterio Gómez Muñoz and Fausto Pedro García Marquez performed the experiments; Carlos Quiterio Gómez Muñoz and Fausto Pedro García Marquez analyzed the data; Carlos Quiterio Gómez Muñoz and Fausto Pedro García Marquez contributed reagents/materials/analysis tools; Carlos Quiterio Gómez Muñoz and Fausto Pedro García Marquez wrote the paper.

Conflicts of Interest: The authors declare no conflict of interest.

References

1. Dai, D.; He, Q. Structure damage localization with ultrasonic guided waves based on a time–frequency method. *Signal Process.* **2014**, *96*, 21–28. [CrossRef]
2. Spinato, F.; Tavner, P.J.; van Bussel, G.; Koutoulakos, E. Reliability of wind turbine subassemblies. *IET Renew. Power Gen.* **2009**, *3*, 387–401. [CrossRef]
3. Ajayi, O.O.; Fagbenle, R.O.; Katende, J.; Ndambuki, J.M.; Omole, D.O.; Badejo, A.A. Wind energy study and energy cost of wind electricity generation in nigeria: Past and recent results and a case study for south west nigeria. *Energies* **2014**, *7*, 8508–8534. [CrossRef]
4. Marugán, A.P.; Márquez, F.P.G. A novel approach to diagnostic and prognostic evaluations applied to railways: A real case study. *Proc. Inst. Mech. Eng. F J. Rail Rapid Transit* **2015**. [CrossRef]
5. Chen, X.; Zhao, W.; Zhao, X.L.; Xu, J.Z. Failure test and finite element simulation of a large wind turbine composite blade under static loading. *Energies* **2014**, *7*, 2274–2297. [CrossRef]
6. Márquez, F.G.; Roberts, C.; Tobias, A.M. Railway point mechanisms: Condition monitoring and fault detection. *Proc. Inst. Mech. Eng. F J. Rail Rapid Transit* **2010**, *224*, 35–44. [CrossRef]
7. Pliego Marugán, A.; García Márquez, F.P.; Lorente, J. Decision making process via binary decision diagram. *Int. J. Manag. Sci. Eng. Manag.* **2015**, *10*, 3–8. [CrossRef]
8. Marquez, F.P.G. An approach to remote condition monitoring systems management. In Proceedings of the Institution of Engineering and Technology International Conference on Railway Condition Monitoring, Birmingham, UK, 29–30 Novemmber 2006; pp. 156–160.
9. Light-Marquez, A.; Sobin, A.; Park, G.; Farinholt, K. Structural damage identification in wind turbine blades using piezoelectric active sensing. In *Structural Dynamics and Renewable Energy*; Springer: New York, NY, USA, 2011; pp. 55–65.
10. García, F.P.; Pinar, J.M.; Papaelias, M.; Ruiz de la Hermosa, R. Wind turbines maintenance management based on FTA and BDD. *Renew. Energy Power Qual. J.* **2012**. Available online: http://icrepq.com/icrepq'12/699-garcia.pdf (accessed on 7 January 2016).
11. Pedregal, D.J.; García, F.P.; Roberts, C. An algorithmic approach for maintenance management based on advanced state space systems and harmonic regressions. *Ann. Oper. Res.* **2009**, *166*, 109–124. [CrossRef]
12. Yang, H.-H.; Huang, M.-L.; Yang, S.-W. Integrating auto-associative neural networks with hotelling T^2 control charts for wind turbine fault detection. *Energies* **2015**, *8*, 12100–12115. [CrossRef]
13. Chen, X.; Qin, Z.W.; Zhao, X.L.; Xu, J.Z. Structural performance of a glass/polyester composite wind turbine blade with flatback and thick airfoils. In Proceedings of the American Society of Mechanical Engineers (ASME) 2014 International Mechanical Engineering Congress and Exposition, Montreal, QC, Canada, 14–20 November 2014.
14. De la Hermosa González, R.R.; Márquez, F.P.G.; Dimlaye, V. Maintenance management of wind turbines structures via mfcs and wavelet transforms. *Renew. Sustain. Energy Rev.* **2015**, *48*, 472–482. [CrossRef]
15. Márquez, F.P.G.; Pedregal, D.J.; Roberts, C. New methods for the condition monitoring of level crossings. *Int. J. Syst. Sci.* **2015**, *46*, 878–884. [CrossRef]
16. García, F.P.; Pedregal, D.J.; Roberts, C. Time series methods applied to failure prediction and detection. *Reliab. Eng. Syst. Saf.* **2010**, *95*, 698–703. [CrossRef]
17. García Márquez, F.P.; García-Pardo, I.P. Principal component analysis applied to filtered signals for maintenance management. *Qual. Reliab. Eng. Int.* **2010**, *26*, 523–527. [CrossRef]

18. Márquez, F.P.G.; Muñoz, J.M.C. A pattern recognition and data analysis method for maintenance management. *Int. J. Syst. Sci.* **2012**, *43*, 1014–1028. [CrossRef]

19. Michaels, J.E. Detection, localization and characterization of damage in plates with an *in situ* array of spatially distributed ultrasonic sensors. *Smart Mater. Struct.* **2008**, *17*, 035035. [CrossRef]

20. Chen, H.; Yan, Y.; Chen, W.; Jiang, J.; Yu, L.; Wu, Z. Early damage detection in composite wingbox structures using hilbert-huang transform and genetic algorithm. *Struct. Health Monit.* **2007**, *6*, 281–297. [CrossRef]

21. Eftekharnejad, B.; Carrasco, M.; Charnley, B.; Mba, D. The application of spectral kurtosis on acoustic emission and vibrations from a defective bearing. *Mech. Syst. Signal Process.* **2011**, *25*, 266–284. [CrossRef]

22. Gómez, C.Q.; Villegas, M.A.; García, F.P.; Pedregal, D.J. Big data and web intelligence for condition monitoring: A case study on wind turbines. In *Handbook of Research on Trends and Future Directions in Big Data and Web Intelligence*; Information Science Reference, IGI Global: Hershey, PA, USA, 2015.

23. Bohse, J. Acoustic emission characteristics of micro-failure processes in polymer blends and composites. *Compos. Sci. Technol.* **2000**, *60*, 1213–1226. [CrossRef]

24. Márquez, F.P.G.; Pérez, J.M.P.; Marugán, A.P.; Papaelias, M. Identification of critical components of wind turbines using FTA over the time. *Renew. Energy* **2015**, *87*, 869–883. [CrossRef]

25. Gorman, M.R. Plate wave acoustic emission. *J. Acoust. Soc. Am.* **1991**, *90*, 358–364. [CrossRef]

26. Ruiz de la Hermosa, R.; García Márquez, F.P.; Dimlaye, V.; Ruiz-Hernández, D. Pattern recognition by wavelet transforms using macro fibre composites transducers. *Mechan. Syst. Signal Process.* **2014**, *48*, 339–350. [CrossRef]

27. Betz, D.C.; Staszewski, W.J.; Thursby, G.; Culshaw, B. Structural damage identification using multifunctional bragg grating sensors: II. Damage detection results and analysis. *Smart Mater. Struct.* **2006**, *15*, 1313–1322. [CrossRef]

28. Coverley, P.; Staszewski, W. Impact damage location in composite structures using optimized sensor triangulation procedure. *Smart Mater. Struct.* **2003**, *12*, 795–803. [CrossRef]

29. Gómez, C.Q.; Ruiz de la Hermosa, R.; Trapero, J.R.; Garcia, F.P. A novel approach to fault detection and diagnosis on wind turbines. *Glob. Nest J.* **2014**, *16*, 1029–1037.

11

Design and Control of a 3 kW Wireless Power Transfer System for Electric Vehicles

Zhenshi Wang [1,2], Xuezhe Wei [2,*] and Haifeng Dai [1,2]

Academic Editor: K. T. Chau

[1] Clean Energy Automotive Engineering Center, Tongji University, No. 4800, Caoan Road, Shanghai 201804, China; 1022wangzhenshi@tongji.edu.cn

[2] School of Automotive Studies, Tongji University, No. 4800, Caoan Road, Shanghai 201804, China; tongjidai@tongji.edu.cn

[*] Correspondence: weixzh@tongji.edu.cn

Abstract: This paper aims to study a 3 kW wireless power transfer system for electric vehicles. First, the LCL-LCL topology and LC-LC series topology are analyzed, and their transfer efficiencies under the same transfer power are compared. The LC-LC series topology is validated to be more efficient than the LCL-LCL topology and thus is more suitable for the system design. Then a novel q-Zsource-based online power regulation method which employs a unique impedance network (two pairs of inductors and capacitors) to couple the cascaded H Bridge to the power source is proposed. By controlling the shoot-through state of the H Bridge, the charging current can be adjusted, and hence, transfer power. Finally, a prototype is implemented, which can transfer 3 kW wirelessly with ~95% efficiency over a 20 cm transfer distance.

Keywords: wireless power transfer (WPT); topology analysis; power regulation; electric vehicle

1. Introduction

Research on wireless power transfer began soon thereafter the famous Tesla coils were invented by Nikola Tesla in 1889 [1,2], and many good results have been achieved [3–6]. In 2007, researchers at MIT proposed strongly coupled magnetic resonances (SCMR), by which they were able to transfer 60 watts wirelessly with ~40% efficiency over distances in excess of 2 m [7]. Various research hot spots, including system architectures, optimization design, frequency splitting, impedance matching and special applications, have been investigated [8–14]. Wireless power transfer is very suitable for charging electric vehicles [15–17], as it can avoid the troublesome plug-in process, provide an inherent electrical isolation and adapt to harsh environments. However, SCMR is not appropriate for automotive applications, as its operating frequency is very high, which goes beyond the limitation of SAE J2954 (work in progress). As another kind of wireless power transfer techniques, inductive power transfer (IPT) has developed for more than twenty years [5], and it mainly focuses on the high power level applications, where the issues of concern normally include power conversion and control [18,19], magnetic structure design [20], control algorithm and strategy [21,22] as well as circuit topology [23]. Basically, both SCMR and IPT conforms to Faraday's and Ampere's laws, and their differences primarily include the design approaches, system architectures, parameter selection and transfer characteristics [6,24].

This paper aims to study a 3 kW vehicle-mounted wireless power transfer system, on which two key parts, the resonant topology analysis and comparison, and the online power regulation, are elaborated. Many resonant topologies are available for wireless power transfer system, but the most basic ones are only series-series, series-parallel, parallel-series and parallel-parallel [23], and the others are all derived from these ones. A wireless power transfer system for electric vehicles requires a resonant topology that should have a unity-power-factor and a current source characteristic [18], no matter whether the distance or angle between the chassis and the ground changes or not. In this paper the characteristics of the LC-LC series and LCL-LCL topologies are analyzed first, and we prove that they both have the required unity-power-factor and current source characteristics. Then the transfer efficiencies of the LC-LC series and LCL-LCL topologies are compared under the same transfer power conditions, and the comparison results validate that the LC-LC series topology is more suitable for the system design due to its higher transfer efficiency. In practice, the distance or angle between vehicle chassis and ground often changes [25], as drivers cannot park their cars at the same location every time, and naturally, online power regulation is indispensable in the battery charging process. Traditional power regulation methods include cascade DC/DC, dead time modulation and phase shifting control [26]. However, the cascade DC/DC may increase the number of devices, decrease the power density and lower the transfer efficiency, while dead time modulation may distort the output waves produced by the H Bridge, and the phase shifting control cannot boost the transfer power, so a novel q-Zsource-based power regulation method is proposed in this paper, which employs two pairs of inductors and capacitors as a unique boost network. The power regulation is realized by controlling the shoot-through state of the H Bridge, and there are no extra power switch devices. By combining the phase shifting control and shoot-through state control, the square-wave voltage produced by the H Bridge can be adjusted arbitrarily, and hence, the transfer power.

2. Comparative Analysis of Resonant Topologies

2.1. LC-LC Series Topology

Wireless power transfer systems normally consist of a power source, a H Bridge, a resonant topology, a rectifier as well as a load. With the classical frequency-domain analysis, we can easily get the amplitude-frequency characteristics of resonant topologies, which are steep spikes, and the maximal point just emerges at the resonant frequency. For example, the LC-LC series topology has a four order transfer function, and the LCL-LCL topology has a six order transfer function. These two resonant topologies only allow the fundamental components of the square-wave voltages produced by the H Bridge to pass through, thus the resonant topology input can be substituted by a quasi-sine voltage source. As the power batteries have strong voltage source characteristics, the rectifier and battery pack can be also simplified into a quasi-sine voltage source, the amplitude of which depends on the battery pack voltage multiplied by $4/\pi$. The simplified LC-LC topology [27,28] is shown in Figure 1.

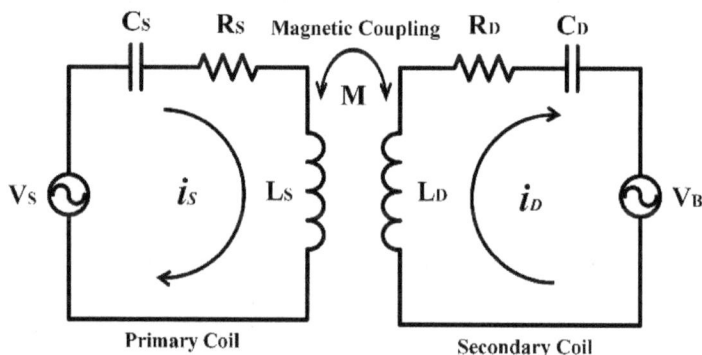

Figure 1. LC-LC series topology.

In Figure 1, L_S and L_D represent the magnetic coils, C_S and C_D represent the resonant capacitors, R_S and R_D represent the parasitic resistances, and M is the mutual inductance between L_S and L_D. When the system works in the resonant state, one has:

$$\begin{cases} V_S = i_S R_S - j\omega_r M i_D \\ j\omega_r M i_S = i_D R_D + jV_B \end{cases} \tag{1}$$

where i_S and i_D are the resonant currents in the primary and secondary coils, and ω_r is the system resonant frequency. By solving Equation (1), we have:

$$\begin{cases} |i_S| = \dfrac{V_S + \dfrac{\omega_r M V_B}{R_D}}{R_S + \dfrac{(\omega_r M)^2}{R_D}} \\ |i_D| = \dfrac{\dfrac{\omega_r M V_S}{R_S} - V_B}{R_D + \dfrac{(\omega_r M)^2}{R_S}} \end{cases} \tag{2}$$

Equation (2) shows that the charging current i_D depends on ω_r, M, R_S, R_D, V_S and V_B. Normally, V_B may change while charging, as it increases with the state of charge (SOC) of the battery. Because $\omega_r = 2\pi f_r$ is up to 100 kHz, M is from 20 µH to 100 µH, R_S is in the milliohm level, and V_S is usually larger than V_B, we have:

$$\frac{\omega_r M V_S}{R_S} \gg V_B \tag{3}$$

By substituting Equation (3) into Equation (2), we can find that i_D remains unchanged during the whole charging process, which realizes a constant-current charging function. As the product of R_S and R_D is smaller than either of them, Equation (2) is further simplified into:

$$\begin{cases} |i_S| = \dfrac{V_B}{\omega_r M} \\ |i_D| = \dfrac{V_S}{\omega_r M} \end{cases} \tag{4}$$

It is worth mentioning that the current source characteristic of LC-LC series topology is tenable only when the charging objects are batteries or some other capacitive load. Based on Equation (2), the transfer power can be expressed as:

$$P_{LC-LC} = |i_D| \cdot V_B = \frac{\omega_r M V_S - V_B R_S}{R_D R_S + \omega_r^2 M^2} \cdot V_B \tag{5}$$

2.2. LCL-LCL Topology

Similarly, the simplified LCL-LCL topology [18,29–31] is shown in Figure 2.

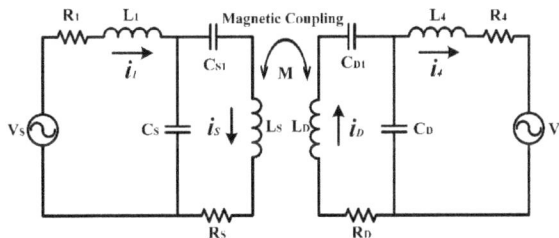

Figure 2. LCL-LCL topology.

In this figure, L_1 and L_4 are the matching inductors of L_S and L_D, R_1 and R_4 are their parasitic resistances. For high power applications, L_S and L_D are normally bulky, which make L_1 and L_4 bulky, and this is not beneficial for the objectives of miniaturization and lightness. Thus the compensating capacitors C_{S1} and C_{D1} are introduced to decrease L_1 and L_4. Still, they should satisfy the following equations:

$$C_{S1} = \frac{1}{\omega_r^2(L_S - L_1)}, \quad C_{D1} = \frac{1}{\omega_r^2(L_D - L_4)} \tag{6}$$

$$\omega_r = \frac{1}{\sqrt{L_1 C_S}} = \frac{1}{\sqrt{L_4 C_D}} \tag{7}$$

Actually, LCL is a transformation of the LC parallel topology [18]. It is well-known that the reflected impedance of the LC parallel topology contains an imaginary part [23], especially when the mutual inductance and the load change online, which makes the tuning process very cumbersome. The additional inductor of the LCL topology can just eliminate this imaginary part whether the mutual inductance and the load change or not. Using a method similar to that in Section 2.1, we can conclude that the LCL-LCL topology also has constant-current charging characteristics. The parasitic resistances are usually small due to the use of Litz wires, thus neglecting the parasitic resistances will not affect the system efficiency sharply, and in practice that loss is very small compared with the loss caused by the H Bridge and rectifier, so we can get the simplified calculation formulas of the LCL-LCL topology as follows:

$$i_S = \frac{V_S}{j\omega_r L_1}, \quad i_4 = j\frac{MV_S}{\omega_r L_1 L_4}, \quad i_D = \frac{V_B}{j\omega_r L_4}, \quad i_1 = j\frac{MV_B}{\omega_r L_1 L_4} \tag{8}$$

The transfer power can be written as:

$$P_{LCL-LCL} = \frac{MV_S V_B}{2\omega_r L_1 L_4} \tag{9}$$

Equation (9) shows that the charging power can be adjusted by V_S. Unlike Equation (5), there are two additional power regulation freedoms L_1 and L_4.

2.3. Comparison between the LC-LC Series Topology and LCL-LCL Topology

The LC-LC series topology and LCL-LCL topology are widely used in practice, as both can realize the constant-current charging characteristics, the unity-power-factor characteristics and even bidirectional power transfer characteristics. Their transfer power characteristics are however different, for instance, the transfer power of the LC-LC series topology increases with the increasing transfer distances according to Equation (5), and the transfer power of the LCL-LCL topology decreases with the increasing transfer distance according to Equation (9). However, their transfer efficiency characteristics have not been compared before, thus this section aims to compare them to provide some suggestions for practical engineering design. The charging power and magnetic coils of the two topologies must be identical, as only then can the efficiency comparison be meaningful. The charging power of the LC-LC series topology equals to that of LCL-LCL topology, if their charging currents are designed to be the same, as they both have the constant-current characteristic. Based on Equations (4) and (8), we can write:

$$\frac{V_S}{\omega_r M} = \frac{MV_S}{\omega_r L_1 L_4} \tag{10}$$

From Equation (10), one has $M^2 = L_1 L_4$. This means that the charging power of the two topologies are the same if the product of two compensating inductors in the LCL-LCL topology equals the mutual inductance M. When the transfer distances are 10, 15, 20, 25 and 30 cm, the measured mutual inductances between two magnetic coils (L_S and L_D) are 107.155 μH, 66.66 μH, 42.538 μH, 28.125 μH, 18.888 μH, respectively. Normally, the distance between the chassis and ground is around 20 cm, thus the corresponding mutual inductance M is around 42.538 µH. Assuming that L_1 equals L_4, one has $L_1 = L_4 = 42.538$ µH. Because the magnetic coils of the LCL-LCL topology are the same as those of the LC-LC series topology, the electric parameters of the LC-LC series topology and LCL-LCL topology can be summarized as shown in Table 1.

Table 1. Detailed parameters of the LC-LC series topology and LCL-LCL topology.

LC-LC Series Topology		LCL-LCL Topology	
Electric parameters	Value	Electric parameters	Value
Primary inductor L_S	290.18 µH	Compensating inductor L_1	43.2 µH
Parasitic resistance R_S	193 mΩ	Parasitic resistance R_1	53 mΩ
Primary capacitor C_S	34.83 nF	Primary resonant capacitor C_S	234.75 nF
Secondary inductor L_D	329.4 µH	Primary inductor L_S	290.18 µH
Parasitic resistance R_D	213 mΩ	Parasitic resistance R_S	193 mΩ
Secondary capacitor C_D	30.89 nF	Compensating capacitor C_{S1}	40.68 nF
		Secondary inductor L_D	329.4 µH
		Parasitic resistance R_D	213 mΩ
		Compensating capacitor C_{D1}	35.07 nF
		Compensating inductor L_4	42.5 µH
		Parasitic resistance R_4	42.5 mΩ
		Secondary resonant capacitor C_D	238.63 nF

The detailed efficiency expressions of LC-LC topology and LCL-LCL topology are given by references [27,30–32], and can be also deduced using Maple. Then we substitute the data of Table 1 into the power and efficiency expressions of the LC-LC series and LCL-LCL topologies, and their resulting transfer characteristics are as shown in Figure 3.

Figure 3a shows that the charging power of these two topologies are the same, despite the different battery voltages, and Figure 3b shows that the efficiency of the LC-LC series topology is higher than that of the LCL-LCL topology. Note that the theoretical results ignore the losses caused by the H Bridge and rectifier, so the efficiency losses are mainly due to the parasitic resistances of the inductors and capacitors. The parasitic resistances of compensating capacitors are usually ignored, for they are far smaller than those of magnetic coils. Compared with the LC-LC series topology, the LCL-LCL topology has another two compensating inductors, the parasitic resistances of which further cause a drop in the transfer efficiency. In order to verify the correctness of theoretical calculations, the experiments are performed, and the results are shown in Figure 4.

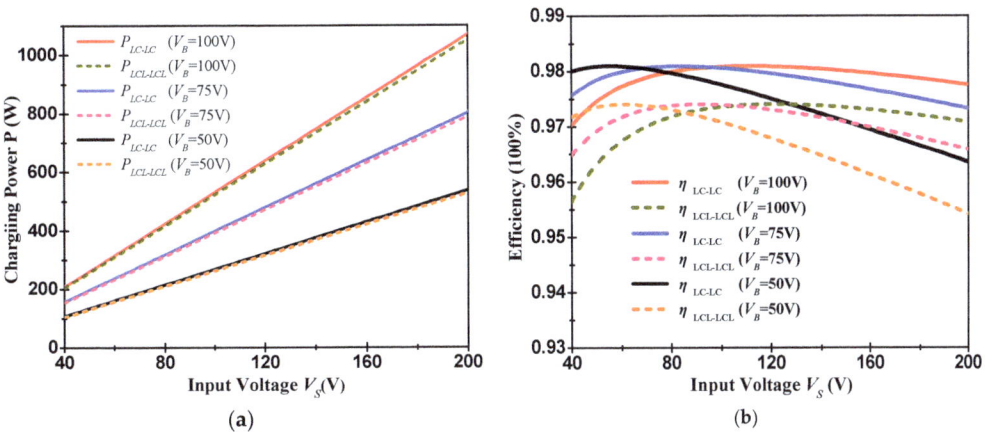

Figure 3. Theoretical transfer power (**a**) and efficiencies (**b**) of the LC-LC series and LCL-LCL topologies.

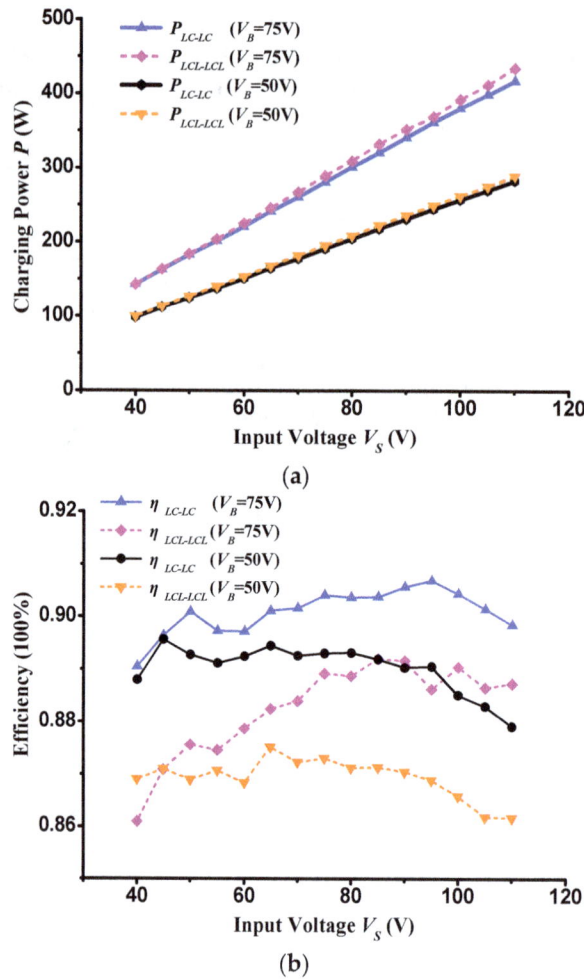

Figure 4. Experimental transfer power (**a**) and efficiencies (**b**) of the LC-LC series and LCL-LCL topologies.

The results in Figure 4a show a good agreement with those in Figure 3a. However, the experimental transfer efficiencies decline sharply compared with those in Figure 3b, which is mainly because the H Bridge and rectifier introduce additional losses. Still, it is obvious that the efficiency of the LC-LC series topology is superior to that of the LCL-LCL topology, so this paper adopts the LC-LC series topology as the power transfer carrier.

3. Online Power Regulation

3.1. Principles of q-Zsource

Z-source is a unique impedance network with two pairs of inductors and capacitors connected in an X shape [33] as shown in Figure 5a, and it was initially used for stabilizing the widely changing voltage produced by fuel cell stacks. Compared with the Z-source, the q-Zsource shown in Figure 5b has better performance [34]. The most obvious two virtues are as follows: first, there is an input inductor L_1, which enables the input current to be continuous and thus limits the transient peak loss. Second, the withstand voltage for C_1 is lowered, for C_1 always combines with power source to charge or discharge L_1. This allows the volume and weight of C_1 to be reduced, and improves system power density.

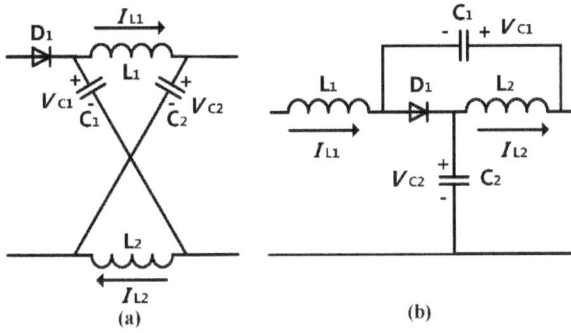

Figure 5. Topologies of Z-source (**a**) and q-Zsource (**b**).

This paper introduces the q-Zsource, not for stabilizing the output, but to produce a changeable output voltage, which can adjust the charging current, and hence, the charging power. The overall system schematic is depicted in Figure 6.

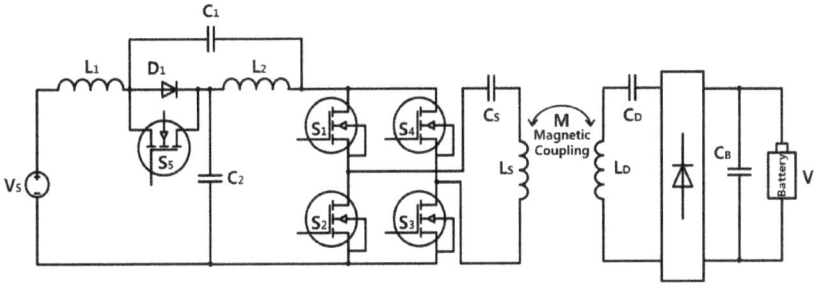

Figure 6. Overall schematic of the wireless power transfer system.

It clearly shows that the whole system consists of the power source, the q-Zsource, the H Bridge, the primary and secondary resonators, the rectifier, the filter and the battery pack. Unlike other system structures, the q-Zsource between the power source and H Bridge is first employed to boost the wireless charging power. By controlling the shoot-through state of the H Bridge, the input voltage to the H Bridge can be boosted through the q-Zsource, so the transmitter current can be adjusted, and hence, the charging current and power for the battery pack. Compared with some typical primary unit current control methods [35], the proposed method has two major merits: first, the digital control of the q-Zsource can be integrated into the primary microprocessor, and there are no active power switches in the proposed method, which can lower the system cost. Secondly, all the current or power regulation devices are included in the primary unit, and this design is beneficial to the vehicle miniaturization and lightness, as the secondary unit can be small and light. There is an additional MOSFET, S_5, connected in parallel with D_1 in Figure 6, and it is used to avoid discontinuous operation conditions when the load is light. To further demonstrate it, assume that there is a light load and the q-Zsource only consists of D_1. The diode D_1 will turn off when the current flowing through it decreases to zero, thus the connection between the q-Zsource and power source is disconnected, and the relationship between them will not be tenable. This abnormal state makes the output voltage of q-Zsource change freely, and further induces a decline of the system transfer characteristics. This unwanted phenomenon will disappear if S_5 is turned on or off actively. S_5 can be also removed if the system always works at the rated state. The q-Zsource has two typical operating states, and the equivalent circuits are depicted in Figure 7.

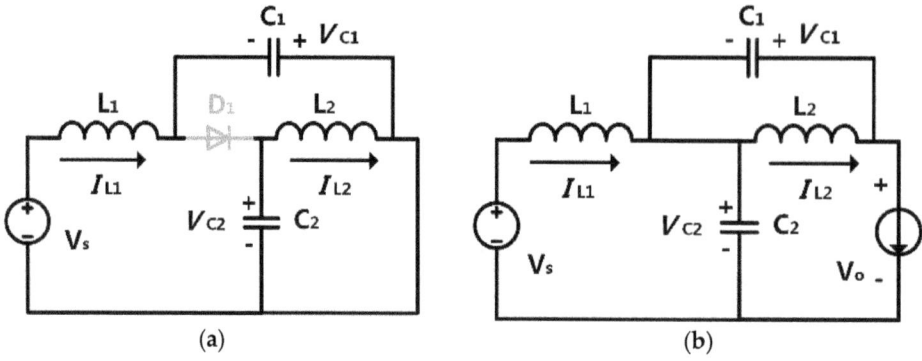

Figure 7. Equivalent circuits of q-Zsource in shoot-through state (a) and non-shoot-through state (b).

The operating mode of the q-Zsource is controlled by the cascade H Bridge. When the H Bridge works in the traditional phase shifting state, no voltage boosting phenomenon happens, but when the system charging power needs to be increased, making either of the bridge arms to be shoot-through, the voltage boosting phenomenon happens. The output of the q-Zsource is shorted by the H Bridge in the shoot-through state, and L_2 is charged by C_2, L_1 is charged by C_1 and V_S, whereas the H Bridge becomes an equivalent current source in the non-shoot-through state as shown in Figure 7b, C_2 and L_2 provide the output voltage together, C_1, L_1 and V_S provide the output voltage together. This also explains why the withstand voltage across C_1 reduces. By controlling the ratio between shoot-through time and non-shoot-through time, the output voltage of the q-Zsource can be adjusted. From Figure 7a, we have:

$$V_S + V_{C1} = V_{L1}, \ V_{C2} = V_{L2} \tag{11}$$

Similarly, from Figure 7b, we have:

$$V_S + V_{L1} + V_{C1} = V_O, \ V_S + V_{L1} = V_{C2}, \ V_{C1} = V_{L2}, \ V_{C1} + V_{C2} = V_O \tag{12}$$

Normally, L_1 equals L_2, and the currents flowing through them are the same, thus one has:

$$V_{L1} = V_{L2} \tag{13}$$

Since the volt-seconds of the inductor should be identical in the steady state, we can get:

$$(V_{C1} + V_S) \cdot T_S = (V_O - V_{C1} - V_S) \cdot T_N \tag{14}$$

T_S represents the shoot-through time and T_N represents the non-shoot-through time, the sum of those is the whole cycle time T. Then substituting Equations (11)–(13) into Equation (14), we have:

$$V_O = \frac{T}{T_N - T_S} \cdot V_S = \frac{1}{1 - 2D} \cdot V_S = BF \cdot V_S \tag{15}$$

$D = T_S/T$ is the duty cycle, BF is the boost factor produced by the shoot-through state, and it is always greater than or equal to 1. Additionally, we can also obtain the voltages across the capacitor C_1 and C_2:

$$V_{C1} = \frac{V_O - V_S}{2} = \frac{T_S}{T_N - T_S} \cdot V_S = \frac{D}{1 - 2D} \cdot V_S \tag{16}$$

$$V_{C2} = \frac{V_O + V_S}{2} = \frac{T_N}{T_N - T_S} \cdot V_S = \frac{1 - D}{1 - 2D} \cdot V_S \tag{17}$$

3.2. Control Method

Unlike the traditional buck or boost converter, the duty cycle D of the q-Zsource cannot reach 50% according to Equation (15). The voltage gain curve of the q-Zsource is shown in Figure 8, and it clearly shows that there are two operating regions.

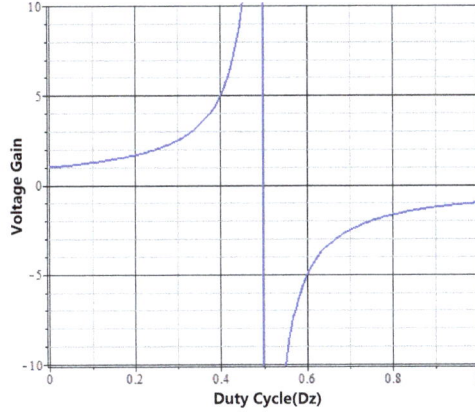

Figure 8. Voltage gain of the q-Zsource.

When D is greater than 0.5, the q-Zsource enters the negative gain region, and produces a negative output voltage, which is hardly used in practice. When D is less than 0.5, it produces a positive output voltage, thus we should limit the duty cycle D to below 0.5. All the traditional control strategies [26] can be used to control the q-Zsource and their theoretical input-output relationships still hold, the only difference is that the shoot-through time is added. The traditional phase shifting control is widely used to produce the square-wave voltages and realize the soft-switching conditions. However, it will not be discussed in this paper, as this control method has already been explained before [36]. It is worth mentioning that the q-Zsource has no effect on the output waves of the H Bridge in this mode, and only acts as a kind of filter.

When the charging power needs to be boosted, the H Bridge enters the new operating mode shown in Figure 9, which supplies the shoot-through state for the q-Zsource to boost the output voltage. Unlike the traditional phase shifting control, an additional shoot-through time $T_{shoot-through}$ is added into the control sequences, The dead time T_{dead}, shifting time $T_{shifting}$ and shoot-through time $T_{shoot-through}$ influence the output waves together. The shoot-through state in Figure 9 is realized by turning on S_3 and S_4 simultaneously, or it can be also realized by simultaneously turning on S_1 and S_2, which depends on the practical situations. The interval between t_0 and t_7 is the whole control cycle, as it is symmetrical, only the operating mode among $t_0 \sim t_3$ needs to be demonstrated. S_1 is turned off at t_0, while I_H is still positive, thus it is forced to flow through the free-wheeling diode of S_2. Before I_H changes, S_2 should be turned on at t_1, which can realize its soft switching. These two steps are similar to the control of a phase-shift-full-bridge, but not exactly the same, as the cascade loads are different. Before S_3 is turned off, S_4 is turned on at t_2, and this state is forbidden in the traditional control. However, precisely because of that, the shoot-through state is provided, which allows the q-Zsource to boost voltage, and different boost factors can be achieved by adjusting the interval between t_2 and t_3. It is noticeable here that the equivalent switching frequency viewed from the q-Zsource is two times the operating frequency of the H Bridge, which greatly reduces the volume and weight of the inductors and capacitors existed in q-Zsource. In addition, the lagging leg (S_3, S_4) is turned off with soft switching, but turned on with hard switching, which lowers the efficiency and needs to be further studied.

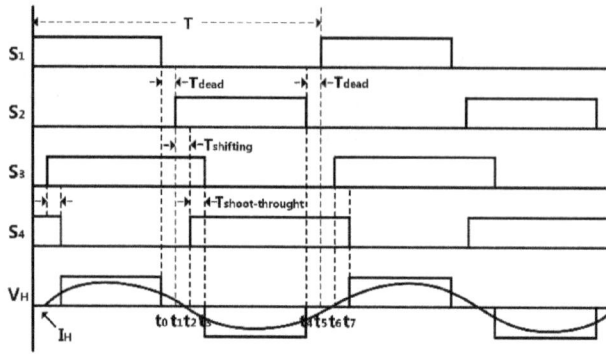

Figure 9. q-Zsource based voltage boost control sequences.

The transmitter and receiver in Figure 6 are high order resonant filters, which only allow the power signals at resonant frequency to pass through. Thus we should evaluate the fundamental components of the waves produced by the H Bridge. The FFT series of square waves depicted in Figure 10 is given as:

$$V = \sum_{n=1,2,3...}^{\infty} \frac{2A}{n\pi}\left[\cos\frac{n\pi T_{non-effective}}{T} - \cos n\pi(1 - \frac{T_{non-effecitve}}{T})\right]\sin n\omega_r t \qquad (18)$$

where A, n, $T_{non-effective}$ and T represent the amplitude of the square wave, harmonic order, non-effective time and cycle time, respectively, and $\omega_r = 2\pi/T$ is the system angular frequency. According to Equation (15), A is actually determined by BF, and $T_{non-effective}$ is determined by T_{dead}, $T_{shifting}$ and $T_{shoot-through}$. Thus the output voltage produced by the H Bridge can be regulated by controlling these parameters appropriately, whereas the voltage stress across the power switches needs to be considered before designing the BF parameter.

The degree of approximation between the square wave shown in Figure 10 and a quasi-sinusoidal wave can be quantified by THD, and the lower the THD, the less the harmonic loss. The THD of the square wave consisted of different non-effective times can be calculated according to Equation (19), where V_n represents the different harmonic amplitudes:

$$THD = \frac{\sqrt{\sum_{n=2,3,4...}^{\infty} V_n^2}}{V_1} \qquad (19)$$

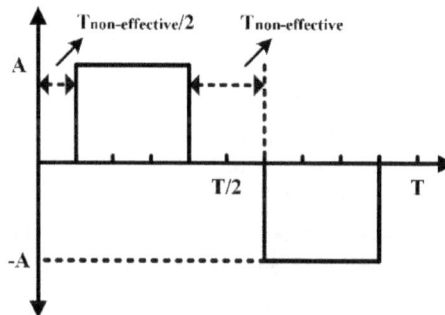

Figure 10. Typical square wave produced by the H Bridge.

The corresponding THD calculation results at different frequencies are shown in Figure 11, which clearly indicates that the lowest THD happens around 2 µs~3 µs non-effective times at 50 kHz and 1 µs~2 µs non-effective times at 80 kHz, rather than 0 µs non-effective time. The optimal non-effective time is where THD has the smallest decrease with the increasing frequencies. Because we adopt 80 kHz as the system operating frequency, the non-effective time should be designed around 1 µs~2 µs to reduce THD as well as harmonic loss.

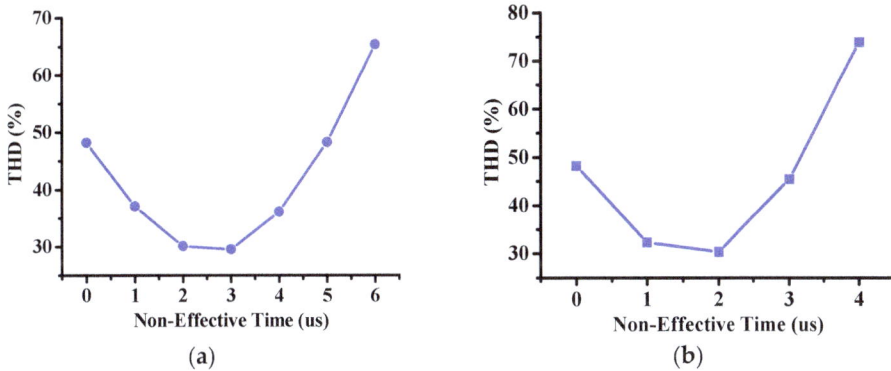

Figure 11. THDs of square waves with different non-effective times at 50 kHz (**a**) and 80 kHz (**b**).

4. Experiments

A prototype was implemented to validate our research results, as shown in Figure 12, and the magnetic coils are designed based on the nested three-layer optimization method, which will be discussed in our other papers. The prototype is fabricated according to the schematic shown in Figure 6, where the power source adopts a 62100H-600 high-voltage DC power supply (Chroma, Taoyuan, Taiwan), the H Bridge employs four SPW47N60C3 MOSFETs (Infineon, Neubiberg, Germany), the resonant capacitors adopt B32672L thin-film series (TDK-EPCOS, Tokyo, Japan) ,the rectifier consists of four IDW30E65D1 fast recovery diodes (Infineon, Neubiberg, Germany) and the battery pack consists of 24 lead-acid battery units.

Figure 13 shows the transfer characteristics of the wireless power transfer system at 20 cm transfer distance, and it is worth mentioning that the q-Zsource does not work, and is only present as a filter. In Figure 13a, the transfer power increases with rising input voltages or battery pack voltages. In practice, the power factor correction (PFC) with 400 V output voltage is employed to enhance AC power quality, and the 300 V battery pack is widely used for many production-ready vehicles, like Toyota Prius, Chevrolet Volt, Mitsubishi i-MiEV as well as Nissan Leaf, thus we define this situation as the system rated operating state. In the rated state, the charging power is 3220 W as shown in Figure 13a, which is a little bigger than 3 kW, since the battery pack voltage increases from 300 V to 309.7 V when the charging current (RMS value is 10.4 A as shown in Figure 14) flows through the battery resistance, causing an extra voltage drop. Figure 13b shows that the transfer efficiencies are nearly unchanged despite the increasing input voltages or battery pack voltages, and the rated efficiency where the input voltage equals 400 V and the battery pack voltage equals 300 V is around 95%.

(a)   (b)

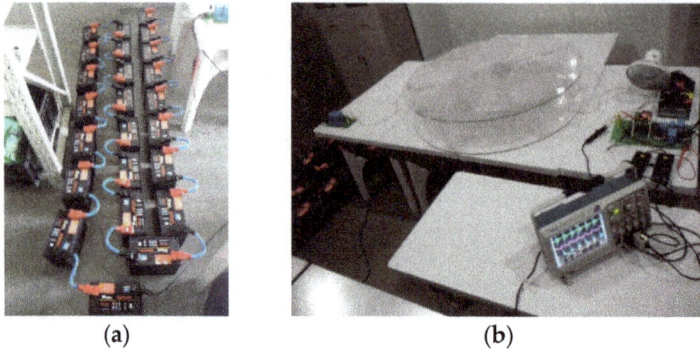

Figure 12. Prototype of wireless power transfer system. (a) 300V lead-acid battery pack. (b) Transmitter and receiver of wireless power transfer system. The magnetic coil is placed on the top of a perspex plate and fixed by eight perspex bars.

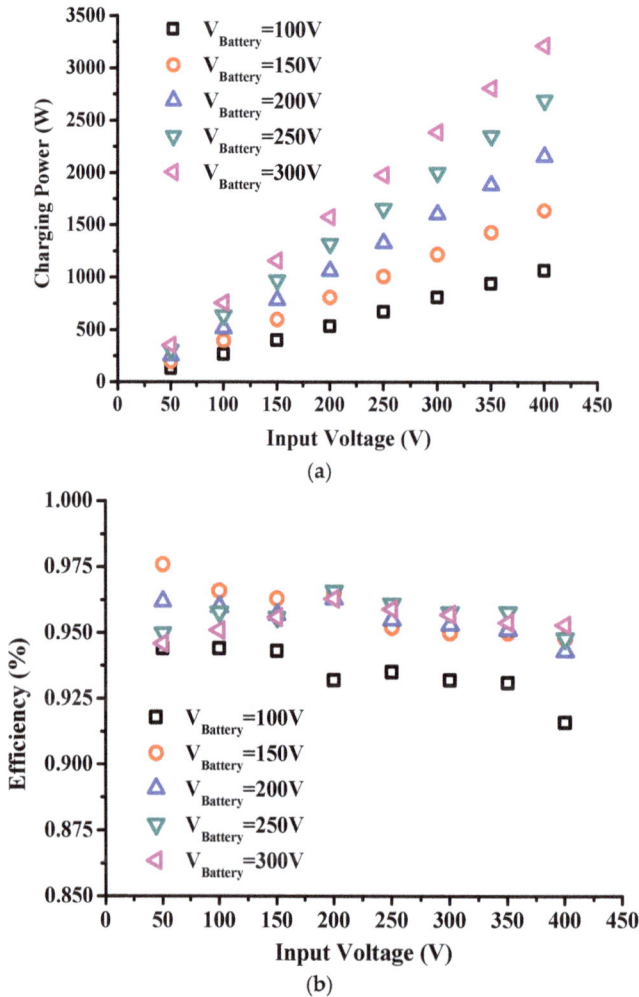

(a)

(b)

Figure 13. Wireless transfer power (a) and efficiency (b) at different input voltages and battery pack voltages at 20 cm transfer distance.

Figure 14 shows some critical waveforms of the wireless power transfer system, where the cyan curve shows the output voltage produced by the H Bridge, the green curve shows the voltage across the resonant capacitor in the transmitter and the purple curve shows the charging current for the battery pack. Because the reduction scale of high voltage probe is 200, the measured voltages need to be multiplied by 200. The amplitude of the square-wave voltage produced by the H Bridge is 400 V. The RMS value of the charging current is 10.4 A, and there are some ripple waves, the amplitude of which depends on the filter capacitors. The bigger the filter capacitors, the smaller the ripple waves.

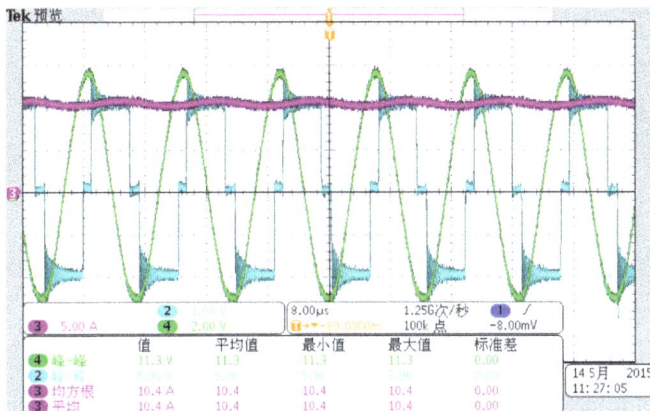

Figure 14. Purple curve: the charging current for the battery pack; cyan curve: the output voltage produced by the H Bridge; green curve: the voltage across the resonant capacitor in the transmitter. They are measured when the system works in the rated state (V_{in} = 400 V, V_B = 309.7 V, $P_{charging}$ = 3220 W, $D_{transfer}$ = 20 cm).

The following experimental results are measured when the q-Zsource works, and two different shoot-through times are shown to clearly demonstrate the q-Zsource principle. Figure 15 indicates that different shoot-through times determine different boost factors. Although the input voltages (green curve) of both Figure 15a,b are identical (200 V), their output voltages (cyan curve) are different. When the shoot-through time is 1 µs, the boost factor is around 1.5, thus the output voltage is around 300 V, and the charging current is 8.26 A. When the shoot-through time is 1.5 µs, the boost factor is around 2, thus the output voltage is around 400 V, and the charging current is 10.2 A. Summarily, the charging currents can be adjusted by controlling the shoot-through times.

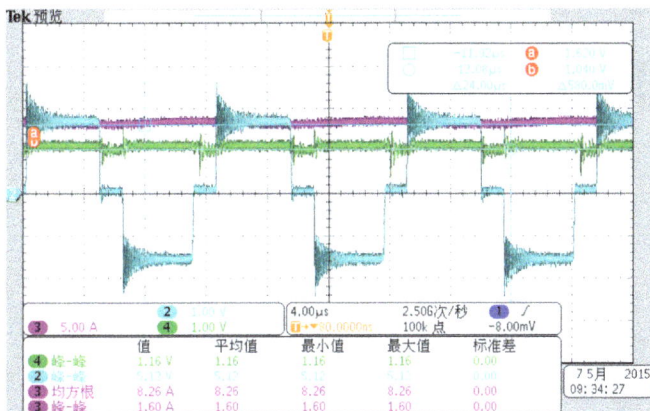

(a)

Figure 15. *Cont.*

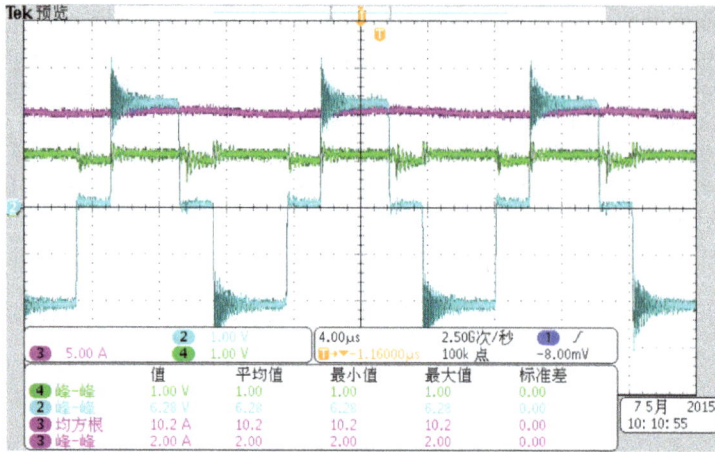

(b)

Figure 15. Purple curve: the charging currents for the battery pack; green curve: the input voltage of the q-Zsource; cyan curve: the output voltage produced by the H Bridge measured (**a**) at the 1 μs shoot-through time; and (**b**) at the 1.5 μs shoot-through time.

In Figure 16a,b the q-Zsource input voltages (green curve) are the same, but their output voltages (cyan curve) are different due to the different boost factors. Unlike the square-wave voltages produced by the H Bridge, the q-Zsource output voltages are pulsatile and always positive. In the shoot-through state, where the q-Zsource output voltage equals zero, the q-Zsource input current (purple curve) increases with a positive slope. However, it decreases with a negative slope when the q-Zsource enters into the non-shoot-through state.

Figure 17 shows the output currents (purple curve) and voltages (cyan curve) produced by the H Bridge, which validate that the system is basically in the resonant state. However, the quasi-resonant state is not beneficial for soft-switching, and hence, efficiency improvement. In practice, the current produced by the H Bridge should lag the voltage to a certain degree, thus an additional 1 nF capacitor is added into the transmitter capacitor array in this paper. The green curve represents the voltage across the q-Zsource capacitor (C_2 in Figure 6), and it is smaller than the output voltage produced by the H Bridge, but it is bigger than the input voltage of the q-Zsource shown in Figure 16.

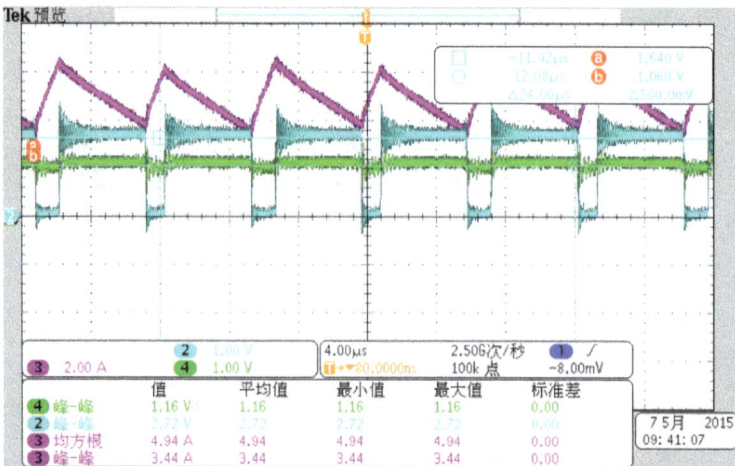

(a)

Figure 16. *Cont.*

(b)

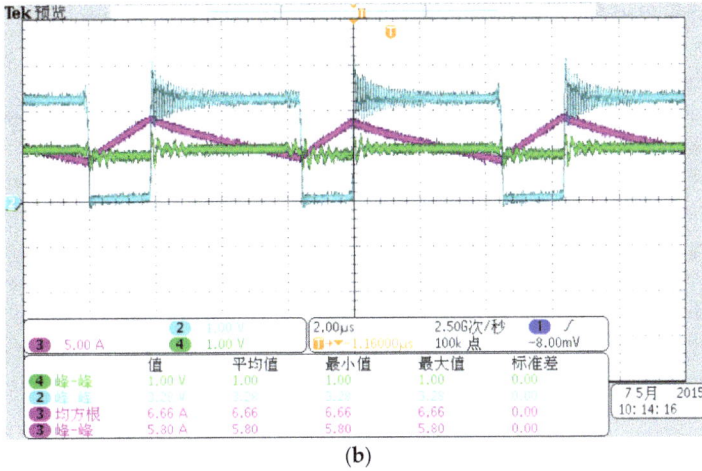

Figure 16. Purple curve: the input current of q-Zsource, green curve shows the input voltage of q-Zsource, cyan curve shows the output voltage of q-Zsource measured (**a**) at the 1 µs shoot-through time; and (**b**) at the 1.5 µs shoot-through time.

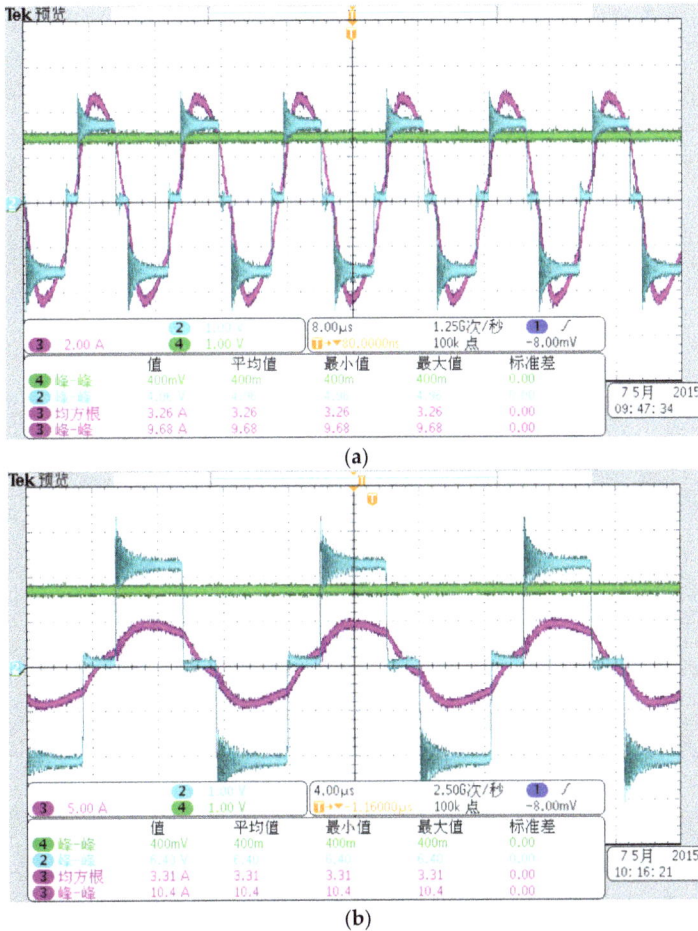

(a)

(b)

Figure 17. Purple curve: the currents produced by the H Bridge; cyan curve: the output voltages produced by the H Bridge; green curve: the capacitor voltages of the q-Zsource measured (**a**) at the 1 µs shoot-through time; and (**b**) at the 1.5 µs shoot-through time.

Figure 18a indicates that the charging power or current can be adjusted by changing the shoot-through times. Take the 200 V input voltage as an example, the charging current is 5.26 A when there is no shoot-through time, and it is increased to 8.26 A with 1 μs shoot-through time, and it is further increased to 10.2 A when the shoot-through time is 1.5 μs. Because the boost factor is close to 2 at 1.5 μs shoot-through time, the input voltages above 200 V are not allowed, which may damage the resonate capacitors. In Figure 18b, the efficiencies decline with the increasing shoot-through times, for the shoot-through state makes the MOSFETs of latter bridge arm lost their soft-switching conditions. Additionally, the operating frequency (160 kHz) of q-Zsource doubles that (80 kHz) of wireless power transfer system, which further causes the decline in the transfer efficiency. This phenomenon can be suppressed by reducing the operating frequency of q-Zsouce. If it is decreased to 40 kHz, the loss can be theoretically reduced as much as four times. However, the reduction of the frequency requires bigger inductors and capacitors than before, thus the q-Zsource parameters need to be re-optimized, which is our future work.

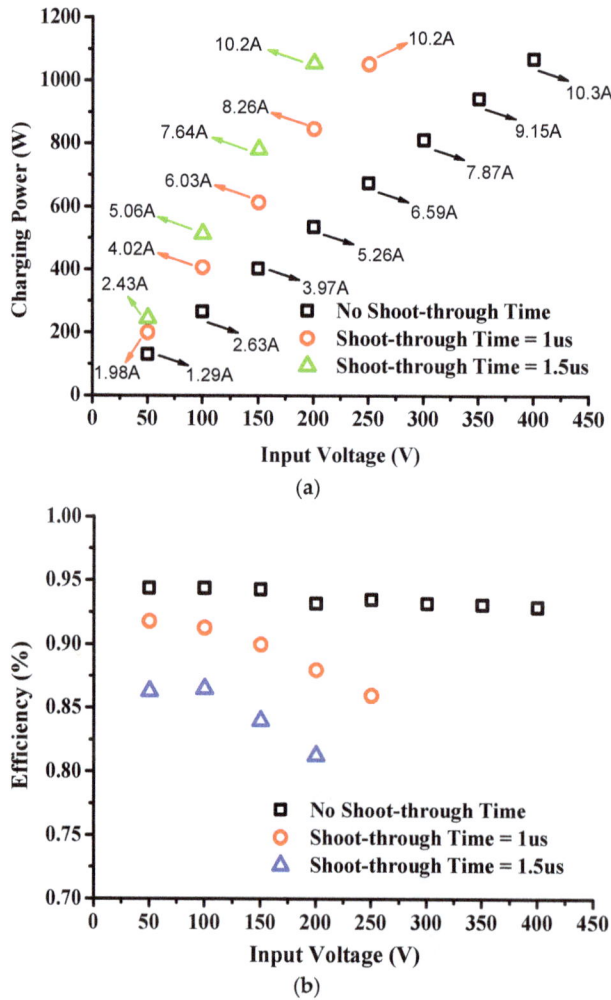

Figure 18. Wireless transfer power (**a**) and efficiency (**b**) characteristics based on the q-Zsource voltage boosting method at 20 cm transfer distance.

5. Conclusions

This paper studies a 3 kW vehicle-mounted wireless power transfer system. First, the efficiency of the LC-LC series topology is verified to be higher than that of the LCL-LCL topology when their transfer power are the same. Then a q-Zsource-based power regulation method is proposed to adjust the charging current online. At last, a 3 kW prototype with ~95% efficiency over a 20 cm transfer distance is implemented to validate our research results. Different shoot-through time durations determine different charging currents despite the same input voltage. When the input voltage is set to be 200 V, a 1 μs shoot-through time can boost the charging current from 5.26 A to 8.26 A, and a 1.5 μs shoot-through time can boost the charging current from 5.26 A to 10.2 A. We hope the work presented in this paper is beneficial to the development of wireless power transfer systems.

Acknowledgments: This work is financially supported by the Major State Basic Research Development Program of China (973 Program, Grant No. 2011CB711202) and the National Natural Science Foundation of China (NSFC, Grant No. 51576142).

Author Contributions: Zhenshi Wang designed the system and analyzed the results, Xuezhe Wei and Haifeng Dai provided guidance and key suggestions.

Conflicts of Interest: The authors declare no conflict of interest.

References

1. Brown, W.C. The history of power transmission by radio waves. *IEEE Trans. Microwave Theory Tech.* **1984**, *32*, 1230–1242. [CrossRef]
2. Tucker, C.A.; Warwick, K.; Holderbaum, W. A contribution to the wireless transmission of power. *Int. J. Electr. Power Energy Syst.* **2013**, *47*, 235–242. [CrossRef]
3. Brown, W.C. Status of the microwave power transmission components for the solar power satellite (SPS). *IEEE MTT-S Int. Microwave Symp. Dig.* **1981**, *81*, 270–272.
4. Glaser, P.E. Power from the Sun: Its future. *Science* **1968**, *162*, 857–861. [CrossRef] [PubMed]
5. Covic, G.A.; Boys, J.T. Modern trends in inductive power transfer for transportation applications. *IEEE J. Emerg. Sel. Top. Power Electr.* **2013**, *1*, 28–41. [CrossRef]
6. Wei, X.Z.; Wang, Z.S.; Dai, H.F. A critical review of wireless power transfer via strongly coupled magnetic resonances. *Energies* **2014**, *7*, 4316–4341. [CrossRef]
7. Kurs, A.; Karalis, A.; Moffatt, R.; Joannopoulos, J.D.; Fisher, P.; Soljacic, M. Wireless power transfer via strongly coupled magnetic resonances. *Science* **2007**, *317*, 83–86. [CrossRef] [PubMed]
8. Sample, A.P.; Meyer, D.A.; Smith, J.R. Analysis, experimental results, and range adaptation of magnetically coupled resonators for wireless power transfer. *IEEE Trans. Ind. Electron.* **2011**, *58*, 544–554. [CrossRef]
9. Li, X.H.; Zhang, H.R.; Peng, F.; Li, Y.; Yang, T.Y.; Wang, B.; Fang, D. A wireless magnetic resonance energy transfer system for micro implantable medical sensors. *Sensors* **2012**, *12*, 10292–10308. [CrossRef] [PubMed]
10. Puccetti, G.; Stevens, C.J.; Reggiani, U.; Sandrolini, L. Experimental and numerical investigation of termination impedance effects in wireless power transfer via metamaterial. *Energies* **2015**, *8*, 1882–1895. [CrossRef]
11. Sun, L.; Tang, H.; Zhang, Y. Determining the frequency for load-independent output current in three-coil wireless power transfer system. *Energies* **2015**, *8*, 9719–9730. [CrossRef]
12. Sanghoon, C.; Yong-Hae, K.; Kang, S.-Y.; Myung-Lae, L.; Jong-Moo, L.; Zyung, T. Circuit-model-based analysis of a wireless energy-transfer system via coupled magnetic resonances. *IEEE Trans. Ind. Electron.* **2011**, *58*, 2906–2914.
13. Kiani, M.; Uei-Ming, J.; Ghovanloo, M. Design and optimization of a 3-coil inductive link for efficient wireless power transmission. *IEEE Trans. Biomed. Circuits Syst.* **2011**, *5*, 579–591. [CrossRef] [PubMed]
14. Dukju, A.; Songcheol, H. A study on magnetic field repeater in wireless power transfer. *IEEE Trans. Ind. Electron.* **2013**, *60*, 360–371.
15. Musavi, F.; Eberle, W. Overview of wireless power transfer technologies for electric vehicle battery charging. *IET Power Electron.* **2014**, *7*, 60–66. [CrossRef]
16. Del Toro, T.G.X.; Vázquez, J.; Roncero-Sanchez, P. Design, implementation issues and performance of an inductive power transfer system for electric vehicle chargers with series-series compensation. *IET Power Electron.* **2015**, *8*, 1920–1930.
17. Siqi, L.; Mi, C.C. Wireless power transfer for electric vehicle applications. *IEEE J. Emerg. Sel. Top. Power Electron.* **2015**, *3*, 4–17. [CrossRef]

18. Keeling, N.A.; Covic, G.A.; Boys, J.T. A unity-power-factor IPT pickup for high-power applications. *IEEE Trans. Ind. Electron.* **2010**, *57*, 744–751. [CrossRef]

19. Hwang, S.-H.; Kang, C.G.; Son, Y.-H.; Jang, B.-J. Software-based wireless power transfer platform for various power control experiments. *Energies* **2015**, *8*, 7677–7689. [CrossRef]

20. Zaheer, A.; Covic, G.A.; Kacprzak, D. A bipolar pad in a 10-kHz 300-W distributed IPT system for AGV applications. *IEEE Trans. Ind. Electron.* **2014**, *61*, 3288–3301. [CrossRef]

21. Madawala, U.K.; Thrimawithana, D.J. New technique for inductive power transfer using a single controller. *IET Power Electron.* **2012**, *5*, 248–256. [CrossRef]

22. Gao, Y.; Farley, K.; Tse, Z. A uniform voltage gain control for alignment robustness in wireless EV charging. *Energies* **2015**, *8*, 8355–8370. [CrossRef]

23. Chwei-Sen, W.; Covic, G.A.; Stielau, O.H. Power transfer capability and bifurcation phenomena of loosely coupled inductive power transfer systems. *IEEE Trans. Ind. Electron.* **2004**, *51*, 148–157.

24. Ricketts, D.S.; Chabalko, M.J.; Hillenius, A. Experimental demonstration of the equivalence of inductive and strongly coupled magnetic resonance wireless power transfer. *Appl. Phys. Lett.* **2013**, *102*, 053904. [CrossRef]

25. Birrell, S.A.; Wilson, D.; Yang, C.P.; Dhadyalla, G.; Jennings, P. How driver behaviour and parking alignment affects inductive charging systems for electric vehicles. *Transp. Res. C Emerg. Technol.* **2015**, *58*, 721–731. [CrossRef]

26. Choi, W.P.; Ho, W.C.; Liu, X.; Hui, S.Y.R. Comparative study on power conversion methods for wireless battery charging platform. In Proceedings of the 2010 14th International Power Electronics and Motion Control Conference (EPE/PEMC), Ohrid, Macedonia, 6–8 September 2010; pp. S15:9–S15:16.

27. Xuan, N.B.; Vilathgamuwa, D.M.; Foo, G.H.B.; Peng, W.; Ong, A.; Madawala, U.K.; Trong, D.N. An Efficiency Optimization Scheme for Bidirectional Inductive Power Transfer Systems. *IEEE Trans. Power Electron.* **2015**, *30*, 6310–6319.

28. Jun-Young, L.; Byung-Moon, H. A Bidirectional Wireless Power Transfer EV Charger Using Self-Resonant PWM. *IEEE Trans. Power Electron.* **2015**, *30*, 1784–1787.

29. Madawala, U.K.; Thrimawithana, D.J. A Bidirectional Inductive Power Interface for Electric Vehicles in V2G Systems. *IEEE Trans. Ind. Electron.* **2011**, *58*, 4789–4796. [CrossRef]

30. Madawala, U.K.; Thrimawithana, D.J. Modular-based inductive power transfer system for high-power applications. *IET Power Electron.* **2012**, *5*, 1119–1126. [CrossRef]

31. Xuan, N.B.; Foo, G.; Ong, A.; Vilathgamuwa, D.M.; Madawala, U.K. Efficiency optimization for bidirectional IPT system. In Proceedings of the 2014 IEEE Transportation Electrification Conference and Expo (ITEC), Dearborn, MI, USA, 15–18 June 2014; pp. 1–5.

32. Madawala, U.K.; Thrimawithana, D.J. Current sourced bi-directional inductive power transfer system. *IET Power Electron.* **2011**, *4*, 471–480. [CrossRef]

33. Zheng, P.F. Z-source inverter. *IEEE Trans.Ind. Appl.* **2003**, *39*, 504–510. [CrossRef]

34. Peng, F.X.; Guang, W.X.; Qiao, C.Z. A single-phase AC power supply based on modified Quasi-Z-Source Inverter. *IEEE Trans. Appl. Supercond.* **2014**, *24*, 1–5. [CrossRef]

35. Boys, J.T.; Covic, G.A. Decoupling Circuits. U.S. 7279850B2, 9 October 2005.

36. Hothongkham, P.; Kongkachat, S.; Thodsaporn, N. Performance comparison of PWM and phase-shifted PWM inverter fed high-voltage high-frequency ozone generator. In Proceedings of the 2011 IEEE Region 10 Conference on Convergent Technologies for the Asia-Pacific Region, Bali, Indonesia, 21–24 November 2011; pp. 976–980.

12

An Algorithm to Translate Building Topology in Building Information Modeling into Object-Oriented Physical Modeling-Based Building Energy Modeling

WoonSeong Jeong and JeongWook Son *

Academic Editor: Chi-Ming Lai

Department of Architectural Engineering, Ewha Womans University, Seoul 120-750, Korea; wsjeong@ewha.ac.kr
* Correspondence: jwson@ewha.ac.kr

Abstract: This paper presents an algorithm to translate building topology in an object-oriented architectural building model (Building Information Modeling, BIM) into an object-oriented physical-based energy performance simulation by using an object-oriented programming approach. Our algorithm demonstrates efficient mapping of building components in a BIM model into space boundary conditions in an object-oriented physical modeling (OOPM)-based building energy model, and the translation of building topology into space boundary conditions to create an OOPM model. The implemented command, *TranslatingBuildingTopology*, using an object-oriented programming approach, enables graphical representation of the building topology of BIM models and the automatic generation of space boundaries information for OOPM models. The algorithm and its implementation allow coherent object-mapping from BIM to OOPM and facilitate the definition of space boundaries information during model translation for building thermal simulation. In order to demonstrate our algorithm and its implementation, we conducted experiments with three test cases using the BESTEST 600 model. Our experiments show that our algorithm and its implementation enable building topology information to be automatically translated into space boundary information, and facilitates the reuse of BIM data into building thermal simulations without additional export or import processes.

Keywords: Building Information Modeling; object-oriented physical modeling; building energy modeling; building topology

1. Introduction

The complex data exchange between architectural building models and energy performance simulation models has been a critical issue at the early design phase, and results in preventing the efficient use of building energy performance simulation in the design process [1]. Specifically, the data exchange process of the building's geometry from a building design model to an energy simulation model as a part of an input file is labor-intensive, tedious, error-prone, and cumbersome, and the model translation process usually requires a significant workload allocated into the building energy simulation project [1–4]. In addition, the decision-making inherent in translating the initial architectural design model into an energy simulation engine is usually subjective and non-reproducible [5].

Since 1996, over four hundreds building energy simulation tools have been listed in the "Building Energy Software Tools Directory" provided by U.S. Department of Energy (DOE) [6], and the tools differ in their simulation engines and graphical user interfaces (GUIs) [7,8]. The GUIs can reduce the amount of time that must be spent in preparing geometry input data from architectural models, but architects and designers have many difficulties in using them since they require extensive knowledge regarding thermal processes and energy simulation [9,10]. We presume that most of the difficulties originate from the implementation concept for the building energy simulation tools' engine; the majority of the tools' engines do not exploit object-oriented programming (OOP) and impede natural object mapping from object-oriented design models.

Recently, many studies have been conducted on the geometry data exchange between Building Information Modeling (BIM) and Building Energy Modeling (BEM) using standard data schemas such as Industry Foundation Classes (IFC) and Green Building XML (gbXML). Some existing simulation tools for BEM were modified for BIM adoption, and other tools were developed to give compatibility with BIM authoring tools [11]. While the BIM-based BEM approach facilitates data translation between design models and energy models, additional efforts such as a manual model check need to be conducted in order to create a reliable energy model from a design model [12]. For example, a manual model check to define a building space is necessary in order to transfer that building space into the space boundaries required for a whole-building energy performance simulation engine. Obtaining space boundaries from the building geometry defined in object-based building design models, including BIM models, is a crucial task [13]. Previous research work [14] demonstrated that the use of object-oriented constructs within BIM and BEM facilitated efficient and reliable translation and improved maintainability. In addition, several researches [14–17] employed a BIM-based energy modeling and simulation approach using BIM and OOPM which provides a number of benefits. First, it enables more reliable OOPM-based BEM model generation from a BIM model. Second, it facilitates the reuse of original BIM data in an OOPM-based BEM without an import/export process. Finally, the approach facilitates further development via object encapsulation and provision.

In this study to improve and enhance the translation of space boundaries between BIM and BEM, we investigated the development of an algorithm to convert BIM models without defining thermal space boundaries into object-oriented physical models for thermal simulation. We utilized C# programming to implement our algorithm within a BIM authoring system, and Modelica, an object-oriented physical modeling language, to create a BEM model and execute the thermal simulation. We conducted the following phases to develop the algorithm and implement it.

(1) Develop an algorithm to translate building spaces into space boundaries conditions between BIM and OOPM.
(2) Implement the algorithm using the targeted object-oriented programming language.
(3) Conduct test cases to demonstrate and validate the algorithm and its implementation.

The objectives of the algorithm development are: (1) to enable building spaces in BIM to be automatically translated into an OOPM-based BEM; (2) to enhance the usability of the original BIM data in building thermal simulation without human interference; and (3) to develop a more reliable OOPM-based BEM model from BIM models.

The research scope is confined to translating the building topology of a BIM model into the space boundary conditions of an OOPM-based BEM model. In this paper, we adopted the terminology of *ModelicaBEM* from [14] representing the OOPM-based BEM models using a Modelica library (Lawrence Berkeley National Laboratory (LBNL) Modelica Buildings Library) and translated from BIM models. Additionally, the terms "Building Information Modeling" or "Building Information Model(s)" are referred to as BIM interchangeably in different contexts. Likewise, the term OOPM implies "Object-Oriented Physical Modeling" or "Object-Oriented Physical Model(s)".

2. Background

2.1. Translating a Building's Geometry Information between Architectural Models and Building Energy Simulation Models

Current building energy simulation tools have been developed to evaluate building energy performance and thermal comfort for the whole building life cycle. Such simulation tools utilize computer programming languages including FORTRAN, C, and C++ to create the simulation engines and the graphical user interfaces (GUIs) [7,8]. The GUIs facilitate rapid input processing by generating the building's geometry information from the user's graphical input data. Although the GUIs support the generation of the building's geometry information as a part of energy model creation from a design model, transition of a building's geometry data produced by architects or building designers is considered a labor-intensive and time-consuming process [1,2,13,18,19]. In order to match the definition of the building's geometry between architectural models and building thermal simulation models, building performance simulation professionals, rather than the building's original designers, still need to reproduce the geometric information. In addition, differing object semantics between the models demand practitioners' decisions, are mostly arbitrary and non-reproducible [5].

In the meantime, various research prototypes [1,2] and energy simulation tools, such as Ecotect, Hevacomp, and eQUEST, have been developed to reduce the redundant process and increase the usability of the original building design information. Recently, BIM-based energy simulation tools have enhanced the efficiency of reusing the building design data by integrating BIM.

However, the translation issue still exists; a certain level of data translation needs to be performed due to incongruent information between BIM and thermal simulation models. Especially, information on the condition of space boundaries is required only in the thermal simulation model, and the data translation to match the space boundary conditions from a design model is considered a significant part of thermal simulation modeling [20].

The effective and efficient building geometry translation of space boundaries can be achieved when the thermal and building design modeling are conducted based on the same modeling concept (object-oriented modeling concept), and the implementation of the translation is performed systematically, under a reliable algorithm.

2.2. Building Topology and Space Boundary Representation

2.2.1. Building Topology Representation in Building Information Modeling (BIM)

Object-oriented building design models such as BIMs consist of building components such as walls, floors, roofs, doors, windows, *etc.*, representing them as three-dimensional objects. The building components in such a model represent their surfaces as an attribute in a child object with height and length. A series of surfaces with a depth attribute composes a building component. In addition, connectivity between building components can represent building topology via object relationships.

The relationships can be defined as classes in BIM, and construction objects from the classes enable the building component connectivity through defined functions. For example, a Revit BIM model represents the connectivity between a room element and walls enclosing the room through *SpatialElementBoundaryOptions* and *BoundarySegment* classes.

2.2.2. Space Boundary Representation in Building Energy Performance Simulation Tools

In contrast, most building energy performance (BEP) simulation tools define the building components as systems of surfaces for whole-building energy simulation models [13,20]. Typically, the surfaces are represented by length and height, and a thickness value can be added to depict a three-dimensional space. The BEP tools depict spaces and/or thermal zones through surfaces called "space boundaries". Defining the space boundaries from building geometry with non-object-oriented modeling tools is a time-consuming task.

Space boundaries are defined by two surfaces; one is the inside and the other is the outside of building components. Only exterior building components are exempted in the space boundaries definition, due to the characteristics of BEP simulation tools: (1) most BEP simulation tools deal with exterior building components consisting of a building envelope; and (2) the BEP tools calculate energy transmission and flow through building components only when they are perpendicular to the components. In other words, the exterior components are not defined as space boundaries in current

BEP tools [20]. Typically, zones in a building model consist of individual rooms or spaces presented as having the same behavioral characteristics as those of the zones.

Currently, energy practitioners model building geometry for BEP simulation by manual re-creation of the original building design model as a part of the input data without recognizing the object mapping process from the building geometry to the space boundaries. The manual transition of building geometry to define space boundaries is usually ad hoc, inconsistent, and arbitrary [20].

Previous research works [20] demonstrated the clarifications and systematized definitions for building geometry translation as five levels or types of space boundaries. The definitions facilitate standardization in preparing building geometry input data, establishing a basis for semi-automation of building geometry data transformation, and limiting further misunderstandings and misrepresentations in the building geometry translation. However, the definition established the rule between model-based Computer Aided Design (CAD) tools and a non-object-oriented BEP simulation engine. While the definition provides such benefits in conducting BEP simulations, the original issue still exists in the translation between the models: object mismatching. The inconsistent model view of buildings causes object mismatching. When the modeling approach in creating architectural models and building energy models is applied with the same modeling concept as for the object-oriented modeling approach, object matching can be conducted intuitively.

2.3. Creating Building Topology from BIM to Translate into Space Boundary Conditions

Building geometry representations from BIM for BEP simulation must be transformed through a series of modifications including simplifying, reducing, translating, or interpreting, due to the inconsistent data structure between BIM and BEP simulation [20,21]. Incongruent object semantics and behavior mismatches cause such modification or abstraction. As described in the previous section, the surface representation is simplified when building components in BIM are translated into BEP simulation models. The semantic and behavior mismatches originate from the distinct model view of the same building model and initiate the manual input process in creating a BEP simulation model. For example, a BIM model represents a building envelope as consisting of building components, whereas a BEP simulation model recognizes the building components as exterior or interior surfaces. To prepare a simulation model, the building components should be transformed into surface elements.

To apply a coherent object relationship between BIM and a building energy model (BEM) for BEP simulation, an object-based modeling approach and consistent model view definition of building models can facilitate an efficient, natural and intuitive model translation. Recently, a BIM-based BEM approach using object-oriented physical modeling (OOPM) has been investigated [14–17]. The approach demonstrated the integration of BIM and OOPM-based energy modeling in: (1) enhancing the interoperability between BIM and BEM; (2) enabling reliable BEM generation from BIM; (3) enabling multi-domain BEP simulations from BIM; and (4) enabling BIM to be utilized as a common user interface for multi-domain BEP simulations. To implement the integration, the research team utilized BIM for an architectural modeling and an OOPM engine to create BEMs and execute BEP simulation.

The previous research work developed a comprehensive data exchange model and framework, facilitating direct mapping between BIM and OOPM-based BEM and supporting an easy-to-use user interface. The framework implemented object mapping between BIM and OOPM-based BEM by automatically creating Modelica (an OOPM language)-based BEM (*ModelicaBEM*) models. A *ModelicaBEM* model represents building geometry from BIM in an OOPM view using Modelica. For example, a room object in BIM can be represented as a thermal zone in *ModelicaBEM*.

One of the main features of the framework generates building topology from created building components. The generated building topology is translated into space boundary conditions defined in the OOPM-based simulation engine, the Lawrence Berkeley National Laboratory (LBNL) Modelica Buildings Library [22]. To enhance the efficiency and enable more natural mapping of room-to-thermal zone translation, a well-organized algorithm associated with BIM and OOPM-BEM should be developed and implemented by demonstrating the automatic generation of space boundary conditions from a complex building in BIM. The previous works demonstrated the overall building topology translation from BIM to OOPM-BEM and identified required information; however, our research work investigated the algorithm to enhance the efficiency of natural mapping between multi-zone BIM models and OOPM-BEM.

3. Research Objectives

This section describes research objectives by defining challenges and tasks for the development of an algorithm. In order to achieve our research objectives, we demonstrated specific application (*TranslatingBuildingTopology*) development. *TranslatingBuildingTopology* is intended to handle the room-to-thermal zone translation from the building topology of BIM to the space boundary conditions of OOPM-BEM to facilitating the creation of *ModelicaBEM* models.

The main challenge is to facilitate object-mapping processes between the two object-oriented models due to the different geometry relationship and semantics. Specifically, the main objective is to enhance a seamless building geometry translation, requiring automatic building topology conversion into the space boundary conditions between BIM and *ModelicaBEM*. To achieve an effective and efficient data transformation, we conducted the following tasks:

- Develop an algorithm to represent the building topology translation based on the investigated object mapping process and identifying required datasets.
- Implement the algorithm by visualizing building components connections and generating specific *ModelicaBEM* model codes.
- Execute thermal simulations by using *ModelicaBEM* models including the translated space boundary conditions from a BIM model.

4. Methodology

This section describes detailed methodology to achieve the research objectives and utilized tools to implement the methodology. In order to develop the room-to-thermal zone translation algorithm, we defined distinctive methodology as follows: (1) identifying the information required for the translation in BIM and *ModelicaBEM*; (2) developing pseudocodes to represent building topology in BIM; (3) implementing the pseudocodes by creating a command using the Revit Application Programming Interface (API); and (4) conducting experiments to validate our methodology, creating *ModelicaBEM* models for validation, using the LBNL Modelica Buildings Library.

To support the development of the algorithm, we investigated the required information in terms of room-to-thermal zone translation; and developed pseudocodes to represent the required information for the translation. The *TranslatingBuildingTopology* command implementation demonstrated the building components connection visualization and *ModelicaBEM* model codes generation as well. For the validation of our algorithm development and its implementation, we conducted a case study by applying the implemented commands to a multi-zone BIM models. Following paragraphs clarify each mentioned step.

4.1. Identifying Required Information

Mismatched objects' semantics and behavior investigation enables us to identify the required information in room-to-thermal zone translation. We indicated the impediment in building geometry translation between BIM and BEP simulation tools due to object semantics and behavior mismatches. However, in order to account for the mismatches clearly we defined them as follows in this study: (1) a semantic mismatch between room objects in Revit and zone objects in LBNL Modelica library; and (2) a behavior mismatch between BIM and *ModelicaBEM*. While we demonstrate the object mapping between Revit and LBNL Modelica library, the semantic mismatches are identified based on the general concepts for the translation. Once the definition of the mismatches is cleared, identified data structures from Revit and the library will be demonstrated for setting the required information. We explain the detailed development process of the definition required information in the Development section.

4.2. Developing Pseudocodes

The pseudocodes description allows us to describe the operation of the room-to-thermal zone translation from the identified required information. The developed pseudocodes enables executable commands to be implemented in the BIM environment. We will describe the development, focusing on function description to be directly implemented using Revit API.

4.3. Implementing the Pseudocodes

Functions in the pseudocodes description facilitates the implementation of executable commands using a specific programming language. We implemented the pseudocode functions as an executable add-in command in the Revit environment using a C# programming language and Revit API. The implemented command, *TranslatingBuildingTopology*, allows retrieval of building components connectivity corresponding to building topology and visualizing the topology as a building components connectivity diagram. Once the topology representation diagram is generated, the command generates the specific values of space boundary conditions as a Modelica code preparation.

4.4. Conducting Experiments

Experiments including three test cases were conducted to demonstrate and validate the room-to-thermal translation algorithm. We utilized the BESTEST Case 600 building model for the experiments, which is one of standard ASHRAE simulation testing models and LBNL's research validated the model using Modelica. For demonstration, a prototype presents how room objects in multi-zone BIM models can be automatically translated into thermal zones incorporated with the space boundary conditions. For validation, thermal simulation executions are conducted by creating fully-prepared *ModelicaBEM* models following LBNL's BEM structure.

4.5. Tools

To develop the algorithm and then implement it, we utilized the BIM authoring tool (Autodesk Revit) and its application programming interface (API), and an OOPM-based BEP simulation engine (LBNL Modelica Buildings Library [22]).

4.5.1. BIM Authoring Tool and Its API

BIM authoring tools, including Revit Architecture [23], ArchiCAD [24], and AECOsim Building Designer V8i [25], facilitate a BIM model as three-dimensional, semantically-rich, and parametrically-modeled during a building's lifecycle [26,27]. Such BIM tools allow building components to be represented by geometry and non-geometry attributes through parameters. In addition the tools provide APIs, enabling software developers to easily access specific building component data and retrieve desired information in a BIM model. The APIs extend the functionality of the tools. For instance, the Revit and ArchiCAD tools support API programming with the C# and C++ programming languages, respectively [28,29]. In our project, we adopted the BIM API capability to access BIM data directly in order to capture building topology and visualize building components' connectivity. The BIM API approach inside the BIM authoring programs also facilitate the generation of specific Modelica values in *ModelicaBEM* models.

4.5.2. Object-Oriented Physical Modeling (OOPM) and Modelica

The OOPM approach has been developed to provide a structured and equation-based modeling method and then enhance the multi-domain simulations capability using a single model [30–32]. Modelica is an object-oriented language for the modeling of large, complex and heterogeneous physical systems [33], developed to represent OOPM including dynamic behaviors using differential algebraic equation (DAE)-based calculations [30]. Component-connection diagrams using Modelica can represent the physical system topology of simulation models, including energy simulation models [30,34]. Such capabilities can support an intuitive object mapping between BIM and *ModelicaBEM* structures naturally compared to the mapping approach between traditional design model and BPS tools. Among domain specific Modelica libraries, LBNL Modelica Buildings Library [22] allows thermal simulations by offering building energy components and solvers. In our project, we adopted the library to comprehend space boundary condition definitions and incorporate with BIM in creating *ModelicaBEM*.

4.5.3. LBNL Modelica Buildings Library

The LBNL Modelica Buildings library consists of dynamic simulation models to demonstrate building energy and control systems [35]. The simulation models include a number of models for building energy analyses such as air-based HVAC (Heating, Ventilation, Air Conditioning) systems, chilled water plants, water-based heating systems, controls, heat transfer between rooms and the outside, multi-zone airflow, single-zone computational fluid dynamics, data-driven load prediction, and electrical DC and AC systems with two or three phases [35]. The library has issued version 2.0 [35], which includes the computational fluid dynamics (CFD) model and the electrical package enabling buildings to be integrated into an electrical grid.

The *HeatTransfer* and *Room* packages in the library are among the major models for thermal analysis of buildings, and have been validated by simulation results comparisons with benchmarked simulation models [36,37]. While the library is still under development and likely to be more extensively used through projects including IEA Annex 60 [38], which has the objective of developing and demonstrating new generation computational tools for building and community energy systems using Modelica [37], the use of it is still new and only a few demonstrated samples currently exist. However, the validation clarifies the capability of whole-building energy simulation [37] and can support object mapping between object-based models due to the object-oriented modeling concept.

In this research, we used the library as a building thermal simulation solver and incorporated it with BIM to create Modelica codes of space boundary conditions. The developed algorithm, by investigating the data structures of Revit and the Modelica library, can prevent both the designers' subjective interpretations of building data and human errors in translating building topology into space boundary conditions.

5. Development of the Methodology

5.1. *Required Information Definition*

While BIM and LBNL's BEM follow an object-oriented modeling concept, object mismatches including object semantics and behaviors exist, resulting in data structures configuration in accordance with related classes. In this project, we demonstrate the two object mismatches, and then describe the required information for room-to-thermal zone translation.

Semantic mismatches of building components impede suitable data exchange of building objects and their attributes between the domains. For example, a building envelope in a BIM model consists of building objects such as walls, floors, roofs, doors, windows, and so on, while an LBNL's BEM model decomposes them as exterior/interior surfaces. In addition, BIM recognizes an interior space as a room object, whereas LBNL's BEM identifies the space as a thermal zone. Three-dimensional geometry information corresponding with building components is stored in a BIM model, but only the wetted surface area is calculated for one-dimensional heat transfer computation in LBNL's BEM.

Behavior mismatches are the other critical aspect of data exchange in the room-to-thermal zone translation. For example, adding an interior wall in a room object to divide the room can be translated into LBNL's BEM as adding two surfaces to define two thermal zones performing heat transfer between the zones. Adding a window in a wall can be interpreted as defining an opaque surface with windows to applying a shading occurrence in a thermal simulation.

Comprehending the two mismatches enables us to identify required datasets. Among the major classes and parameters in LBNL Modelica Buildings Library, the *MixedAir* class models a thermal zone with completely mixed air for heat transfer through space boundaries. In addition, the class defines parameters for thermal simulations including convection, conduction, infrared radiation and solar radiation. Validation of the application using the *MixedAir* class has been conducted [22,36,37,39]. Table 1 shows the required parameters in creating an enclosed room with surfaces and constructions. *MixedAir* class defines the space boundary conditions as five construction types (*datConExt*, *datConExtWin*, *datConBou* or *surBou*, and *datConPar*), which are the major parameters for the room-to-thermal zone translation. The construction types describe the different statuses of space boundary conditions; if a thermal zone consists of three surfaces without windows and an interior surface shared with another thermal zone, the number of exterior boundaries (opaque surfaces) is five. The detailed description will be given in the algorithm development section.

Table 1. Class and parameters in creating enclosed rooms for the computation of heat transfer from LBNL Modelica Buildings library [22,35].

Class	Object Properties	
	Name	**Description**
MixedAir	*datConExt*	Opaque surfaces.
	nConExt	Number of *datConExt*.
	datConExtWin	Opaque surfaces with windows.
	nConExtWin	Number of *datConExtWin*.
	datConPar	Interior partitions in a thermal zone.
	nConPar	Number of *datConPar*.
	datConBou	Opaque surfaces on interior walls between thermal zones.
	nConBou	Number of *datConBou*.
	surBou	Opaque surfaces on the same interior walls between thermal zones.
	nSurBou	Number of *SurBou*.

The investigated parameters' information enables us to identify behaviors for the object mapping from the building components information in BIM into the space boundary construction types in *ModelicaBEM*. Table 2 shows the inspected object-mapping behaviors. The behavior description allows investigation into the kind of building components and properties that must be retrieved in a BIM model. For example, retrieving the wall, floor, and roof components and inspecting their wetted surfaces are the corresponding activity to the mapping of opaque surfaces into a *ModelicaBEM* model.

Table 2. Object mapping description between a room object and a thermal zone.

Space Boundary Conditions	Behaviors for Room-to-Thermal Zone Mapping in BIM
Opaque surfaces (*datConExt*)	Retrieving the wetted surfaces attached into a room object in building components including walls, floors, and roofs.
Number of opaque surfaces (*nConExt*)	Computing the number of building components attached into a room object.
Opaque surfaces with windows (*datConExtWin*)	Retrieving the wetted surface of a wall object with windows attached into a room object.
Number of opaque surfaces with windows (*nConExtWin*)	Computing the number of walls including with windows attached into a room object.
Interior partitions (*datConPar*)	Retrieving the wetted surfaces of an interior wall object. Both surfaces of the interior wall are retrieved.
Number of interior partitions (*nConPar*)	Computing the number of interior wall components in a room object.

Table 2. *Cont.*

Space Boundary Conditions	Behaviors for Room-to-Thermal Zone Mapping in BIM
Opaque surfaces on interior walls between thermal zones (*datConBou*)	Retrieving the surfaces on interior walls shared with other room objects.
Number of *datConBou* between thermal zones (*nConBou*)	Computing the number of surfaces of interior walls attached into the selected room object.
Opaque surfaces on the same interior walls between the thermal zones (*surBou*)	Retrieving the other surfaces on the same interior walls where the *datConBou* construction type is computed.
Number of surBou between the thermal zones (*nSurBou*)	Computing the number of surfaces of the other sides of the interior walls of retrieving the *datConBou* construction type information.

In addition, the behavior description facilitates the development of the algorithm. For instance, retrieving a room object's wetted surfaces information is a process that traverses a room object to inspect which building components enclose the room, and then traverses each building component again to retrieve the wetted surface information. The following algorithm development section explains the details of the development, including data flow description and pseudocode development.

5.2. Algorithm Development

The algorithm development follows a diagramming-describing-implementing approach: (1) developing a flow chart to represent the identified behavior description; (2) describing pseudocodes based on the flow chart; and (3) implementing a *TranslatingBuildingTopology* command based on the pseudocodes description.

5.2.1. Develop a Flow Chart Diagram

The investigated behavior description allows us to recognize the translation process between BIM and *ModelciaBEM*. Such an identified translation process can be represented as a set of flow charts consisting of: (1) filtering room objects; (2) retrieving and storing building components; and (3) assigning types of space boundary condition, which explains a data flow for object mapping (Figure 1). Each step of the flow chart can be described as a function description in pseudocodes.

(1) Filter room objects (in Figure 1A)

The created BIM model contains building components, represented as elements in Revit. Such building components have relationships with each other and follow their own data structure. For example, a room object should be enclosed with building objects and instantiated from the Room class. The Room class has super-subclass relationships with the *SpatialElement* class, which provides boundary segments in accordance with the building elements enclosing the room object. In order to search and store room objects in a BIM model, we defined a series of processes, including: (1) traversing the created BIM model; (2) collecting all building elements into a variable; (3) retrieving spatial elements from among the building elements; (4) filtering room objects in the spatial elements; and then (5) storing the room objects into a room-arraylist variable.

(2) Retrieve and store building components (in Figure 1B)

The room-arraylist variable enables the created building components to be categorized in the wall, floor, and roof categories. As shown in the B section in Figure 1, each index in the arraylist allows the filtering of building components in each room object by traversing the room-arraylist variable, using an iterator, until the number of iterations is the same as the size of the room-arraylist variable. Each room object facilitates inspection of the building components categorization due to the relationship with other classes, including the *SpatialElementBoundaryOptions* and *BoundarySegment* classes, which provide enveloping wall components. The following pseudocodes description section explains the details of the building components categorization using the related classes.

(3) Assign space boundary construction types (in Figure 1C)

The arraylists consisting of wall-arraylist, floor-arraylist, and roof-arraylist include room object information and enveloping building components information. For example, a wall-arraylist contains the identification information of a room object in the first index and the walls' identification information from the second index, sequentially. Therefore, comparison of two wall-arraylists enables the identification of which walls are shared between room objects. If there is the same wall identification information in two wall-arraylists, we can assign one of the walls as an opaque surface on an interior wall (*datConBou* in the space boundary condition) and the other wall can be the other opaque surface on the same interior wall (*surBou* in the space boundary condition).

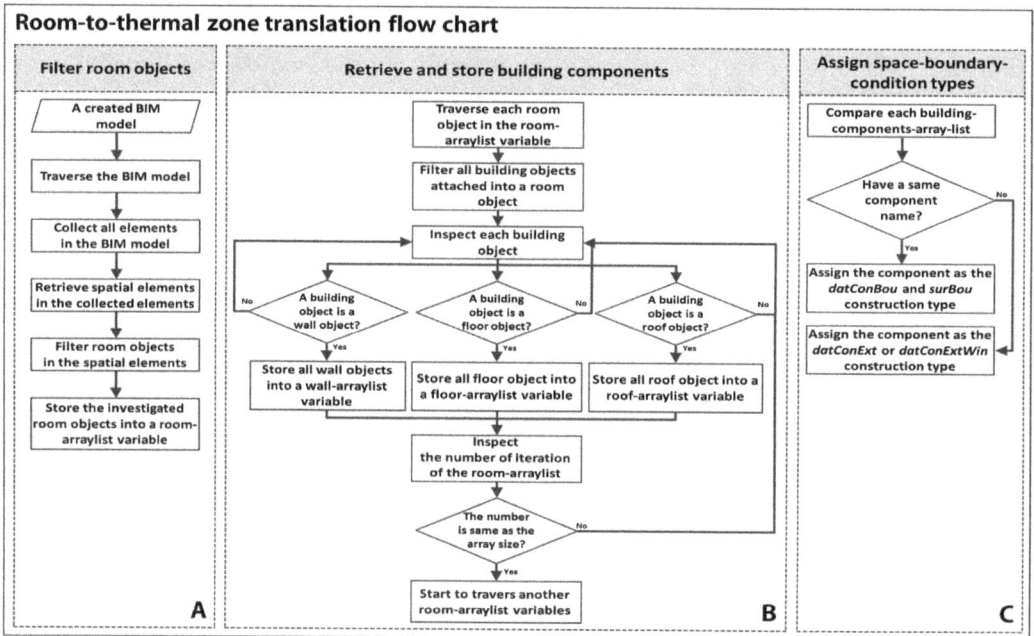

Figure 1. Room-to-Thermal zone translation flow chart describing a data flow corresponding with the identified behavior description. (**A**) Filter room objects; (**B**) Retrieve and store building components; (**C**) Assign space-boundary-condition types.

The developed flow chart diagram (Figure 1) allows us to describe a series of functions through pseudocodes. The following describe pseudocodes section explains the description of the pseudocodes.

5.2.2. Describe Pseudocodes

The main purpose of the pseudocodes description is to support the implementation of the flow chart as a specific command in the BIM environment. In order to facilitate implementation in a specific BIM authoring tool (Revit), we explored the Revit API, which allows access to the BIM data structure and the retrieval of the created building components information, including geometry and non-geometry information. The predefined classes in the API enable us to describe the flow chart as pseudocodes. For example, we defined the *WallCollector* function corresponding with the process of retrieving and storing the wall components from a room object in the flow chart (Figure 1B), and utilize the *BoundarySegment* class to access the enveloping wall components from a room object. The pseudocodes in Figure 2 show the process of assigning space boundary condition types using a room object.

Pseudocodes description for generating space boundary conditions information from building topology

---

Require: The room object(R) from a BIM model is a defined as a zone

function *WallCollector*(Room *R*)

{The return *wc* is an arraylist containing of all wall objects enclosing a room object}

    List *boundarySegement* ← containing all building components composing of a room

        for *j* from 1 to size of *boundarySegement*

           if(current object's type == wall object type

                wc[*j*] ← Extend(*wc*[*boundarySegementElement*])

return *wc*

function *RetreiveSharedWall*(Room *R1*, Room *R2*)

{The return *nameOfSharedWall* array is utilized to inform which wall is shared between adjacency rooms and the

values of *datConBou* and *surBou*}

    List *roomWallConnectivity* ← containing the information of the name of room objects such as *R1* or *R2* and

    walls consisting of the room

        for *j* from 1 to size of *roomWallConnectivity* for *R1*

           if *roomWallConnectivity* for *R2* has same name of a wall object in the *roomWallConnectivity* for

           R1 then *nameOfSharedWall*[*j*] ← Extend(string WallName)

return *nameOfSharedWall*

function *SetSpaceBoundary*(String *sharedWallInfo*, Array *WallComponents*)

{The return is assigning the values incorporating with *ConBou* and *SurBou* in space-boundary-condition

variables}

    String array *eachSharedWallInfor* ← containing the information of the shared wall objects between *R1* and *R2*.

        for *j* from 1 to size of *eachSharedWallInfor* for *R1*

           if *eachSharedWallInfor* for *R2* set the value of *ConBou*, then *eachSharedWallInfor* for *R2* set the

           value of *SurBou*

return *void*

Figure 2. A pseudocode description to match building topology into space boundary conditions.

The defined functions in Figure 2 enable the algorithm to detect shared wall objects between room objects. The functions describe only the investigation process of space boundary condition type from a BIM model. Other required processes defined in the flow chart will be directly implemented using Revit API; we used the *FilteredElementCollector* class to collect all building elements in a BIM model. The detailed implementation explanation will be discussed in the Implement the Pseudocode Description section.

5.2.3. Implement the Pseudocode Description

We implemented the pseudocodes using the C# language incorporated with Revit API and developed a *TranslatingBuildingTopology* command. We developed a chain of functions to compose the command as shown in Figures 3–5. *TranslatingBuildingTopology* enables the building topology in a BIM model: (1) to be graphically represented as a connectivity diagram; and (2) to be translated into specific values for space boundary condition variables in a *ModelicaBEM* model. The following code blocks in the figures demonstrate the room-to-thermal zone translation, focusing on the shared wall information retrieval process.

(1) Room Objects Retrieval Process Implementation

We created an element-collect instance from the Revit API class (*FilteredElementCollector* class) to retrieve all elements in a Revit BIM model. Then we filtered the collected elements into spatial elements, as shown in Figure 3a. The *RoomFilter* construction enables the filtered spatial element to sort room objects. Finally, we created an array-list instance to collect room objects from the room filtered element using the *RoomFilter* construction (Figure 3b).

```
#region Make a filter to store room objects from the current document

FilteredElementCollector roomCollector = new FilteredElementCollector(doc);
roomCollector.OfClass(typeof(SpatialElement));
                                                    (a) Element collector declaration

roomCollector.WherePasses(new RoomFilter());
IList<Element> roomElements = roomCollector.ToElements();
                                                    (b) Room object collector declaration
#endregion
```

Figure 3. A code block to describe room objects retrieval process from a Revit BIM model.

(2) Enveloping Wall Retrial Function Implementation

After completing the collection of room objects in an array-list, we can explore the array-list to investigate the enveloping walls of each room object. We conducted the investigation with the *WallCollector* function implementation shown in Figure 4. The function consists of two declaration parts: (1) enveloping element collector declaration (Figure 4a); and (2) wall object collector declaration (Figure 4b). The envelope element collector declaration part demonstrates the process of collecting boundary elements which enclose a room object. We collected the enveloping elements in an arraylist instance, and then we explored the instance until all the elements are investigated, whether they are wall objects or not. Once an element in the arraylist meets the condition, we stored it in a wall arraylist as a wall object. Figure 4 shows the implementation of the two declarations using provided classes from Revit API; we utilized the *SpatialElementBoundaryOptions* and *BoundarySegment* classes to define the declarations.

```
#region Function: Collecting enveloping wall objects from each room element
public static ArrayList WallCollector(Room room)
{
    ArrayList WallCollectArray = new ArrayList();

    SpatialElementBoundaryOptions options = new SpatialElementBoundaryOptions();

    IList<IList<Autodesk.Revit.DB.BoundarySegment>> boundary = room.GetBoundarySegments(options);
                                                    (a) Enveloping element collector declaration
    foreach (IList<Autodesk.Revit.DB.BoundarySegment> boundarySS in boundary)
    {

        int j = 1;
        String roomNumber = room.Number.ToString();
        WallCollectArray.Add(roomNumber.ToCharArray());

        foreach (Autodesk.Revit.DB.BoundarySegment boundarySegment in boundarySS)
        {
            if (boundarySegment.Element != null)
            {
                WallCollectArray.Add(boundarySegment.Element);
            }
            j = j + 1;
        }
    }

    return WallCollectArray;
                                                    (b) Wall object collector declaration
}
#endregion
```

Figure 4. A code block to describe enveloping wall objects retrieval process in each room object.

(3) Shared Wall Information Retrieval Function Implementation

The shared wall information between rooms supports the defining of the space boundary condition, especially the conditions of both surfaces of the interior wall shared by room objects (named as *datConBou* and *surBou* in *ModelicaBEM*). Once shared wall objects are identified, we can assign the *datConBou* and *surBou* conditions based on the object information, including the number of *datConBou* and *surBou*. We developed a *RetrieveSharedWallInformation* function as shown in Figure 5. The implementation is mainly focused on the retrieval of the shared wall's name. Once the name of the shared wall is identified, we can retrieve other walls' information, such as area, tilt, and azimuth, by modifying parts of the wall information provided by the Wall class in Revit API. To collect the shared wall's name, we identified the connectivity between room and wall objects, as shown in Figure 5a. The connectivity represents that the relationships between a room and enveloping walls attached to the room by storing the name of the room and walls into an arraylist. Based on the rooms' connectivity information, we implemented a comparison of the rooms' connectivity as shown in Figure 5b. After conducting the comparison, we can acquire the name of the shared wall with the rooms' names. For example, if a wall whose name is *FrontWall* is shared with two rooms (named *RightRoom* and *LeftRoom*), the function provides a series of string-type values separated by comma and semi-colon such as "*FrontWall, RightRoom; FrontWall, LeftRoom*". In addition, the *RetrieveSharedWallInformation* function enables the connectivity to be visualized through a series of nodes. We will demonstrate the visualization in the following experiments section.

```
#region Function: Return the shared wall's name with the room's name.
public static String RetrieveSharedWallInformation(ArrayList code_Connection)
{
    String result = null;
    ArrayList sharedWall = new ArrayList();
    ArrayList RoomWallConnectivity = new ArrayList();

    RoomWallConnectivity = GetSepeartedArray(code_Connection, IOLibrary.FileStream.Room_ModelicaBEM,
        IOLibrary.FileStream.Wall_ModelicaBEM);
                                        (a) Collecting the connectivity information
    int j = RoomWallConnectivity.Capacity;
    int k = RoomWallConnectivity.Count;

    Object[] AllRoomConnectivity = RoomWallConnectivity.ToArray();
    Object[] AllRoomConnectivity_Comparison = RoomWallConnectivity.ToArray();
    String store = null;
    for (int i = 0; i < AllRoomConnectivity.Length; i++)
    {
        String[] temp = AllRoomConnectivity[i] as String[];

        for (int p = 0; p < AllRoomConnectivity_Comparison.Length; p++)
        {
            String[] temp2 = AllRoomConnectivity_Comparison[p] as String[];

            if (CompareArrays(temp, temp2) == CompareArrays(temp, temp))
            {
                String nothing = null;
            }
            else
            {
                store= CompareArrays(temp, temp2);
                sharedWall.Add(store);
            }
        }
    }
    Null value setting
    }                                   (b) Retrieval of the shared wall's name with room's name
    IEnumerator ie = sharedWall.GetEnumerator();
    Store shared wall's name into a string variable
    return result;
}
#endregion
```

Figure 5. A code block to describe the shared wall information retrieval process between room objects.

The implemented functions enable building topology, represented by the object connectivity in the implementation, to be translated into space boundary condition types such as *datConBou* and *surBou* retrieved by the shared wall information. In the experiments section, we will demonstrate the implemented functions by applying them into BIM models, including a validated energy model (BESTEST 600), and providing energy simulation results to validate the translation.

6. Experiments

We conducted: (1) experiments by applying the implemented *TranslatingBuildingTopology* command to a multi-zone BIM; and (2) simulation result comparisons between two *ModeliaBEM* models. One model includes the translated space boundary conditions generated from the command and the other is manually created following the OOPM simulation engine specifications (following LBNL Modelica Buildings library and ModelicaBIM library [16]). For the experiments, we conducted three test cases (a one-room model, a two-room model, and a three-room model). These cases are based on a one-room model adopted from the BESTEST Case 600 building: evenly dividing the room in the one-room model to create two rooms, and evenly dividing the right room in the two-room model to create three rooms. We hypothesized that if our algorithm represented the translation, and if the implementation of the flow chart was correct, the *ModelicaBEM* models of each case could produce close or identical simulation results.

6.1. Test Cases

For the test cases, we created equivalent three BIM models incorporated with Autodesk Revit Architecture: a one-room model (Test Case 1), a two-room model having two windows (Test Case 2), and a three-room model having a window (Test Case 3). Based on the one-room model, the other models were created to present more variation of building topology. The following sections discuss each test case, focusing on: (1) building topology representation focusing on the wall-room connectivity; (2) space boundary condition representation in a *ModelicaBEM* model; and (3) simulation result comparisons.

6.1.1. Building Topology Representation

(1) Test Case 1: Creating a Basic Building Model

To conduct Test Case 1, we created a one-room building model using Revit Architecture corresponding to the BESTEST Case 600 building model description. The definition of the BESTEST Case 600 model is given in [40,41] and is as follows.

- One room of a single thermal zone, 8.0 m × 6.0 m × 2.7 m in length, width and height, respectively.
- The building model consists of four exterior walls, a roof, a floor, and two south-oriented 6 m^2 windows, but no doors, as shown in Figure 6a.
- The building components have building envelope material properties as described in Table 3.
- No shade and no internal heat gains from occupancy and equipment occur inside the building.
- The floor is located above ground level and is not attached to the ground.
- The location of the building is Denver, CO, U.S.
- The building is lightweight.

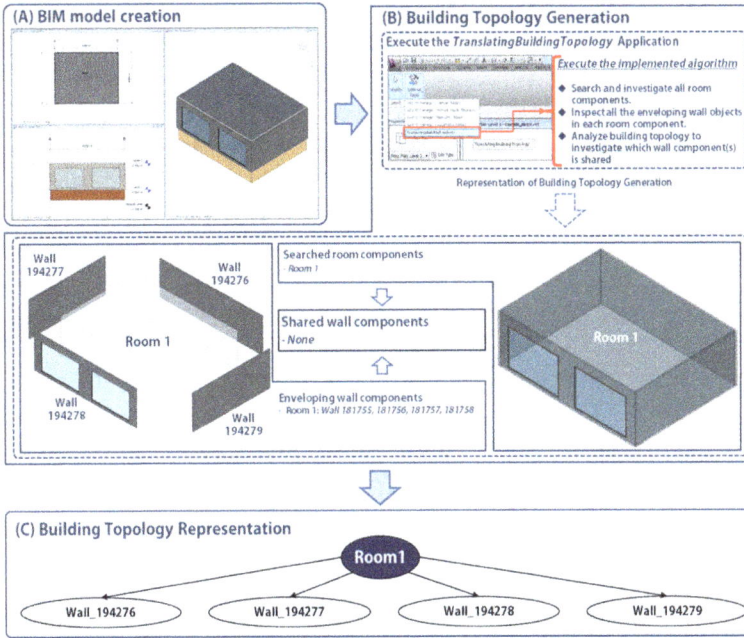

Figure 6. Building topology representation process diagram for Test Case 1.

Table 3. Material specification for the BESTEST Case 600 building model adapted from [40,42].

Material	Thickness (m)	Thermal Conductivity (W/m· K)	Specific Heat Capacity (J/kg· K)	Mass Density (kg/m^3)
Exterior Wall (Inside to Outside)				
Plasterboard	0.012	0.160	840	950
Fiberglass quilt	0.066	0.040	840	12
Wood siding	0.009	0.140	900	530
Floor (Inside to Outside)				
Timber flooring	0.025	0.140	1200	650
Insulation	1.003	0.040	0	0
Roof (Inside to Outside)				
Plasterboard	0.010	0.160	840	950
Fiberglass quit	0.1118	0.040	840	12
Wood siding	0.019	0.140	900	530

We created a BIM model based on the description shown in Figure 6a. The BIM model contains the material information shown in Table 3 by adopting the adding material command from [15]. The BIM model consists of four wall objects, a roof object, and a floor object. The south-facing wall object has two 6 m^2 windows. The room object is enclosed by the building objects and the room has no interior wall object.

The *TranslatingBuildingTopology* command translates the room object in the BIM model into the building topology representation. Figure 6 shows the translation (Figure 6b) and the building topology representation (Figure 6c).

We adopted the Microsoft automatic graph layout (MSAGL) tool [43] to implement the graphical representation of the building topology. The *WallCollector* and *RetrieveSharedWallInformation* functions enable the *TranslatingBuildingTopology* command to retrieve the wall object information attached to the room object shown in Figures 6b and 6c. The command collects the wall's identification number from a wall object in the BIM model and then generates a new wall's name by combining the wall-id number with the predefined identification word (Wall_), e.g., Wall_194276. Each new wall's name is represented in an oval shape node and is connected into a room node with an arrow (Figure 6c).

(2) Test Case 2: Creating a Two-Room Building from the One-Room Building

We modified the one-room building model by installing an interior wall to evenly divide the single zone into a two-thermal zone, removing the left window object in the south-facing wall of the one-room model, and installing an east-facing 6 m^2 window as in Figure 7a.

After the command explored all the wall objects enveloping the two room objects, it retrieved all walls' id numbers (Figure 7b). The walls' id numbers enable the command to demonstrate which wall object is shared between the rooms, using the comparison function in the *TranslatingBuildingTopology* command. Test Case 2 shows that the installed interior wall is the shared wall object (Wall_181758) through node connectivity representation, which means that the behavior of adding an interior wall is correctly translated into the behavior of retrieving a shared wall object having two surfaces between two thermal zones.

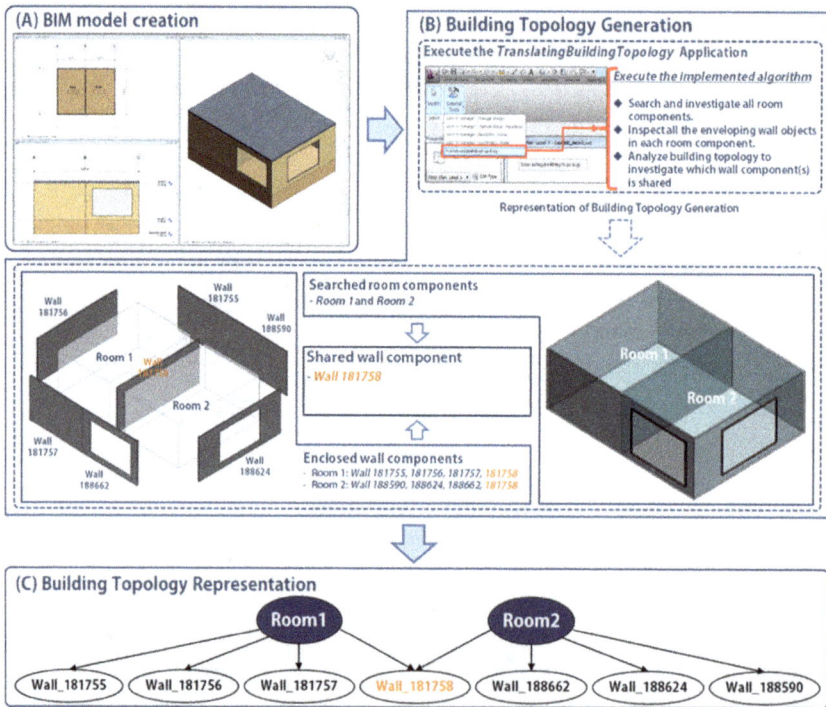

Figure 7. Building topology representation process diagram for Test Case 2.

(3) Test Case 3: Creating a Three-Room Building from the Two-Thermal-Zone Building

To demonstrate a more diverse building topology translation, we installed another interior wall in the right room object of Test Case 2 (Figure 8a). The updated BIM model has three rooms; the west room has the same dimension as in Test Case 2, but the east room in Test Case 2 is divided into a south room and a north room. The south and north room measure 4.0 m × 3.0 m × 2.7 m in length, width, and height, respectively. In adding another interior wall to create south and north rooms, we removed the east-facing window. Note that we referred to the west room, the south room and the north room as Rooms 1–3 respectively, as shown in Figure 8b. The only difference between Room 1 and Room 3 is the size. We will demonstrate how the difference can affect the simulation result in the following simulation result comparisons section.

The *TranslatingBuildingTopology* command shows the wall-room connectivity based on the shared wall-id numbers retrieval process (Figure 8c). The connectivity diagram implies that the heat transfer is conducted through Wall_194014 between Rooms 1 and 2, through Wall_181758 between Rooms 1 and 3, and through Wall_192844 between Rooms 2 and 3.

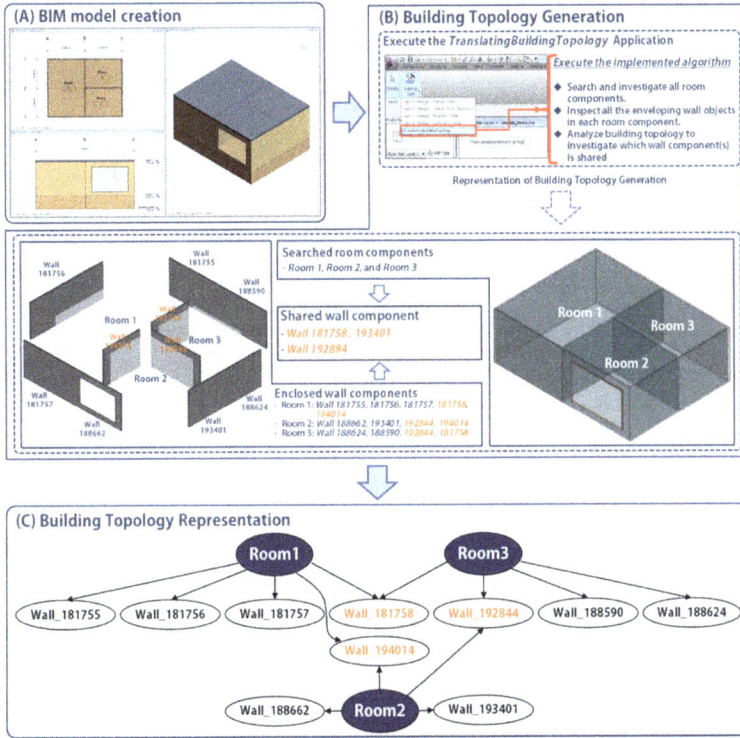

Figure 8. Building topology representation process diagram for Test Case 3.

6.1.2. Space Boundary Condition Representation in a *ModelicaBEM* Model

To represent the space boundary condition by following the object-oriented concept and the architecture viewpoint instead of an engineer's, we adopted the LBNL Modelica Buildings library (as an OOPM simulation engine) and *ModelicaBIM* library (to wrapper the simulation engine as the architecture viewpoint) [16]. The specification of the two libraries enables the created building topology to be represented as instances in a *ModelicaBEM* model.

(1) Test Case 1: Single-Thermal-Zone *ModelicaBEM* Model

Once the building topology representation is executed, the *TranslatingBuildingTopology* command generates the values of the space boundary conditions. As shown in Figure 9, the room has three exterior walls (translated into the number of *datConExt*), excepting the exterior wall with windows (translated into the number of *datConExtWin*). The Room 1 object is translated into the *ThermalZone* object in a *ModelicaBEM* model. The *ThermalZone* object includes two space boundary condition types (*datConExt* and *datConExtWin*) and the values of the types are translated from the wall-room connectivity in the building topology. The number of the *datConExt* in the *ModelicaBEM* model can be five (Figure 9A) due to the calculation of the total number of opaque surfaces in the room model corresponding with the surfaces of three exterior walls, the floor, and the roof object. The variables in *datConExt* (Figure 9A) and *datConExtWin* (Figure 9B), such as layers, area (A), tilt (til), and azimuth (azi), have the values from the instantiated building component objects, such as Wall_194276.*structure* in the layers variable. The wall-room connectivity enables the instantiation to be conducted by following the specification in the *ModelicaBIM* library.

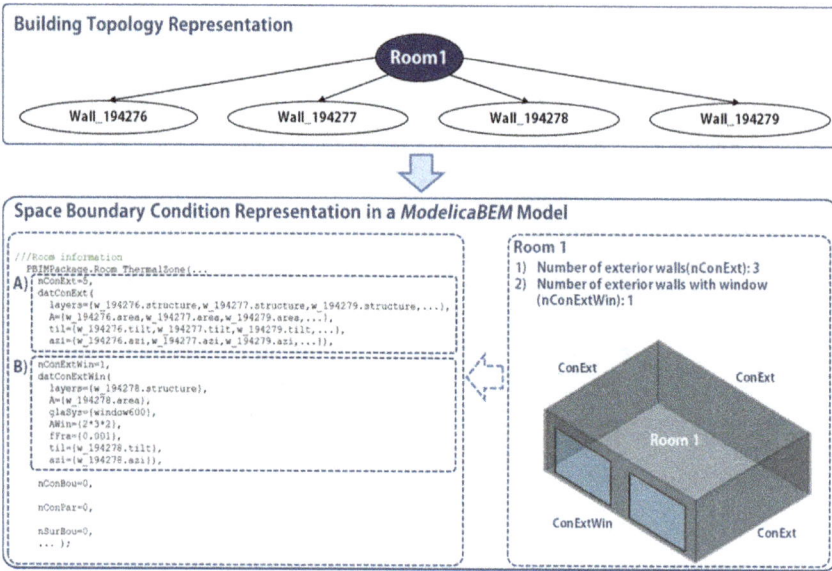

Figure 9. Space boundary condition representation of the one-thermal-zone *ModelicaBEM* model.

(2) Test Case 2: Two Thermal Zones *ModelicaBEM* Model

The main modification between the single thermal zone model and the two thermal zones model is whether the conduction heat transfer is performed through the interior wall's surface where the interior wall is shared by the two zones. Figure 10 represents the differentiation by identifying the translation of the interior wall object into the *datConBou* and *surBou* variables.

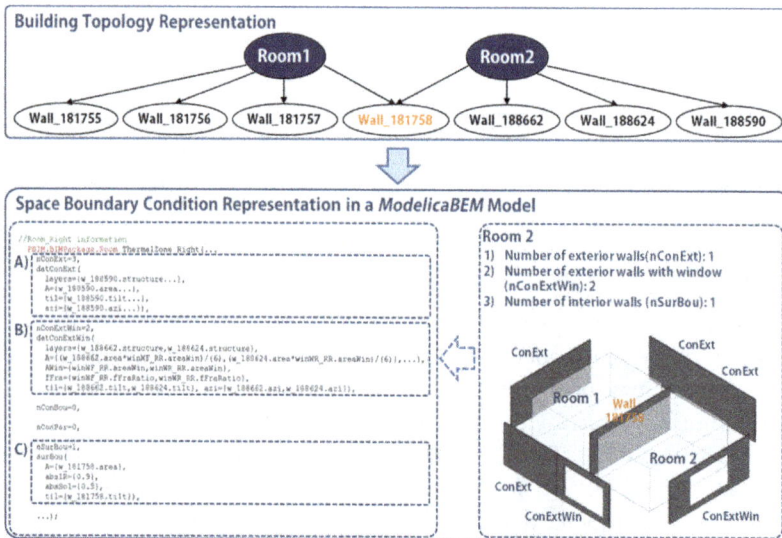

Figure 10. Space boundary condition representation of the two-thermal-zones *ModelicaBEM* model.

The LBNL library defines two construction types: opaque surfaces on interior walls between thermal zones as *datConBou* and the other opaque surfaces on the same interior walls between the thermal zones as *surBou*. The Wall_181758 interior wall is identified as the shared interior wall by the *TranslatingBuildingTopology* command. Therefore, we can identify one surface of the wall as the *datConBou* type and the opposite surface as the *surBou*. We implemented the room-filter process in the flow chart as following the room number sequence, and assigned the *datConBou* type to the surface attaching to the first room object. Therefore, we can assign the *datConBou* type to the surface oriented into the Room 1 object and the *surBou* type into Room 2, as shown in Figure 10C. The Room 2 object has two exterior walls with windows, the same as in Test Case 1; therefore, the *datConExt* and the *datConExtWin* types can follow same process as for Room 1 in Test Case 1 (Figure 10A,B).

(3) Test Case 3: Three Thermal Zones *ModelicaBEM* Model

In order to demonstrate the variable building topology translation, we expanded the model with two thermal zones by installing one more interior wall (the interior Wall_192844 object in Figure 11) to split Room 2 as two thermal zones.

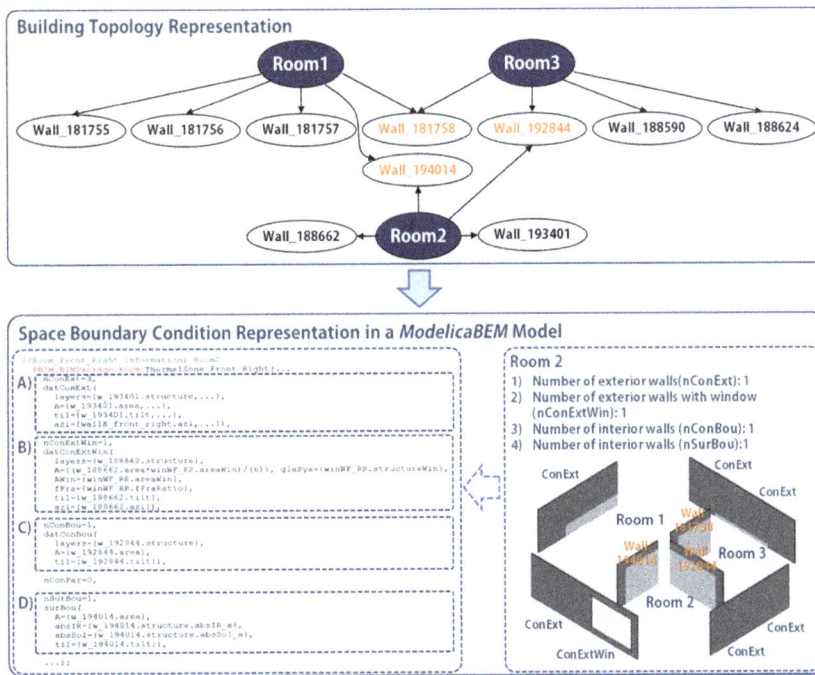

Figure 11. Space boundary condition representation of the three-thermal-zones *ModelicaBEM* model.

The installation of the interior wall between Rooms 2 and 3 enables the interior surface of the Wall_192844 object oriented into Room 2 to be assigned as *datConBou* (Figure 11C) and the surface of the interior wall oriented into Room 3 as *surBou* based on our implementation. The interior wall between Rooms 1 and 2 identified as the same condition in Test Case 2; therefore, the *surBou* construction type can be assigned into the interior surface of the Wall_194014 object oriented into Room 2 (Figure 11D).

The three test cases verify our algorithm development and its implementation by representing: (1) the building topology as building object connectivity; and (2) the translation of the connectivity into the space boundary condition types in each *ModelicaBEM* model.

6.1.3. Simulation Result Comparisons

We performed thermal simulation from the created *ModelicaBEM* models using a Dymola 2012 simulation program as a Modelica development environment, LBNL Modelica Buildings library version 1.3 as a simulation solver, and a simulation interval of 3600 s for a one-year period.

In order to validate our approach, we utilized the same BESTEST Case 600 building from Test Case 1 because the simulation results from the LBNL library for the BESTEST Case 600 have been validated [37]. Note that the LBNL's building model is created by manually writing the building description code, following the specification from the library, while our algorithm's *ModelicaBEM* model is created by including the automatically-translated building topology data in the *ModelciaBEM* model. We apply the coherent model conditions (described in the previous building topology representation section) to all the building models except the building location: Denver, Colorado for Test Case 1 and Chicago, Illinois for Test Cases 2 and 3. The building dimensions of all the test cases are the same, but room dimensions differ based on the test cases' description, incorporating the split of thermal zones. We created two *ModelicaBEM* models for all the test cases, the algorithm's version and LBNL's version, to demonstrate an annual indoor air temperature change without an HVAC system module.

In summary, the simulation results of all the test case models created by using the developed algorithm (algorithm version) agree with the results from the model created with the LBNL Modelica Buildings library (LBNL version). Two models of each test case show almost identical (Test Case 1) or close (Test Cases 2 and 3) simulation results of annual indoor air temperature of all three test cases. In Table 4, the dates when the highest and lowest temperatures are found in each room of the algorithm version model are the same as the dates in the LBNL version model. For example, in Test Case 1, as shown in Table 4, the highest temperature occurred at 3:00 p.m. on 17 October and the lowest temperature at 7:00 a.m. on 4 January in both *ModelicaBEM* models.

Overall, the *ModelicaBEM* models including the translated building topology data produce very similar simulation results to LBNL's models. This is expected because we applied the same thermal simulation algorithm from the LBNL Modelica Buildings library to all the *ModelicaBEM* models.

However, two major differences exist between the *ModelicaBEM* models: the modeling approach and the model structures. The simulation result comparison experiments can have unpredictable simulation results due to the differences if the algorithm is not developed or implemented correctly.

Regarding the modeling approach, our implemented algorithm allows a more comprehensive energy model creation to reflect the actual building semantics. In the examples of Test Cases 2 and 3, the LBNL's model represents the interior wall object as different surface objects to follow energy semantics, e.g., boundary condition. However, our approach automatically recognizes both surfaces from the interior wall object and identifies the orientation of each surface to represent the actual building configuration, e.g., building component connectivity: one is for a room object and the other is for the other room object.

Regarding the model structure, the *ModelicaBEM* models, by applying our algorithm implementation, use architecture semantics, such as rooms instead of engineering-based semantics, such as *MixedAir* objects. To demonstrate the architectural modeling structure, we adopted the ModelicaBIM library, allowing wrapping the *MixedAir* class in LBNL Modelica Buildings library as the Room class in the ModelicaBIM library.

Table 4. Annual peak temperatures of the Test Cases.

Test Case	Room Name	Highest Temperature (°C)/Date, Time	Lowest Temperature (°C)/Date, Time
1	Room 1	Algorithm version: 65.7 °C/17 October, 3:00 p.m. LBNL version: 65.9 °C/17 October, 3:00 p.m.	Algorithm version: −19.8 °C/4 January, 7:00 a.m. LBNL version: −19.8 °C/4 January, 7:00 a.m.
2	Room 1	Algorithm version: 34.9 °C/19 July, 9:00 p.m. LBNL version: 35.4 °C/19 July, 9:00 p.m.	Algorithm version: −12.3 °C/8 January, 11:00 a.m. LBNL version: −12.2 °C/8 January, 11:00 a.m.
	Room 2	Algorithm version: 42.2 °C/18 July, 10:00 a.m. LBNL version: 43.6 °C/18 July, 10:00 a.m.	Algorithm version: −13.2 °C/8 January, 7:00 a.m. LBNL version: −13.2 °C/8 January, 7:00 a.m.
3	Room 1	Algorithm version: 34.2 °C/19 July, 9:00 p.m. LBNL version: 34.6 °C/19 July, 9:00 p.m.	Algorithm version: −12.5 °C/8 January, 11:00 a.m. LBNL version: −12.5 °C/8 January, 11:00 a.m.
	Room 2	Algorithm version: 40.8 °C/18 July, 1:00 p.m. LBNL version: 41.7 °C/18 July, 1:00 p.m.	Algorithm version: −10.5 °C/31 January, 2:00 a.m. LBNL version: −10.5 °C/31 January, 2:00 a.m.
	Room 3	Algorithm version: 34.0 °C/19 July, 9:00 p.m. LBNL version: 34.4 °C/19 July, 9:00 p.m.	Algorithm version: −11.2 °C/8 January, 11:00 a.m. LBNL version: −11.2 °C/8 January, 11:00 a.m.

7. Conclusions and Future Work

This paper presents a new translation algorithm integrating BIM and Modelica-based OOPM for building thermal simulation. The implemented command from the algorithm enables inter-disciplinary data exchange between the two object-oriented models (BIM models and *ModelicaBEM* models). The algorithm can leverage the consistent use of the original building data including building topology in the energy simulation without manually rewriting them in energy models. The building data in BIM already created by architects or designers can significantly eliminate the overhead in identifying the required building description data for input data in building energy modeling. The suggested algorithm and implementation in this paper have the capability to improve error-prone manual energy modeling processes.

Our algorithm and its implementation utilized BIM API instead of the standard data schemas, such as IFC and gbXML, for natural mapping processes. We acknowledge the benefits of adopting the standard schemas to enhance interoperability for model translation among various BIM tools and energy simulation tools. However, the use of BIM API enables direct access to BIM data in developing specific commands and preserves parametric modeling capability. We will expand the feasibility of the algorithm by applying it to other BIM tools to enhance the interoperability.

Our suggested algorithm facilitates the development of a system interface, enabling multi-zone BIM models to be automatically translated into *ModelicaBEM* models with high efficiency and accuracy. The advance of the algorithm is to enhance the efficiency of natural mapping between two object-oriented models, which can facilitate efficient model translation from object-based design models to diverse building performance simulation. We will expand our algorithm and the implementation of it to cover more complicated building models, including the vertical stacking of rooms and the addition of doors. Additionally, a future project will investigate the use of standard schemas such as IFC for building topology translation into OOPM-BEM. Moreover, the current version is focused on thermal simulation; we will expand the coverage of the algorithm into more simulation domains, including daylight and photovoltaics.

Acknowledgments: The authors are grateful to Wei Yan at Texas A & M University for valuable advises and contribution to this research. This research was supported by Basic Science Research Program through the National Research Foundation of Korea (NRF) funded by the Ministry of Science, ICT (Information and Communication Technology) and Future Planning (No. NRF-2013R1A1A1010562 and No. NRF-2013R1A2A2A04014772).

Author Contributions: All authors read and approved the manuscript. All authors contributed to this research work, discussed the results and implications and involved in the manuscript development at all stages. WoonSeong Jeong provided precious ideas on the establishment of algorithm and its implementation as well as experiments. JeongWook Son discussed the main idea to develop this research work and reviewed and revised the manuscript as well as led the development of the paper.

Conflicts of Interest: The authors declare no conflict of interest.

References

1. Bazjanac, V. *IFC BIM-Based Methodology for Semi-Automated Building Energy Performance Simulation*; Lawrence Berkeley National Lab.: Berkeley, CA, USA, 2008.

2. Bazjanac, V. Implementation of semi-automated energy performance simulation: Building geometry. In Proceedings of the CIB, Istanbul, Turkey, 8–11 October 2009; pp. 595–602.

3. Hand, J.W.; Crawley, D.B.; Donn, M.; Lawrie, L.K. Improving the data available to simulation programs. In Proceedings of the Building Simulation, Montréal, QC, Canada, 7–9 August 2005; pp. 373–380.

4. Bazjanac, V.; Maile, T.; Rose, C.; O'Donnell, J.T.; Mrazović, N.; Morrissey, E.; Welle, B.R. An assessment of the use of building energy performance simulation in early design. In Proceedings of the Building Simulation, Sydney, Australia, 14–16 November 2011; pp. 1579–1585.

5. Hitchcock, R.J.; Wong, J. Transforming IFC architectural view BIMs for energy simulation: 2011. In Proceedings of the Building Simulation, Sydney, Australia, 14–16 November 2011; pp. 1089–1095.

6. U.S. Department of Energy (U.S. DOE). Best Directory | Building Energy Software Tools. Available online: http://www.buildingenergysoftwaretools.com (accessed on 4 August 2015).

7. Clarke, J. *Energy Simulation in Building Design*, 2nd ed.; Routledge: London, UK, 2001.
8. Crawley, D.B.; Hand, J.W.; Kummert, M.; Griffith, B.T. Contrasting the capabilities of building energy performance simulation programs. *Build. Environ.* **2008**, *43*, 661–673. [CrossRef]
9. Maile, T.; Fischer, M.; Bazjanac, V. *Building Energy Performance Simulation Tools—A Life-Cycle and Interoperable Perspective*; Center for Integrated Facility Engineering (CIFE) Working Paper 107; Lawrence Berkely National Lab.: Berkeley, CA, USA, 2007; pp. 1–49.
10. Attia, S. *State of the Art of Existing Early Design Simulation Tools for Net Zero Energy Buildings: A Comparison of Ten Tools*; Université Catholique de Louvain: Louvain La Neuve, Belgium, 2011.
11. Aksamija, A. BIM-based building performance analysis: Evaluation and simulation of design decisions. In Proceedings of the 2012 ACEEE Summer Study on Energy Efficiency in Buildings, Pacific Grove, CA, USA, 12–17 August 2012.
12. General Services Administration. 3D-4D Building Information Modeling. Available online: http://www.gsa.gov/portal/content/105075?utm_source=PBS&utm_medium=print-radio&utm_term=bim&utm_campaign=shortcuts (accessed on 4 August 2015).
13. Rose, C.M.; Bazjanac, V. An algorithm to generate space boundaries for building energy simulation. *Eng. Comput.* **2013**, *31*, 271–280. [CrossRef]
14. Jeong, W.; Kim, J.B.; Clayton, M.J.; Haberl, J.S.; Yan, W. Translating building information modeling to building energy modeling using Model View Definition. *Sci. World J.* **2014**, *2014*, 1–21. [CrossRef] [PubMed]
15. Jeong, W.; Kim, J.B.; Clayton, M.J.; Haberl, J.S.; Yan, W. A framework to integrate object-oriented physical modelling with building information modelling for building thermal simulation. *J. Build. Perform. Simul.* **2015**. [CrossRef]
16. Kim, J.B.; Jeong, W.; Clayton, M.J.; Haberl, J.S.; Yan, W. Developing a physical BIM library for building thermal energy simulation. *Autom. Constr.* **2015**, *50*, 16–28. [CrossRef]
17. Yan, W.; Clayton, M.; Haberl, J.; Jeong, W.; Kim, J.; Kota, S.; Bermudez Alcocer, J.; Dixit, M. Interfacing BIM with building thermal and daylighting modeling. In Proceedings of the 13th International Conference of the International Building Performance Simulation Association, Chambery, France, 26–28 August 2013; pp. 3521–3528.
18. Bazjanac, V.; Maile, T. *IFC HVAC Interface to EnergyPlus—A Case of Expanded Interoperability for Energy Simulation*; Lawrence Berkeley Natl. Lab: Berkeley, CA, USA, 2004.
19. O'Donnell, J.; See, R.; Rose, C.; Maile, T.; Bazjanac, V.; Haves, P. SimModel: A domain data model for whole building energy simulation. In Proceedings of the Building Simulation, Sydney, Australia, 14–16 November 2011; pp. 382–398.
20. Bazjanac, V. Space boundary requirements for modeling of building geometry for energy and other performance simulation. In Proceedings of the CIB W78: 27th International Conference, Cairo, Egypt, 16–18 November 2010.
21. Bazjanac, V.; Kiviniemi, A. Reduction, simplification, translation and interpretation in the exchange of model data. In Proceedings of the 24th W78 Conference Maribor 2007, Maribor, Slovenia, 26–29 June 2007; pp. 163–168.
22. Wetter, M.; Zuo, W.; Nouidui, T.S.; Pang, X. Modelica buildings library. *J. Build. Perform. Simul.* **2014**, *7*, 253–270. [CrossRef]
23. Revit Architecture–Building Design Software—Autodesk. Available online: http://www.autodesk.com/products/revit-family/overview (accessed on 15 January 2016).
24. About ArchiCAD—A 3D CAD Software for Architectural Design & Modeling. Available online: http://www.graphisoft.com/products/archicad (accessed on 15 January 2016).
25. Bentley System. Building Information Modeling—Bentley Architecture. Available online: https://www.bentley.com/en/products/brands/aecosim (accessed on 15 January 2016).
26. Eastman, C.M.; Teicholz, P.; Sacks, R.; Liston, K. *BIM Handbook: A Guide to Building Information Modeling for Owners, Managers, Designers, Engineers and Contractors*; John Wiley and Sons: Hoboken, NJ, USA, 2008.
27. Lee, G.; Sacks, R.; Eastman, C.M. Specifying parametric building object behavior (BOB) for a building information modeling system. *Autom. Constr.* **2006**, *15*, 758–776. [CrossRef]
28. Autodesk Developer Center—Autodesk® Revit® Architecture, Autodesk® Revit® Structure and Autodesk® Revit® MEP. Available online: http://usa.autodesk.com/adsk/servlet/index?siteID=123112&id=2484975 (accessed on 4 August 2015).

29. GRAPHISOFT. Graphisoft Developer Center. Available online: http://www.graphisoft.com/support/developer (accessed on 4 August 2015).

30. Fritzson, P. *Principles of Object-Oriented Modeling and Simulation with Modelica 2.1*; John Wiley and Sons: Hoboken, NJ, USA, 2010.

31. Tiller, M. *Introduction to Physical Modeling with Modelica*; Kluwer Academic Publishers: Boston, MA, USA, 2001.

32. Wetter, M. Modelica-based modeling and simulation to support research and development in building energy and control systems. *J. Build. Perform. Simul.* **2009**, *2*, 143–161. [CrossRef]

33. Tummescheit, H. *Design and Implementation of Object-Oriented Model Libraries Using MODELICA*; Lund University: Lund, Sweden, 2002.

34. Fritzson, P.; Bunus, P. Modelica—A general object-oriented language for continuous and discrete-event system modeling and simulation. In Proceedings of the 35th Annual Symposium, San Diego, CA, USA, 14–18 April 2002; pp. 365–380.

35. Wetter, M. Modelica Library for Building Energy and Control Systems. Available online: https://simulationresearch.lbl.gov/modelica (accessed on 4 August 2015).

36. Wetter, M.; Zuo, W.; Nouidui, T.S. Modeling of heat transfer in rooms in the MODELICA "Buildings" library. In Proceedings of the Building Simulation, Sydney, Australia, 14 November 2011; pp. 1096–1103.

37. Nouidui, T.S.; Phalak, K.; Zuo, W.; Wetter, M. Validation and application of the room model of the Modelica buildings library. In Proceedings of the 9th International Modelica Conference, Munich, Germany, 3–5 September 2012; pp. 727–736.

38. Energy in Building and Communities Programme. IEA EBC Annex 60. Available online: http://www.iea-annex60.org (accessed on 23 November 2015).

39. Wetter, M. Multizone airflow model in Modelica. In Proceedings of the 5th International Modelica Conference, Vienna, Austria, 4–5 September 2006; pp. 431–440.

40. Judkoff, R.; Neymark, J. *International Energy Agency Building Energy Simulation Test (BESTEST) and Diagnostic Method*; National Renewable Energy Lab.: Golden, CO, USA, 1995.

41. Pedersen, C.O.; Liesen, R.J.; Strand, R.K.; Fisher, D.E. *A Toolkit for Building Load Calculations*; American Society of Heating, Refrigerating and Air-Conditioning Engineers Inc.: Oklahoma City, OK, USA, 2001.

42. ASHRAE. *ANSI/ASHRAE Standard 140-2007 Standard Method of Test for the Evaluation of Building Energy Analysis Computer Programs*; America Society of Heating, Refrigerating and Air-Conditioning Engineers Inc.: Atlanta, GA, USA, 2010.

43. Microsoft Corporation. Microsoft Automatic Graph Layout—Microsoft Research. Available online: http://research.microsoft.com/en-us/projects/msagl (accessed on 4 August 2015).

Permissions

All chapters in this book were first published in Energies, by MDPI; hereby published with permission under the Creative Commons Attribution License or equivalent. Every chapter published in this book has been scrutinized by our experts. Their significance has been extensively debated. The topics covered herein carry significant findings which will fuel the growth of the discipline. They may even be implemented as practical applications or may be referred to as a beginning point for another development.

The contributors of this book come from diverse backgrounds, making this book a truly international effort. This book will bring forth new frontiers with its revolutionizing research information and detailed analysis of the nascent developments around the world.

We would like to thank all the contributing authors for lending their expertise to make the book truly unique. They have played a crucial role in the development of this book. Without their invaluable contributions this book wouldn't have been possible. They have made vital efforts to compile up to date information on the varied aspects of this subject to make this book a valuable addition to the collection of many professionals and students.

This book was conceptualized with the vision of imparting up-to-date information and advanced data in this field. To ensure the same, a matchless editorial board was set up. Every individual on the board went through rigorous rounds of assessment to prove their worth. After which they invested a large part of their time researching and compiling the most relevant data for our readers.

The editorial board has been involved in producing this book since its inception. They have spent rigorous hours researching and exploring the diverse topics which have resulted in the successful publishing of this book. They have passed on their knowledge of decades through this book. To expedite this challenging task, the publisher supported the team at every step. A small team of assistant editors was also appointed to further simplify the editing procedure and attain best results for the readers.

Apart from the editorial board, the designing team has also invested a significant amount of their time in understanding the subject and creating the most relevant covers. They scrutinized every image to scout for the most suitable representation of the subject and create an appropriate cover for the book.

The publishing team has been an ardent support to the editorial, designing and production team. Their endless efforts to recruit the best for this project, has resulted in the accomplishment of this book. They are a veteran in the field of academics and their pool of knowledge is as vast as their experience in printing. Their expertise and guidance has proved useful at every step. Their uncompromising quality standards have made this book an exceptional effort. Their encouragement from time to time has been an inspiration for everyone.

The publisher and the editorial board hope that this book will prove to be a valuable piece of knowledge for researchers, students, practitioners and scholars across the globe.

List of Contributors

Erdong Zhao and, Liwei Liu
School of Economics and Management, North China Electric Power University, Beijing 102206, China

Jing Zhao
School of Mathematics and Statistics, Lanzhou University, Lanzhou 730000, China

Zhongyue Su
College of Atmospheric Sciences, Lanzhou University, Lanzhou 730000, China

Ning An
Gerontechnology Lab, School of Computer and Information, Hefei University of Technology, Hefei 230009, China

James Michael Hooper and James Marco
Warwick Manufacturing Group (WMG), University of Warwick, Coventry CV4 7AL, UK

Gael Henri Chouchelamane and Christopher Lyness
Jaguar Land Rover, Banbury Road,Warwick, Coventry CV35 0XJ, UK

Lan-Rong Dung and Ming-Han Lee
Department of Electrical and Computer Engineering, National Chiao-Tung University, 1001 Ta-Hsueh Rd., Hsinchu 30010, Taiwan

Hsiang-Fu Yuan and Jieh-Hwang Yen
Institute of Electrical Control Engineering, National Chiao-Tung University, 1001 Ta-Hsueh Rd., Hsinchu 30010, Taiwan

Chien-Hua She
MiTAC International Corp., No. 1, R & D 2nd Road, Hsinchu Science-Based Industrial Park, Hsinchu 30010, Taiwan

Zhanfeng Song
Department of Electrical Engineering and Automation, Tianjin University, Tianjin 30072, China

Yanjun Tian and Zhe Chen
Department of Energy Technology, Aalborg University, Aalborg 9220, Denmark

Yanting Hu
Faculty of Science and Engineering, University of Chester, Chester CH1 4BJ, UK

Reza Ahmadi Kordkheili, Mehdi Savaghebi and Josep M. Guerrero
Department of Energy Technology, Aalborg University, Pontoppidanstraede 101, Aalborg 9220, Denmark

Seyyed Ali Pourmousavi
NEC Laboratories America Incorporations, Cupertino, CA 95014, USA

Mohammad Hashem Nehrir
Electrical and computer engineering department, Montana State University, Bozeman, MT 59717, USA

Christopher D. Elvidge
Earth Observation Group, National Centers for Environmental Information, National Oceanic and Atmospheric Administration, 325 Broadway, Boulder, CO 80205, USA

Kimberly Baugh, Feng-Chi Hsu and Tilottama Ghosh
Cooperative Institute for Research in the Environmental Sciences, University of Colorado, Boulder, CO 80303, USA

Mikhail Zhizhin
Cooperative Institute for Research in the Environmental Sciences, University of Colorado, Boulder, CO 80303, USA
Russian Space Research Institute, Moscow 117997, Russia

Qingwu Gong, Jiazhi Lei and Jun Ye
School of Electrical Engineering, Wuhan University, Wuhan 430072, China

Chen Wang
School of Mathematics and Statistics, Lanzhou University, Lanzhou 730000, China

JieWu
School of Mathematics and Computer Science, Northwest University for Nationalities, Lanzhou 730030, China

Jianzhou Wang
School of Statistics, Dongbei University of Finance and Economics, Dalian 116025, China

Weigang Zhao
Center for Energy and Environmental Policy Research, Beijing Institute of Technology, Beijing 100081, China
School of Management and Economics, Beijing Institute of Technology, Beijing 100081, China

Mingxiang Deng, Wei Li and Yan Hu
State Key Laboratory of Water Environment Simulation, School of Environment, Beijing Normal University, Beijing 100875, China

Carlos Quiterio Gómez Muñoz and Fausto Pedro García Márquez
Ingenium Research Group, Department of Business Management, University of Castilla-La Mancha, Ciudad Real 13071, Spain

Zhenshi Wang and Haifeng Dai
Clean Energy Automotive Engineering Center, Tongji University, No. 4800, Caoan Road, Shanghai 201804, China
School of Automotive Studies, Tongji University, No. 4800, Caoan Road, Shanghai 201804, China

XuezheWei
School of Automotive Studies, Tongji University, No. 4800, Caoan Road, Shanghai 201804, China

WoonSeong Jeong and JeongWook Son
Department of Architectural Engineering, Ewha Womans University, Seoul 120-750, Korea

Index

www.ingramcontent.com/pod-product-compliance
Lightning Source LLC
Chambersburg PA
CBHW061946190326
41458CB00009B/2801